World-Systems Evolution and Global Futures

Series Editors

This series seeks to promote understanding of large-scale and long-term processes of social change, in particular the many facets and implications of globalization. It critically explores the factors that affect the historical formation and current evolution of social systems, on both the regional and global level. Processes and factors that are examined include economies, technologies, geopolitics, institutions, conflicts, demographic trends, climate change, global culture, social movements, global inequalities, etc. Building on world-systems analysis, the series addresses topics such as globalization from historical and comparative perspectives, trends in global inequalities, core-periphery relations and the rise and fall of hegemonic core states, transnational institutions, and the long-term energy transition. This ambitious interdisciplinary and international series presents cutting-edge research by social scientists who study whole human systems and is relevant for all readers interested in systems approaches to the emerging world society, especially historians, political scientists, economists, sociologists, geographers and anthropologists.

This book series is indexed in Scopus.

All titles in this series are peer-reviewed.

Barry B. Hughes

Exploring and Shaping International Futures

Fifth Edition

Frederick S. Pardee Institute for International Futures
UNIVERSITY OF DENVER

Springer

Barry B. Hughes
Frederick S. Pardee Institute for
International Futures
Josef Korbel School of Global and Public
Affairs, University of Denver
Denver, CO, USA

ISSN 2522-0985 ISSN 2522-0993 (electronic)
World-Systems Evolution and Global Futures
ISBN 978-3-032-22083-7 ISBN 978-3-032-22084-4 (eBook)
https://doi.org/10.1007/978-3-032-22084-4

This work was supported by University of Denver.

The first three editions (1993, 1996, and 1999), published by Westview Press, were named *International Futures: Choices in the Search for a New World Order*. The fourth edition (2006) was co-authored by Dr. Evan E. Hillebrand and published by Paradigm Publishers.
4th edition: © Barry B. Hughes 2006

This Springer imprint is published by the registered company Springer Nature Switzerland AG
The registered company address is: Gewerbestrasse 11, 6330 Cham, Switzerland

To Frederick S. Pardee
He believed in our ability to "improve the human condition"
and greatly helped us in the effort to do so.

Preface

Life holds a wonder of opportunities and pleasures.

Constant challenges limit the development of life's potential and threaten life itself.

Both statements are true across human lives and within the experience of most individuals. Yet, the balance among the opportunities, pleasures, challenges, and threats varies tremendously within individual life spans and across societies and historical periods. Individual humans and humanity collectively are constantly struggling to shift the balance and improve our futures.

This book immerses you in thinking about longer-term global futures and a wide range of the issues, challenges, and opportunities that we collectively face around human development and sustainability. Its goal is to assist in the extension and refinement of your understanding and vision of alternative possible futures and your thinking about policy strategies that might help move the world toward a humane and environmentally sustainable one.

The book involves you in exploring alternative futures in two ways. The first is via traditional text. The volume argues that we understand the future through extrapolation of trends and through causal understandings of global change. Chapter 1 introduces the questions that the book addresses, Chapter 2 assists you in understanding historical trends, and the rest of the book helps you think about the future and the causal dynamics of global systems.

We are, of course, part of the causal systems we examine. Thus, the second major emphasis in the book is assistance in your own analysis of general leverage points where the more specific actions of policymakers and even individuals might help preferred futures unfold.

That assistance involves your immersion in exploration of alternative futures through use of a highly interactive computer simulation model called International Futures (IFs). IFs is a global model system that simulates long-term population, economic, human development, food, energy, environmental, and sociopolitical change. It represents the world in 188 countries and it aggregates those into country groupings of your choosing. It therefore facilitates quantitative exploration of futures (forecasting) and facilitates thinking broadly about uncertainties and possibilities across the rest of the twenty-first century (foresight).

The home of the International Futures model system is the Frederick S. Pardee Institute for International Futures at the University of Denver's Josef Korbel School of Global and Public Affairs. At https://korbel.du.edu/pardee/ you can use an online version of IFs or you can download a version for your own computer. Using the online version eliminates issues of installation and system compatibility. Such issues should, however, not be significant on computers with Microsoft Windows. The downloadable version runs somewhat faster and is more fully under your control. The Help system associated with both versions of the model is comprehensive with respect to the use of the model and its structure.

Since its inception in 1979, IFs has evolved as a tool for thinking about the future. This volume is based on the eighth generation of the model (Appendix 2 reviews its evolution across the generations); this is effectively the fifth edition of this volume, although the changes in editions over time have been quite dramatic. The most recent and substantially different version of this volume (Hughes & Hillebrand, 2006) documented the fourth generation.

To our new and continuing associates in use and development of the IFs model, thank you for joining in the exploration of alternative futures. Now please jump into the text and model interaction. I hope you can both learn and enjoy.

Denver, USA Barry B. Hughes

Acknowledgments

Over nearly five decades, countless organizations and individuals have contributed to the building of the International Futures (IFs) forecasting system with their use of it, their feedback and suggestions for enhancement, and their direct involvement in that enhancement. Many have also contributed to building and developing its home, the Frederick S. Pardee Institute for International Futures, and to the work of the Institute.

It is impossible to recognize individually all the students, instructors, and analysts who have used the IFs system over its many years and provided essential advice for enhancement or even more directly participated in its development. One of the best decisions ever made in the IFs project was to make the system freely available for users and thereby to create a huge network of participants in its development. Among those important in the first half of those five decades I particularly thank Steve Bankes, Gerald Barney, Donald Borock, Peter Brecke, Stuart Bremer, Jonathan Cave, Richard Chadwick, Mark Crescenzi, Thomas Cusack, Karl Deutsch, William Dixon, Theodore Gordon, Harold Guetzkow, Elizabeth Hanson, Paul Herman, Evan Hillebrand, Ronald Inglehart, Douglas Lemke, Pentti Malaska, Edward Mansfield, Mihajlo Mesarovic, Sergei Parinov, Robert Pestel, Brian Pollins, Peter Rindfuss, Dale Rothman, Phil Schrodt, Paul Senese, Thomas Shook, Dale Smith, Harvey Starr, Douglas Stuart, Donald Sylvan, William Thompson, and Jonathan Wilkenfeld. At the University of Denver, home of IFs since 1980, important graduate student colleagues in very early years include Shannon Brady, James Chung, Warren Christopher, Kay Drucker, Michael Ferrier, Michael Niemann, Padma Padula, and Terrance Peet-Lukes.

Work during the 1970s in support of a Club of Rome project led by Mihajlo Mesarovic and Edward Pestel made possible initial creation of IFs late in the decade. That initial version of IFs led to an invitation by Stuart Bremer for my involvement in the GLOBUS model project at the Science Center of Berlin across most of the 1980s. Work there and interaction with graduate students at the University of Denver led to the first edition of this book in 1993, the second in 1996, and the third in 1999, all directed toward educational use.

For early financial support, I give thanks to the National Science Foundation, the Cleveland Foundation, the Exxon Education Foundation, General Motors, the

Kettering Family Foundation, the Pacific Cultural Foundation, and the United States Institute of Peace. None of these institutions bears any responsibility for the model system or its analysis, but their support made much early development and use possible.

In the more recent half of the IFs history since the late 1990s, recognition of the value provided by its analysis moved it from almost entirely educational use to extensive use by intergovernmental organizations, nongovernmental organizations, and national governments. An early and critically important step on that path in 2000–2003 was funding within, and use by, the TERRA project of the European Commission, made possible by Robert Pestel. The third edition of this book and the work with GLOBUS also led to interest in IFs by Evan Hillebrand and the Strategic Assessments Group at the U.S. Central Intelligence Agency. Collaboration with Hillebrand and with Ronald Inglehart at the University of Michigan contributed to enhanced representation in IFs of both domestic and international sociopolitical systems and ultimately to the fourth edition of this book in 2006, co-authored by Evan Hillebrand.

Hillebrand and Paul Herman also brought IFs to the attention of the U.S. National Intelligence Council (NIC) and its preparation of quadrennial *Global Trends* reports to the President, also publicly available. Funding generated by analytical contributions to Project 2020 of the NIC helped make possible the creation of an openly available version of the model for the web, a huge effort. In total, IFs contributed to four *Global Trends* reports thanks especially to Mathew Burrows (then at the NIC). Further, with encouragement from Ronald Inglehart, the European Union Center at the University of Michigan also provided support for enhancing the user interface and general ease of use of the IFs system.

Over the last 20-plus years, IFs has benefited greatly from, and hopefully contributed substantially to, many projects. For instance, thanks to Dale Rothman, IFs supported the United Nations Environment Programme's fourth *Global Environmental Outlook* and has been involved in several more recently. There has also been a strong partnership with the Institute for Security Studies in Africa and its founding leader, Jakkie Cilliers, leading to collaboration with the African Union Development Agency. Rob Lempert supported cooperation with RAND in the United States and Peter Johnston with RAND in Europe. Brian O'Neill and Detlef van Vuuren have helped connect IFs to the integrated assessment modeling community, as did Paul Lucas.

In a tremendously important step for the IFs project in the first decade of this century, Frederick Pardee began providing long-term sustaining financial support that has undergirded creation of a home for IFs since 2007 in the Frederick S. Pardee Center (now Institute) for International Futures at the University of Denver's Josef Korbel School of Global and Public Affairs. Fred's support began with sponsorship of a five-volume series called Potential Patterns of Human Progress (PPHP), with new modeling in the IFs system supporting annual volumes starting in 2009 on poverty, education, health, infrastructure, and governance.

In the early years of this century, starting even before the support of Fred Pardee, the IFs system advanced substantially due to the involvement of a team of graduate assistants and colleagues. It is only possible to mention some of their many contributions. Anwar Hossain built a strong and useful database and made possible the addition of a social accounting matrix structure to the model. Mohammod Irfan created the scenario tree approach to model input and led model development on education and, with Dale Rothman, on infrastructure. José Solórzano took the lead on conversion of the model to the Web and, with Randall Kuhn, led modeling on health. Keith Gehring was a leader in incorporating IFs into the curriculum of the Korbel School and in teaching with and about it elsewhere around the world. He has continued over many years to provide ideas for its improvement. Marc Sydnor provided great support for the fourth edition of this volume. Elizabeth Bremer substantially supported the production of a volume in 2019 documenting IFs and placing it in the context of integrated assessment models. Jonathan Moyer led badly needed development of the Help system and provided invaluable feedback from his intensive use across the entire IFs system.

This listing is a small fraction of the organizations and projects that have unfolded over the last two decades and of the people who have assisted in the development and use of IFs and the advances of the Pardee Institute. It is impossible to name them all personally, just as it was for the early decades of the work. My apologies to those omitted here. It is critical to name at least some of those now at the Institute. Jonathan Moyer (Associate Professor and Director), José Solórzano (Lead System Developer), and Mohammod Irfan (Senior Scientist) were already identified as critical, long-time members of the team. Collin Meisel is Director of Analysis and Associate Director for Geopolitical Analysis. Taylor Hanna is Associate Director for Development Analysis. Pam Hoberman is Director of Operations and Peter Stansbery is Grants and Business Manager. Yutang Xiong is Lead Data Operations Manager. Keith Gehring is Teaching Associate Professor. Alphabetically by first name, Abdelrahman (Bido) Ibrahim, Caleb Petry, Cecilia Peterson, Deva Sahadevan, Devin Joshi, Divya Bandi, Eli Margolese-Malin, Haider Khan, Mickey Rafa, Randall Kuhn, Samantha Kunz, Stellah Kwasi, and Timothy Sisk also are or recently have been key team members, a list changing over time. Among recent past team members to whom we owe special thanks, both for modeling and analysis, are David Bohl and Steve Hedden. The project is blessed by having both Mohammod Irfan and José Solórzano as still active long-time members of the team and by Jonathan Moyer becoming Director of the Institute a decade ago. Jonathan has been an incredibly dedicated and creative force in giving both the Institute and IFs long-term strength and sustainability.

Jennifer Knerr was a wonderful editor in support of the first four editions of this volume. At Springer, it has been productive and a pleasure to work with Senior Editor Niko Chtouris and Production Supervisor Jayalakshmi Raju on this fifth edition.

Janet Dickson has had many important roles over time. In addition to being my supportive wife, she and I together made the connection to Fred Pardee. Janet wrote and edited across the five PPHP volumes and edited this and an earlier book on global modeling. She has helped guide the development of the Institute, including co-leading the development of the Institute's first strategic plan with Jonathan Moyer, now over five years ago.

For all these people, I am grateful. It's been a long and wonderful ride.

Contents

Acronyms

AIDS	Acquired Immune Deficiency Syndrome
AUDA	African Union Development Agency
B.C.E.	Before the Common Era
C.E.	Common Era
CBR	Crude Birth Rate
CDR	Crude Death Rate
CFCs	Chlorofluorocarbons
CGIAR	Consultative Group on International Agriculture Research
CO_2	Carbon Dioxide
COLT	Country and Organization Leader Travel Database
COVID-19	Coronavirus Disease of 2019
COVID	Coronavirus Disease
DICE	Dynamic Integrated Climate-Economy
DiME	Diplomacy, Military, Economy index
EU	European Union
FAO	UN Food and Agriculture Organization
FBIC	Formal Bilateral Influence Capacity index
FDI	Foreign Direct Investment
GDI	Gender Development Index
GDP	Gross Domestic Product
GEM	Gender Empowerment Measure
GEM	General Equilibrium Model
GEO	Global Environmental Outlook
GHG	Greenhouse Gas
GII	Gender Inequality Index
GIS	Geographic Information System
GLOBUS	Computer Simulation of Worldwide Political and Economic Developments
GNP	Gross National Product
GSG	Global Scenario Group
HDI	Human Development Index
IAMC	Integrated Assessment Model Consortium

IAMs	Integrated Assessment Models
ICPSR	Inter-university Consortium for Political Research
ICT	Information and Communications Technology
IEA	International Energy Agency
IFs	International Futures
IGOs	Intergovernmental Organizations
IIASA	International Institute of Applied Systems Analysis
IMAGE	Integrated Model to Assess the Global Environment
IMF	International Monetary Fund
INGOs	International Nongovernmental Organizations
IOs	International Organizations
IPAT	Impact (Environmental) from Population, Affluence, Technology
IPCC	Intergovernmental Panel on Climate Change
IPDs	Infectious and Parasitical Diseases
ISS	Institute for Security Studies
LDCs	Less Developed Countries (Contemporary Usage Often "Least")
MDGs	Millennium Development Goals
MER	Market Exchange Rate
MFP	Multi-Factor Productivity
MIDs	Militarized Interstate Disputes
NATO	North Atlantic Treaty Organization
NEPAD	New Partnership for Africa's Development
NIC	National Intelligence Organization (U.S.)
OECD	Organization for Economic Cooperation and Development
OPEC	Organization of Petroleum Exporting Countries
PDF	Probability Distribution Function
PID	Proportional-Integral-Derivative controller
PITF	Political Instability Task Force
PPP	Purchasing Power Parity
R&D	Research and Development
RCPs	Representative Concentration Pathways
SAM	Social Accounting Matrix
SDGs	Sustainable Development Goals
SHAPE	Sustainable Development Pathways Achieving Human Wellbeing While Safeguarding the Climate and Planet Earth
SSPs	Shared Socioeconomic Pathways
TFR	Total Fertility Rate
TNT	TriNiTrotoluene Explosive
U.S.	United States
UCDP	Uppsala Conflict Data Program
UN	United Nations
UN DESA	United Nations Department of Economic and Social Affairs
UNCED	United Nations Conference on Environment and Development
UNCTAD	United Nations Conference on Trade and Development
UNDP	United Nations Development Programme

UNEP	United Nations Environment Programme
UNESCO	United Nations Educational, Scientific, and Cultural Organization
UNICEF	United Nations International Children's Emergency Fund
UNIDO	United Nations Industrial Development Organization
UNPD	United Nations Population Division
USAID	U.S. Agency for International Development
V-DEM	Varieties of Democracy Electoral Democracy Index
WASH	Water, Sanitation, and Hygiene
WB	World Bank
WFS	World Values Survey
WHO	World Health Organization
WIC	Wittgenstein Centre for Demography and Global Human Capital
WIM	World Integrated Model

List of Figures

List of Tables

Part I

Foundations: Improving Global Futures

Action in the Face of Uncertainty 1

Abstract

This chapter sets the stage for the book. It stresses the highly significant changes and challenges facing us across this century with respect to human development, environmental sustainability, and sociopolitical efforts to manage and support those. It discusses the global setting of the Sustainable Development Goals and the creation of alternative scenarios concerning the future, illustrated by the Shared Socioeconomic Pathways. A major emphasis is on both the uncertainty of future pathways and the importance of efforts to better understand and shape those. The International Futures forecasting and analysis system is identified as a foresight-enhancing tool to be used across the volume to better understand possible alternative futures and to explore interventions that may help move global change toward humane and environmentally sustainable futures.

1.1 Introduction

In a remarkable period between the fifteenth and eighteenth centuries, Europeans explored, and from their own point of view, discovered the rest of the world. They acted in the face of great uncertainty. On the push outward, in 1522, one ship of five launched in Ferdinand Magellan's expedition returned to Spain from the east with only 18 men, completing the first voyage around the world. Magellan himself died in the Philippines attempting forcibly to convert locals to his religion. The actions of the explorers dramatically reshaped the human world.

The explorers attempted to reduce uncertainty by enhancing their maps of the world. Map makers of the period advanced global mapping from about the level of

Janet Dickson, Pardee Institute Senior Advisor, contributed much to this chapter and edited the entire volume.

B. B. Hughes, *Exploring and Shaping International Futures*, World-Systems Evolution and Global Futures, https://doi.org/10.1007/978-3-032-22084-4_1

the ancient Romans to something much closer to contemporary standards (Shirley, 1984). Perhaps the greatest of all map makers in the period, Gerardus Mercator (1512–1594), developed a navigation-enhancing projection of three dimensions onto two. He also produced the first global atlas. His maps reshaped human understanding of geography and facilitated successful exploration.

Our own era is still one of exploring and reshaping the world, of acting in the face of uncertainty, and of trying to develop the best maps possible to guide us. Today the economic product of the average human has attained levels surpassing any in history. Yet billions of people suffer abject poverty. Global disparities in income and malnutrition levels do (or *should*) challenge the consciences of nearly all who live in relative comfort. We devote about two and one-half percent of total global economic output to the pursuit of military security, but we simultaneously have created an insecure world with thousands of nuclear weapons and several ongoing high casualty conflicts. The technological sophistication of our scientists and engineers has created new marvels in electronics, biology, and other fields; nonetheless, large portions of our shared environment have greatly deteriorated. Many dimensions of uncertainty frame our future.

Some of the choices that we contemporary humans make on human development, economic, environmental, and security issues could well have consequences as important as the decisions made by Magellan and his contemporaries. We will never fully anticipate their consequences for ourselves or others because, like Mercator, we are only mapping the world as we and others discover it.

What might be the future of human developmental, environmental, and political-social systems throughout the twenty-first century? That is the overarching question that motivates this volume. The easy and correct answer is that no one knows. If we were so fatalistic as to believe that we had no control over the future, we might simply accept that response and return our attention to daily life. Most of us believe, however, that knowledge and actions substantially shape our own future and the futures of our descendants. We fear misguided action or inaction. Consequences could include failure to address poverty and hunger, with potential for domestic and international conflicts and environmental despoliation, including rapid climate change. We hope that thoughtful behavior can instead ensure a healthy, peaceful, and prosperous world. We therefore confront that very real challenge: we cannot know the future, but it is essential to act in the face of that uncertainty.

To reduce the challenge to manageable proportions, we can decompose our general question about the future into three more specific ones. First, *what kind of future would we prefer?* Second, *where do current changes appear to be taking us?* Third, *how much and what leverage do we have to bring about a future we prefer?* These questions collectively help us grapple with the necessity of choice in the face of incomplete knowledge. The task of this book is to assist in investigating these three questions and thereby in addressing the challenge. In the process we can organize and expand our own mental maps as guides to action.

Throughout our efforts we need always to recognize the uncertainties in our exploration of the future. Our enterprise needs to wrap forecasting of alternative possible futures in an understanding of our knowledge limitations, to recognize that we are engaged in a broader process of explorative foresight. Also, while we seek to identify leverage that suggests policy orientations, our analysis is at a macro level that helps identify directions for policy efforts, not specific policy recommendations. We seek to enhance our thinking about strategic possibilities.

1.2 The Three Questions Elaborated

Although this volume will guide us deeper into each question, in large part through our use of the International Futures (IFs) model system, it is useful to begin thinking about them now.

1.2.1 What Future Would We Like?

There is an old saying that "the only thing worse than not getting your heart's desire is getting it." We probably have all had the experience of wanting something and then finding that obtaining it did not make us happy. Often the problem is a failure to clarify our own values.

Three sets of values and goals with which few would disagree have increasingly framed images of desired futures (see Table 1.1). First, humans as individuals should be able to develop their capabilities as fully as possible, attaining literacy, securing nutrition and health care that support a reasonable life expectancy, and gaining access to a basic level of economic resources. The broader purposes of these capabilities are to allow individuals substantial freedom of choice in their pursuit of a fulfilling life (Sen, 2000). Second, humans in their interactions with a broader biological and physical environment should be able to live in a sustainable manner so that lifestyles and choices do not jeopardize the life conditions of their own futures and those of subsequent generations (United Nations, 1987). Third, humans in their interactions with one another desire peace and security (Kant, 1795), basic fairness and justice (Rawls, 1971), and good governance systems to help maintain those. These are fundamental to pursuing human well-being and environmental sustainability. Collectively, these goals have increasingly come to be recognized as the pillars of "sustainable human development," the overarching or "metagoal" of most who think about and act to enhance global futures.

It is, however, not just the general statements of values and goals that have come into increasingly sharp focus, but the articulation of more specific goals and

Table 1.1 Human spheres of action and values

Humans as individuals	Human well-being
Humans with the environment	Environmental sustainability
Humans with each other	Cooperation/security/fairness

Source The author

targets for action. In recent decades, so many global conclaves and institutions have enunciated such goals and targets, often not met, that a cynic might conclude that words have substituted for action. Yet, in the global conversation that these meetings, pronouncements, and declarations have punctuated, increasing clarity and resolve have emerged.

Around the environment, the first United Nations Conference on the Human Environment in Stockholm during 1972 and the Club of Rome with its publicity around possible biological and physical *Limits to Growth* (D. H. Meadows et al., 1972) helped set the stage. In 1983, the United Nations charged the Brundtland Commission (officially the World Commission on Environment and Development; see United Nations, 1987) with assessing the world's environmental condition and establishing an agenda for it. That commission defined sustainable development in a manner that has strongly shaped the subsequent discussion: namely, development that meets the needs of the present without compromising the ability of future generations to meet their own needs. A series of international conventions and protocols followed, including very specific and successful agreements to reduce the emissions of chlorofluorohydrocarbons and, to date, less successful ones to reduce greenhouse emissions, especially carbon dioxide.

One might think that global action on human development with attention to issues such as poverty reduction and hunger elimination would have emerged even earlier than attention to the environment, and it often did within higher-income countries, as exemplified by the New Deal in the United States during the Great Depression of the 1930s. Perhaps the extreme poverty in countries like China and India and across an African continent still mostly under colonial control until about 1960 long discouraged belief that global action beyond some foreign aid had any potential to ameliorate it.

Yet, goals concerning poverty and hunger reduction plus enhancement of literacy have appeared regularly over recent decades. In 1992, the United Nations Conference on the Environment and Development in Rio de Janeiro, Brazil, at which the framework convention on greenhouse gases was adopted, explicitly identified development as a global commitment that needs to be integrated with environmental concerns. In 1995, The World Summit for Social Development in Copenhagen called for three dimensions of sustainable development: economic, social, and environmental (taken up again in the second world summit in Qatar in 2025).

Even with respect to security and social fairness, goals have become clearer. Increased and sometimes questionable willingness emerged in the global community to contravene long-established principles of national sovereignty and to intervene, peacefully whenever possible, in countries where order has disintegrated or gross social injustice is obvious. Yet, it is in the areas of human development and environmental sustainability that much domestic action and collaboration have now emerged.

In 2000, the UN Millennium Summit (UNDP, 2001: 21–24; UNDP, 2002: 13–33) accomplished something very remarkable. The global community committed itself to an *integrated set of specific targets* that centered on improving the global human condition by 2015. Among the specific goals that the summit set, as reiterated and extended in 2002 at the World Summit on Sustainable Development in Johannesburg, South Africa, were:

- Reducing by one half the proportion of the world's population surviving on less than $1 a day (at 1993 Purchasing Power Parity). At that time, about 1.2 billion people were attempting to cope at this level, with about 2.8 billion living on less than $2 per day (UNDP, 2001: 9).
- Reducing by one half the proportion of humanity that suffered from hunger.
- Achieving universal completion of primary schooling.
- Achieving gender equality in access to education.
- Reducing under-five mortality rates by two-thirds and reducing maternal mortality rates by three-quarters.
- Halting and beginning to reverse the spread of HIV/AIDS, malaria, and other major diseases.
- Reducing by one half the proportion of the world's people without access to safe drinking water.
- Creating a global compact to mobilize rich and poor in pursuit of all goals.

Known collectively as the Millennium Development Goals (MDGs), the full set of goals for 2015 cut across and integrated eight economic, environmental, and human development issues. During the 2012–2017 period, with the reaching of the 2015 MDG horizon and with some success, but falling far short on most goals, the UN led the global community in developing the Sustainable Development Goals (SDGs) as successors to the MDGs. They were refined in 2020 and officially have a 2030 target horizon. Figure 1.1 shows the 17 goals. Each has multiple associated targets, and the full set has 251 specified indicators.

This volume will return to the SDGs often in subsequent chapters, especially Chap. 8 with its focus on human development goals and Chap. 11 with broad attention to environmental issues. The SDGs are a great help in thinking about the future we want to see, as well as motivators for identifying potential action in the face of uncertainty. They help in studying the global system, setting our own goals based on core values, and identifying possibilities for reaching them.

The SDGs, unfortunately, will largely also not be fully met. Yet, they have mobilized international organizations, non-governmental organizations, governments in both rich and poor countries, and great numbers of individuals in their pursuit. Their horizon extension, or very similar successor goal sets, will continue to shape progress toward sustainable human development.

This volume will indicate throughout where IFs can be useful in exploring actual and potential progress toward the SDGs. Also, IFs has coverage in some areas, especially with respect to human development and sociopolitical representations, that goes beyond the SDG statements and their elaboration. Parts III–V of the volume sequentially give special attention to the three spheres of action and values shown in Table 1.1. For instance, Chap. 8 in Part III gives much attention to the potential for analysis of human development goals with IFs.

Fig. 1.1 The Sustainable Development Goals. *Source* UN Office for Sustainable Development: https://unosd.un.org/content/sustainable-development-goals-sdgs

It would, however, be a very considerable misrepresentation to suggest that goals as widely accepted as those identified above give rise to universal agreement on specific actions to be pursued. Values interact with understandings of the dynamics that characterize global systems, and well-intentioned individuals can disagree quite strongly about those dynamics (and on the relative importance of different values). To address the role of understanding those dynamics, it will be necessary to add investigation of causality and the understanding of world views to our tool kit of forecasting and foresight techniques.

1.2.2 Where Do We Appear to Be Going?

One of the most common techniques for studying change is **extrapolation** of historical patterns. Extrapolation is trend projection. If global population is growing at 0.9 percent this year, a simple extrapolation assumes that it will grow at 0.9 percent each year in the future. A more sophisticated extrapolation would recognize that the annual population growth rate has declined from 2.0 percent each year in the late 1960s to about 0.9 percent now. Therefore, the rate of growth may decline further in the next few decades. But extrapolation is not the only technique for exploring where change is taking us. It is a potential starting point, and sometimes we will use it. But we can very often do better.

Forecasting by **historical analog** is a more qualitative approach. For example, contemporary global change is sometimes said to have much in common with the period at the end of World War II. The current world system, it can easily be claimed, is going through the greatest and most rapid change it has experienced since the 1940s and early 1950s. Such arguments are, however, perhaps too short-sighted. It is possible that the systems of our world are now moving into a period of transformations more nearly like those that characterized the Industrial Revolution, sweeping out from England in the late 1700s and early 1800s, eventually to encompass the entire world. Even further, some have suggested that we are entering a period of transformation as dramatic as the breakdown of feudalism and the emergence of the state system within Europe during the 1600s and eventually globally.

Every era might view its own time as unique and therefore may exaggerate its place in history. So let us review very briefly some of the specific elements of our contemporary transformative era, its similarities and differences with changes in previous eras, and some implications for the future.

The world's increasing population is now moving through the top half of an S-curve of change in total size with the likelihood yet this century of even a decreasing population. This follows a great many millennia of extremely slow population growth and then a rolling series of what might be characterized as demographic explosions over about the last 250 years, first in the more developed world and then over most of the last century throughout the developing world, where the ongoing shocks are still rocking the system. Yet, global fertility decline has definitively gained momentum and most of the world is clearly moving into a phase that we have not seen historically, where issues associated with aging populations supplement and then may largely replace those associated with growing populations.

Some unfortunate countries will grow old before they grow rich, reinforcing the obvious need to think also about economics. The world's economy has roughly followed the same pattern of accelerating growth as demographics, but with markedly faster growth rates, thereby making the average person in most countries rich by any historical standard. In contrast to demographic growth, however, the ongoing patterns of the global economy are even less certain. Across roughly the same period as demographic explosion, the economies of the world have moved from almost entirely agricultural to increasingly manufacturing to currently predominantly service based. What will artificial intelligence, robotics, at least early movement out into the solar system, and bioengineering do to economies the rest of this century? The uncertainties are great, and historical analogs are not obvious.

The last two centuries have been marked also by transformations in the global energy system, from almost entirely animal and wood-based systems to ones driven heavily by coal and then to overwhelming reliance on oil and gas. The current century will, almost without doubt, see another very significant transformation, this time to renewable energy forms. Its implications for systems including employment and the environment will be very different than the transition to fossil fuels.

Peoples in previous eras frequently damaged their local environments, thereby undercutting their social systems and often setting up patterns of malnutrition and warfare. What appears unique about the current era is the global reach of our changes to the environment, actually overfishing the once seemingly endless abundance of the oceans; cutting and burning massive rainforests; emptying in decades aquifers that sometimes filled only over millennia; creating the conditions for multiple inches of topsoil to wash away; and moving steadily toward, and then past, doubling of atmospheric carbon dioxide with attendant rises in global temperature.

On the geopolitical side of global change is a metatrend of technological, economic, and sociopolitical change that has been reshaping the world at least since the days of the European explorations and colonial empire-building and, of course, that could be traced back further to the eras of Marco Polo or Genghis Khan (or even further back through Sumerian, Egyptian, Middle Kingdom [China], Grecian, and Roman periods). That history has been marked by the regular rise and fall of great powers, often suggested in the modern state system to happen roughly every 100 years. The rapid rise of China since the 1980s and, more recently, India, is signaling all but certain transition in global sociopolitical structures.

In contrast to previous periods, however, the last 100 years have seen the creation of the League of Nations, the United Nations, and tremendous numbers of international governmental and non-governmental organizations linking the world geopolitically—as do the nuclear weapons and missile systems of many countries. It is possible to believe that the historically unprecedented global economic growth rates of the twentieth century, the dramatic expansion of global financial flows since the last quarter of that century, and the emergence of information and communications technologies linking the world's populations almost instantaneously augur a potentially significant transformation of geopolitical change.

There are many other changes of seemingly remarkable significance occurring in the systems noted above: the shift that occurred near 2010 of humanity from primarily rural to primarily urban residence patterns; the movement of the world's population from a primarily illiterate condition to mostly literate in the last century and potentially to nearly universal literacy in the current one; a continuing albeit very irregular and perhaps less definitive spread of democracy around the world; the rise of nutrition levels to the point where even developing countries are beginning to see obesity challenge or replace undernutrition as a problem; the loss around the world of thousands of unique languages and of the cultural content carried by them, as well as the growing rebellion of many cultures to the threats of globalization; the increasing ability of small groups and even individuals to mobilize destructive power that threatens thousands and even millions of other humans; and much more.

In short, global change to be experienced during the lifetime of today's young people may well challenge that of nearly any historical period and weaken the ability of both extrapolation and historical analogs to fully understand it.

Expanding our mental models can benefit from augmenting and going beyond extrapolation and historical analogs with **causal analysis**. Extrapolation might sometimes allow its user to make a fortune on the stock market (technical analysts or "chartists" rely on it heavily) and might also provide some very good guesses about global futures. It has, however, significant limitations. Consider the old story about a person falling from the top of the Empire State Building. As she passes the 51st floor, a friend at a window asks how it is going. The response is "so far, so good." The reason that most of us see a little dark humor in this is that we automatically reach beyond extrapolative to causal analysis, a consideration of dynamic cause-and-effect relations.

While causal analysis can be superior to extrapolation and use of historical analogs, in practice the identification of causal variables and elaboration of relationships becomes complex. This book will analyze global change with attention to both extrapolation and history, two key foundations of mental maps. The International Futures system is, however, a representation of dynamic changes within and across all its integrated models. Therefore, most emphasis will be placed on its causal representations.

1.2.3 What Leverage Do We Have?

Analysis of human leverage in this volume is complicated for several reasons. One is that values vary somewhat (and sometimes significantly) across individuals and groups, and we therefore may choose to differentially focus our attention with respect to leverage. Another is that our understandings of the workings of the world tend to vary quite substantially and often to be underdeveloped, so that we frequently identify different courses of action to accomplish even nearly identical goals. Our own explorations need to continuously shape and reshape our mental and computer-based maps of the dynamics of human and other systems. We have a need not just for forecasts, but for insight and therefore foresight. Still another reason is that our analytical tools represent the world at a macro level, while policy interventions and human action are at a more micro level. Thus, discussion and analysis in this volume will focus more on strategic points of leverage for action, less on specific actions.

As we explore potential leverage using IFs, we will be creating **alternative scenarios of the future**. Although "scenario" is traditionally defined as a story unfolding within a play, those who do forecasting refer to it as an unfolding story about the future. As such, a scenario typically has much complexity, cutting across many issues and dynamic processes. In our discussions, however, we will use the same word to refer to individual or small sets of interventions, very often building those into complex stories experimentally and gradually. Chapters 4 and 5, especially the latter, will introduce the process for building scenarios with IFs in steps and with complexity.

That building of scenarios in steps takes us to still another reason that analysis of leverage is complicated. Almost any intervention we make in a complex system will not only affect the variable(s) of our immediate interest, but those proximate effects will also propagate to other variables across interacting human, social, and natural systems. Think, for example, of how any significant changes in total population size or total gross national product size will affect food and energy demand and then environmental variables. As our analysis proceeds, we will always be looking at how a first round of interventions needs to be adjusted and supplemented by other interventions.

The issue of secondary and tertiary consequences of actions will become critical. There are, for instance, considerable disputes over the implications of giving food and other aid to less developed countries. Often those disputes do not center on the primary impact of aid on recipients but, instead, on the secondary implications of the aid for changes in the economic and political systems of the recipients and the tertiary or third-order implications of those changes in turn upon the long-term well-being of the recipients.

Such secondary and tertiary consequences make it almost impossible to study issues in isolation. Everything becomes connected to everything else, and tracing through consequences of action becomes messy for any analyst. This is a major reason for those who study the future to use computers. If we can represent these complex interactions in a computer simulation or model, we can then let the computer trace through the implications of our actions. The International Futures model system will allow you to engage in your own experimentation with human intervention and to undertake your own assessment of the extent of human leverage and of secondary and tertiary consequences. It is meant to be both a modern atlas of mental maps and a tool for helping its users expand and refine their own maps.

As you work with IFs, your building of early and simple scenarios will therefore progressively lead to increasingly complex ones. In the most general terms, those complex scenarios will map out (1) the path(s) that we appear to be on, and/or (2) alternative, more or less desirable paths. Chapter 14 identifies some of the complex scenario sets that have been elaborated in a variety of forecasting projects and will hopefully facilitate integration of your own thinking about the future across multiple issue areas. (A peek ahead could be useful.)

In addition to unique scenarios, there is one set of five widely used scenarios to which some subsequent chapters refer. These are the Shared Socioeconomic Pathways (SSPs). Current versions of IFs include forecast runs of those scenarios immediately ready for analysis. The SSPs grew out of the work of the global integrated assessment model (IAM) community, which creates and uses large-scale computer simulations and directs special attention to environmental issues, especially climate change (see Chap. 11 on the environment). The traditional IAM approach has been to posit five alternative pathways for sociopolitical futures, rather than to analyze their unfolding dynamically, and then turn to forecasting the environmental futures associated with them. The SSPs have therefore most often been used in environmental analyses, such as in 2023 by the sixth assessment report (IPCC, 2023) of the Intergovernmental Panel on Climate Change (IPCC). Presentation of the SSPs is generally organized by two dimensions related to ameliorating negative futures: mitigation and adaptation challenges related to climate change (see Fig. 1.2).

Although the IAM community grew largely from efforts to identify alternative futures with respect to environmental change, it has increasingly looked also to the broader SDG issue set. For instance, the SHAPE project (2019–2023) produced analysis across a wide range of SDGs using two IAMs and two specialized topic

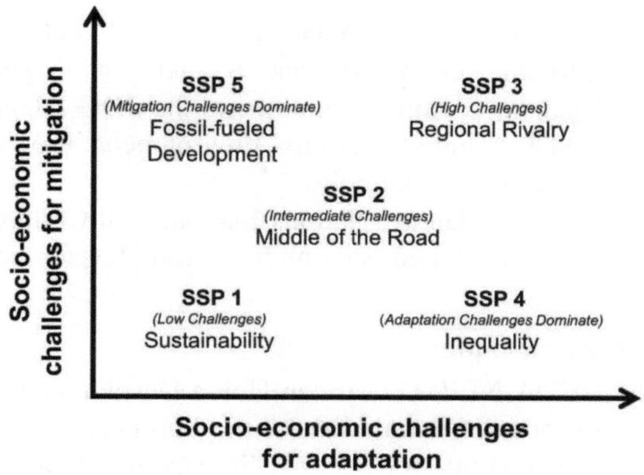

Fig. 1.2 Shared Socioeconomic Pathway scenario space. *Source* O'Neill et al., (2017: 170): http://dx.doi.org/10.1016/j.gloenvcha.2015.01.004. Reprinted with permission

models (Soergel et al., 2024). IFs itself is an IAM with an especially broad and integrated perspective across the SDGs, dynamically building both sociopolitical and environmental model elements (Hughes, 2025).

The SSP representation in IFs draws upon the storylines of the SSPs as initially laid out by O'Neill et al. (2017). But, whereas most IAM use of the SSPs incorporates exogenous representations from specialized models of important variables like GDP, population, and educational attainment of adults, the representations of such variables in the SSP scenarios of IFs compute those endogenously within the dynamic system structure from the many variables that drive them. That endogenous representation in IFs is important in the integrated analysis of potential leverage for advance toward human development and environmental sustainability.

1.2.4 Mobilizing Understandings and Leverage

It is not an exaggeration to suggest that a great many intergovernmental organizations (IGOs), international non-governmental organizations (INGOs), and national governments are heavily engaged in the enterprise of understanding where we are going, thinking about where we want to go, and exercising the leverage we have in shaping our direction. The existence of the SDGs has very considerably helped direct their mid- and long-range efforts. Tools such as the IFs system support these efforts, especially when the tools are open-access and without cost as IFs is.

The Pardee team has undertaken many projects for and with such organizations. Partners have included:

United Nations (UN) Family:

1. UN Secretary-General's Office
2. UN Women
3. UNDP (United Nations Development Programme): Various Human Development Report office teams, UNDP Regional Bureaus and Hubs (project

focus on Central Asia, Africa, the Sahel), UNDP Country Offices (Egypt, Ethiopia, Guinea-Bissau, Kazakhstan, Nigeria, North Macedonia, Sudan, Uganda, Ukraine, Yemen); and the SDG Integration Team
4. UNEP (United Nations Environment Programme), including their Global Environment Outlooks.
5. UNICEF (United Nations International Children's Emergency Fund)
6. UNIDO (United Nations Industrial Development Organization)

Others include:

1. AUDA-NEPAD (African Union Development Agency)
2. European Commission (of the European Union)
3. ISS (Institute for Security Studies in Africa)
4. Octopus [Green] Energy Group (UK registration)
5. USAID (U.S. Agency for International Development): general, also USAID Egypt, USAID Ethiopia, USAID Uganda, USAID Ukraine)
6. Various national governments

Chapter 14 will return to the kind of analyses done with such partners, often developing multiple scenarios. Your own interaction with IFs will help explore some of those. Hopefully it will also help you improve your own mental model of global change and how we might collectively help shape it.

1.3 Conclusion

This chapter has begun to address three questions, namely where global patterns seem to be taking us, what future we prefer, and what leverage we have. As the Danish physicist Niels Bohr put it, "Prediction is very difficult, especially about the future" (cited in Watkins, 1990: 152). Although IFs will assist you in better addressing our three questions, it is no crystal ball. Even with the best computer simulations, the future remains essentially unpredictable. All models are simplifications of reality and therefore inevitably wrong (although another saying points out "but some are useful"). Some forecasts will inevitably be correct, whether based on astrology or on computer simulation. Some of those based on computer simulation may even be correct for the right reasons—they will reflect an accurate causal understanding of the way the world works.

Understanding of that unpredictability is, however, the reason that those of us who develop or use IFs never refer to the alternative scenarios we develop as predictions. We refer to them as alternative scenarios, forecasts, or projections. And it is also why we think of computer models as tools to help us think about and hopefully somewhat improve our own mental models.

There is no alternative to choice and action, even if the choice is inaction. Let's enhance our mental models and explore and shape our desired future as best we can.

References

Hughes, B. B. (2025). Analysis of integrated global SDG pursuit: Challenges and progress. *Sustainability, 17*(15), 6672. https://doi.org/10.3390/su17156672

Intergovernmental Panel on Climate Change (IPCC). (2023). Summary for policymakers. In H. Lee & J. Romero (Eds.), *Climate change 2023: Synthesis report.* Contribution of working groups I, II and III to the sixth assessment report of the Intergovernmental Panel on Climate Change (pp. 1–34). IPCC. https://www.ipcc.ch/report/ar6/syr/downloads/report/IPCC_AR6_SYR_FullVolume.pdf

Kant, I. (1795). *Perpetual peace: A philosophical sketch.*

Meadows, D. H., Meadows, D. L., Randers, J., & Behrens, W. K., III. (1972). *Limits to growth.* Universe Books.

O'Neill, B. C., Kriegler, E., Ebi, K. L., Kemp-Benedict, E., Riahi, K., Rothman, D. S., van Ruijven, B. J., van Vuuren, D. P., Birkmann, J., Kok, K., Levy, M., & Solecki, W. (2017). The roads ahead: Narratives for shared socioeconomic pathways describing world futures in the 21st century. *Global Environmental Change, 42*, 169–180. https://doi.org/10.1016/j.gloenvcha.2015.01.004

Rawls, J. (1971). *A theory of justice.* Harvard University Press.

Sen, A. (2000). *Development as freedom.* Anchor Books.

Shirley, R. W. (1984). *The mapping of the world: Early printed world maps 1472–1700.* Holland Press.

Soergel, B., Rauner, S., Daioglou, V., Weindl, I., Mastrucci, A., Carrer, F., Kikstra, J., Ambrósio, G., Aguiar, A. P. D., Baumstark, L., Bodirsky, B. L., Bos, A., Dietrich, J. P., Dirnaichner, A., Doelman, J. C., Hasse, R., Hernandez, A., Hoppe, J., Humpenöder, F., Kriegler, E., et al. (2024). Multiple pathways towards sustainable development goals and climate targets. *Environmental Research Letters, 19*(12), Article 124009. https://doi.org/10.1088/1748-9326/ad80af

United Nations. (1987). *Our common future: Report of the World Commission on Environment and Development.* Oxford University Press.

United Nations Development Programme (UNDP). (2001). *Human development report 2001: Making new technologies work for human development*. Oxford University Press.

United Nations Development Programme (UNDP). (2002). *Human development report 2002.* Oxford University Press.

Watkins, J. E., Jr. (1990). What may happen in the next hundred years. In E. Cornish (Ed.), *The 1990s and beyond* (pp. 150–155). World Future Society.

Global Change

2

Abstract

A critically important foundation for exploring possible and desired futures is analysis of where global change appears to be taking us. This chapter lays that foundation. It begins by identifying multiple shapes of trends often found in quantitative analyses. It then looks for such patterns of change in long historical data series across issue areas that subsequent chapters will explore in detail, including with attention to their causal drivers and their interaction. The issue areas surveyed here include demographics, economics, and various aspects of human development (poverty reduction, healthy life expectancy extension, increased education, and access to safe water and sanitation). Important change patterns also appear in agriculture that affect access to food and nutrition, in energy that shape its accessibility, and in environmental arenas that often interact closely with agricultural, energy system, and technological evolution. The chapter similarly identifies the character of trends in environmental systems, including the extent of forests, usage of water, and atmospheric accumulation of greenhouse gases. Not least, it identifies some patterns of change in domestic and international sociopolitical structures and character. In combination, these trends lay that foundation for exploring where the complex evolution of global systems appears to be taking us.

Clay Morgan, former Pardee Institute Fellow, is co-author of this chapter.

B. B. Hughes, *Exploring and Shaping International Futures*, World-Systems Evolution and Global Futures, https://doi.org/10.1007/978-3-032-22084-4_2

2.1 Introduction

Where does global change appear to be taking us? Chapter 1 argued that addressing this question is fundamental to exploring what leverage we have in shaping our future. This chapter reviews historical patterns as one foundation of understanding before we turn to forecasting.

At what specific aspects of change should we look? We begin with demographics, because it is fundamental to know how many of us there are and how fast our numbers are growing. Just as the demographic system aggregates us as individuals, our economic system aggregates us as producers and consumers, so we turn to that system next and ask how it might be changing. Then we look at patterns of human development in areas such as poverty reduction, health improvement, and educational advance.

We move next to a look at some areas of change that support human development, namely agriculture and energy. Is the food supply keeping up with our growth in numbers and, in fact, improving nutrition? Energy is a "master resource," and with sufficient energy we could grow food on the top of Mount Everest and extract all necessary raw materials from sea water.

Yet agriculture and energy system changes greatly affect environmental sustainability. How great is that impact and how is it changing in areas such as deforestation, water usage, and greenhouse gas emissions? Technological change is critical also; it not only greatly influences our patterns of resource extraction and waste creation but also our interaction with each other. Are there recognizable patterns of technological innovation?

Finally, our domestic and international political and social structures stand at the peak of human and sustainable development efforts (Mesarovic & Pestel, 1974: 29). What sociopolitical trends and transformations can we recognize?

2.2 Types of Change

Before examining specific trends, however, it is important to understand several common patterns of change, most of which Fig. 2.1 represents.

The first is linear (not shown in Fig. 2.1). A linear growth process adds or subtracts equal increments year after year. The Brazilian Amazon rainforest in 2025 was about 3.3 million square kilometers (down from 4.2 million before 1970 (Wikipedia). If loggers, ranchers, and homesteaders in the Brazilian Amazonian rainforest were to clear 20,000 square kilometers each year (about the average rate early in the century), the forest size would decline linearly and it would completely disappear in about 150 years (fortunately the rate of clearance has somewhat declined). Much analysis of historical patterns and future prospects looks for linear patterns. In reality, change is seldom linear.

Many processes grow **exponentially**. Investors often rely on exponential growth. If you put $1,000 (a stock value in our analytical terms) into a savings account earning 5.0% annual interest and leave the account undisturbed, your

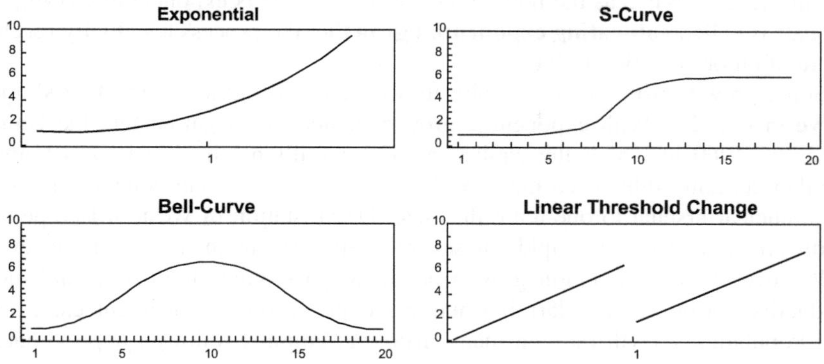

Fig. 2.1 Types of growth. *Source* Author

money will grow exponentially, earning $50.00 the first year and $52.50 the second year (flows added to the stock value). At the end of 14 years the balance will have grown to $1979.90. An exponential growth process of this kind produces an upward-sloping curve (see Fig. 2.1). If the global economy were to grow at 3.0% a year through the twenty-first century, roughly the average rate of the twentieth century, its size would exhibit the same type of upward-sloping growth curve reaching over 1,000 trillion dollars in 2021$ by century end.

In the example of a bank account growing at 5.0%, an initial investment nearly doubled in 14 years. In fact, we can characterize exponential growth processes by their **doubling time**. The **Rule of 72** provides an easy way of estimating that doubling time. Dividing 72 by a percentage growth rate provides a good estimate of the number of periods required for doubling. An economy growing at 3.0% will double in about 24 years.

Sometimes the growth rate of a process accelerates. Annual global economic growth rates quite steadily accelerated after 1700 when, on a per capita basis, growth was barely over 0.0% per year. A process growing at an accelerating percentage rate is **super-exponential**.

Both linear and exponential growth implicitly assume that a process goes on indefinitely. We know, however, that there are often limits that constrain growth processes. The rate of Amazonian deforestation has slowed as the remaining forest area declined, and the pressures to protect it increased. Sometimes such limits cause an abrupt cessation of growth.

Very often, growth processes do not simply reach a limit and then abruptly cease. Instead, they often follow one of two patterns with respect to limits. The first is a pattern of exponential (or even super-exponential) growth up to a turning point, and then a pattern of slowing growth up to a limit, producing an S-curve. The growth of most organisms traces such a curve. When a sapling is very small, it grows only a few inches in a year. As it grows, it adds a greater increment each year throughout its early life. Eventually, however, that growth increment declines and then ceases. This growth pattern traces an **S-curve** like that in Fig. 2.1. Whereas the bottom half of the S-curve is exponential growth, the top half is called **saturating exponential growth**—the process eventually reaches a saturation or cessation point.

Some growth processes eventually turn negative as shown in the **bell-shaped curve** in Fig. 2.1. With exponential growth of nuclear megatonnage, the United States by 1960 had built the equivalent of 1.4 million Hiroshima-sized bombs, and that certainly influenced the U.S. decision to negotiate subsequent reductions of its nuclear arsenal to one-tenth that size. Deforestation in Western Europe and North America was very rapid for several centuries but then slowed and eventually reversed as population growth declined, governance improved, and more productive forms of both farming and wood production/replacement came into use. Population growth rates in many historical empires followed approximately an S-curve pattern to a maximum value and then subsequently often tracked something like an inverted S-curve back down because of environmental destruction and warfare. There is no guarantee that such processes will be natural, smooth, and manageable. Jared Diamond (2004) pointed to the example of Easter Island where a society destroyed itself by outgrowing the ability of the island to provide food and energy—a process called **overshoot and collapse.**

The final basic pattern of growth shown in Fig. 2.1 involves **threshold change.** Global oil prices have often exhibited such behavior with quick jumps in the 1970s, substantial drops in the 1980s, and another substantial rise across the first decade of this century before declining again. Interstate war is another threshold event. Because of its irregularity, threshold change is the most difficult type to forecast, and with the International Futures (IFs) system we do very little of it.

These growth and decline patterns (linear, exponential, S-shaped, bell-shaped, and threshold change) can combine in complex ways. For instance, the repetition of bell-shaped growth over time can create a cyclical pattern (consider the repeated growth and collapse of empires). Or a cyclical pattern might combine with an underlying pattern of long-term linear or exponential growth. For instance, the U.S. economy has exhibited exponential long-term growth, but roughly four-year-long business cycles have often snaked around that underlying trend.

The complexity of possible patterns is, of course, what makes forecasting with **extrapolation** (projecting a growth pattern beyond the period for which we have data) dangerous. As obvious as it is, the most common error in forecasting is simple extrapolation of linear or exponential growth without considering the possibility of changes in the pattern.

For instance, Chap. 1 noted the forecasting flaw of the individual falling from a building and declaring half-way down "so far, so good." Consider a similar story about exponential growth. An airplane with four engines departs Hawaii for a 5-h trip to Los Angeles. When the plane loses power in one engine the captain announces that there is no problem, but the flight will last 6 h. When it loses another engine, the captain extends the forecast of flight time to 8 h. After the third engine burns, a more nervous captain reassures passengers of the crew's continuing ability to fly and to land but forecasts a total trip duration of 11 h. One passenger turns to another and says, "You know, if that fourth engine goes out, we could be up here all night!" Yes, again sick. But illustrative.

Many people lost fortunes in stock markets after 1929 by investing in the belief that the market was following a pattern of exponential growth, only to discover that they had been at the top of those exponential curves and about to undergo a threshold change. Mark Twain poked fun at such extrapolation:

> In the space of 176 years the Lower Mississippi has shortened itself 242 miles. That is an average of a trifle over one mile and a third per year. Therefore, any calm person, who is not blind or idiotic, can see that in the old oölitic Silurian period, just a million years ago next November, the Lower Mississippi River was upward of one million three hundred thousand miles long, and stuck out over the Gulf of Mexico like a fishing-rod. And by the same token any person can see that 742 years from now the Lower Mississippi will be only a mile and three-quarters long, and Cairo and New Orleans will have joined their streets together, and be plodding comfortably along under a single mayor and a mutual board of aldermen. There is something fascinating about science. One gets such wholesale returns of conjecture out of such a trifling investment in fact. (Buchanan, 1974: 17)

How are we to know when extrapolation is useful and when it is ludicrous or dangerous? Linear extrapolation of the percentage of its economy that the U.S. has spent on health care (about 5.0% in 1960, 9.0% in 1980, and 13.0% in 2000) would have led to forecasts of 17% in 2020 and 21% in 2040. But the rates have roughly stabilized in the range of 17–18 percent since about 2009, except for a one-year surge to nearly 20% in the COVID-19 pandemic year of 2020. Longer time horizons, and the fact that health care cannot possibly take 101% of GDP, made continued linear expectations (quite often held by many) unreasonable.

Forecast risk obviously increases with length of time horizon; it also generally increases with the rate at which a process is growing or declining; and it often decreases somewhat with the length of the base period from which we are forecasting. The best rule of all is to use common sense and to combine extrapolation with at least a rudimentary consideration of the **causal dynamics** of the system under consideration. Doing so can help identify the existence of limits on a process and can even help in spotting likely turning points. We will try to keep these rules in mind as we turn to historical data and forecasts as a consistent aid to our thinking about long-term forecasting in the following chapters. As you read about historical patterns, very often trends, ask yourself how you think future years may look.

2.3 Demographic Change

Global population grew exponentially, in fact super-exponentially, over a very long period. Humanity populated most of the globe by 8000 B.C.E. and numbered about 5 million (Ehrlich & Ehrlich, 1972: 12). By 1000 C.E., population had reached approximately 250 million, and it doubled to a level of 500 million in roughly 1600 C.E. The growth rate over that nearly 10,000-year period was less than 0.05%

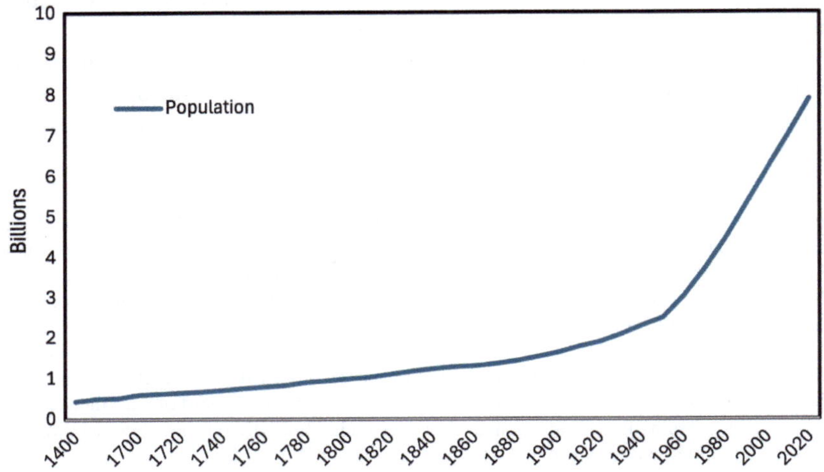

Fig. 2.2 Global population since 1400. *Source* Data from Our World in Data. https://archive.ourworldindata.org/20250825-064920/grapher/population.html

annually. By the end of the eighteenth century, however, population growth rate had accelerated dramatically (to about 0.5%) and in the nineteenth and twentieth centuries the rate continued to rise (see Fig. 2.2).

Simple exponential extrapolations of world population growth made in the mid-1960s created considerable anxiety. Were global population to continue growing at the 2.1% annual rate near the end of that decade, it would double from 3 to 6 billion people by the year 1995, rise to 12 billion by 2030, 24 billion in 2065, and reach 48 billion in about 2100 (the Rule of 72, explained earlier, makes such estimates easy).

The level grew to 6 billion in 1999, not far off such a forecast. Since then, however, population growth has slowed considerably, growing to just over 8.2 billion in 2025 and expected to reach "only" 8.6 billion in 2030 rather than 12 billion. Much evidence supports the current belief by many forecasters that population growth will be exhibiting the pattern of an S-shaped curve rather than continuing in super-exponential form. In fact, we passed the turning point of that S-shaped curve in the late 1960s. By 2025 the global growth rate fell to 0.9%, close to the rates in the 1930s. See Table 2.1, which also shows annual rates by continent in 2025 (note how the growth in Africa stands out, creating issues around forecasting its future growth). Thus, instead of a steady climb to 24 billion and even higher levels, global population will quite possibly not exceed 10 billion people this century. It may well begin to fall before the century's end, an issue we explore in Chap. 6.

Population forecasts, at least over a period of 20–30 years, tend to be more accurate than those in the other issue areas that we will review. Most of the humans

Table 2.1 Global population

	Population (millions)							% Rate
	1750	1850	1950	2000	2025	2050	2100	2025
Africa	106	111	224	812	1558	2595	3860	2.4
Asia	498	801	1369	3676	4837	5217	4326	0.6
Europe	125	208	368	728	752	708	573	−0.1
Former USSR	42	76	180	290	306	311	279	0.6
Latin America	16	38	167	523	666	729	607	0.7
North America	2	26	172	315	524	578	581	0.6
Oceania	2	2	13	31	46	56	66	0.1
World	791	1262	2520	6085	8251	9738	9892	0.9

Sources 1750–1950 from The American Philosophical Society (1967): 137. 1950–2000 from the United Nations Statistics Division, July 2005. 2025–2100 from IFs 8.54

who will have children in the next 20 years have already been born, and fertility and mortality generally change slowly. Nonetheless, there remain uncertainties about longer-term population forecasts, as suggested by the often poor forecasts from the 1960s and 1970s. Some have argued, for instance, that the human population has reached or exceeded the ultimate limits imposed by its food and energy supplies (Brown 1981; Donnella Meadows et al., 1972; Malthus 1798) or its ability to protect against disease. If so, population growth might unfortunately fall for mortality rather than fertility reasons. Such debates correctly take us beyond extrapolation and into causal analysis. We turn to such analysis in later chapters.

2.4 Economic Change

Despite periodic economic downturns (recessions and depressions), global economic growth has been extremely rapid and, like global population prior to 1970, accelerating over the last 200 years. Living standards have improved enormously. The Industrial Revolution began in England, spread elsewhere in Europe, and then moved to the United States, Japan, and more recently to economies in Asia and Latin America. In the process, the world experienced a "great divergence," generally dividing into rich countries and poor ones. The early industrial economies developed primarily in the Northern half of the global land mass, while most of the lower-income countries are in the Southern half. Figure 2.3 traces in constant prices the growth of gross domestic product, the total production of goods and services—Gross Domestic Product or GDP—from 1965 to 2025. It uses the Organisation for Economic Cooperation and Development (OECD) member countries as the proxy for higher-income/developed countries.

Figure 2.3 shows that the absolute gap in total GDP between developing and developed countries shrank rapidly in the last half of the twentieth century, with

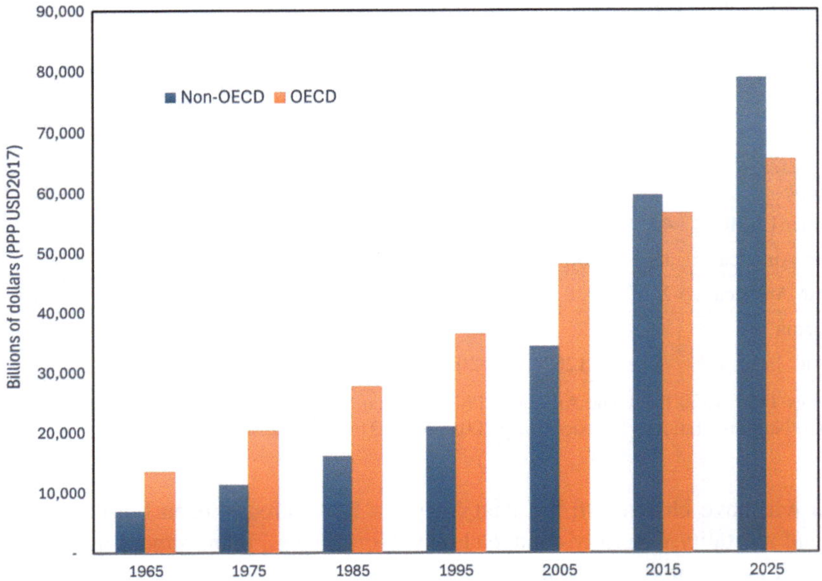

Fig. 2.3 GDP in developing and developed countries. *Sources* Data from World Development Indicators for 1960 to 2015 and the International Monetary Fund for 2025. OECD member countries used as proxy for developed countries

the developing world surpassing the developed world in total economic size by 2015.

Yet, the non-OECD countries contain 6 times as many people as the OECD countries and their population growth rate has been more than twice as fast. Thus, the gap in per capita GDP is huge. Figure 2.4 sketches the trend in GDP per capita in the two sets of countries since 1965 and its ratio. (The data shown in Figs. 2.3 and 2.4 are in purchasing power parity dollars rather than exchange rate dollars, a distinction Chap. 7 elaborates.)

Despite Fig. 2.4 showing continuation of the centuries-long North–South divergence in GDP per capita, the story around the ratio of GDPs per capita is different. In about 1995, the upward trend that long characterized that ratio also began to reverse. The ratio is now lower than it was in 1965 when many countries in the Global South were only a few years removed from being colonies. We appear to have moved into a "great convergence", at least across countries (Milanović, 2023).

This hopeful trend of narrowing the ratio in GDP per capita is, however, not a uniform pattern across the Global South, much less within countries. Economic performance varies enormously among the developing countries. Many countries in East and South Asia are doing very well, while much of sub-Saharan Africa is not. Related to this, world poverty levels remain abysmally high, a topic taken up later in the next section.

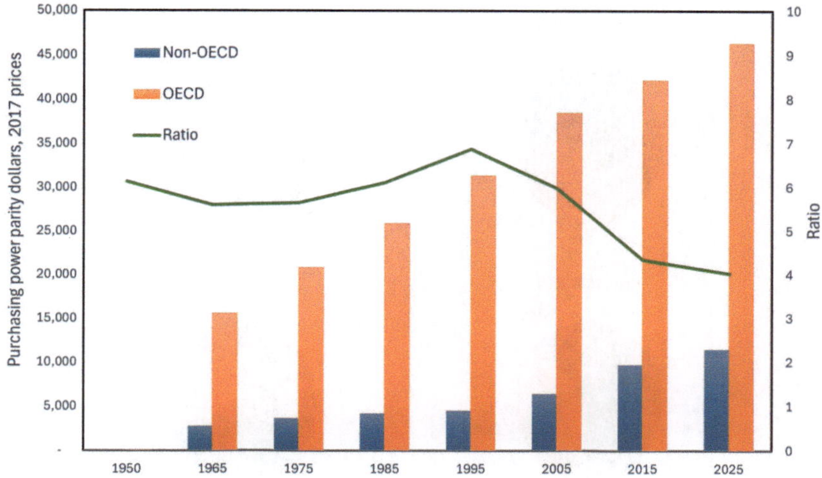

Fig. 2.4 GDP per capita in developing and developed countries. *Sources* Data from World Development Indicators for 1960 to 2015 and the International Monetary Fund for 2025. OECD member countries used as proxy for developed countries

As Figs. 2.3 and 2.4 have begun to do with respect to development levels, forecasting with IFs will supplement such aggregated views of the global economy with two types of disaggregation: by geography and by economic sector. Prior to industrialization, the agricultural sector of an economy dominates it. At the time of the American Revolution, about 80% of the U.S. population was engaged in producing food. Over time (see Fig. 2.5), the industrial (Ind) sector became dominant and the agricultural (Ag) sector experienced a very long relative decline (not an absolute decline because it generally continued to increase its overall production even as it used less labor and became a smaller share of the total economy). The third large portion of an economy is the service (Ser) sector.

Figure 2.5 shows the relative size of these three sectors over time, again across non-OECD and OECD country populations. Note that, in what the world generally considers developing countries, the agricultural sector has continued to shrink compared to the industrial and service sector. The industrial sector has now begun to decline not just globally but also as a share of GDP across the developing world, even as those countries increase their share of global industrial production. And even in what the UN considers the set of countries with the lowest contemporary human development level, the service sector has become more than 60% of the total economy.

Still further, the nature of service economies is changing in rich and poor countries. Ideally in looking at historical patterns we would explicitly identify a fourth sector of the economy and category of workers, much of which most data sources now misleadingly lump with the service sector. That fourth category is the information sector. Despite the importance of the information economy, analysts do not

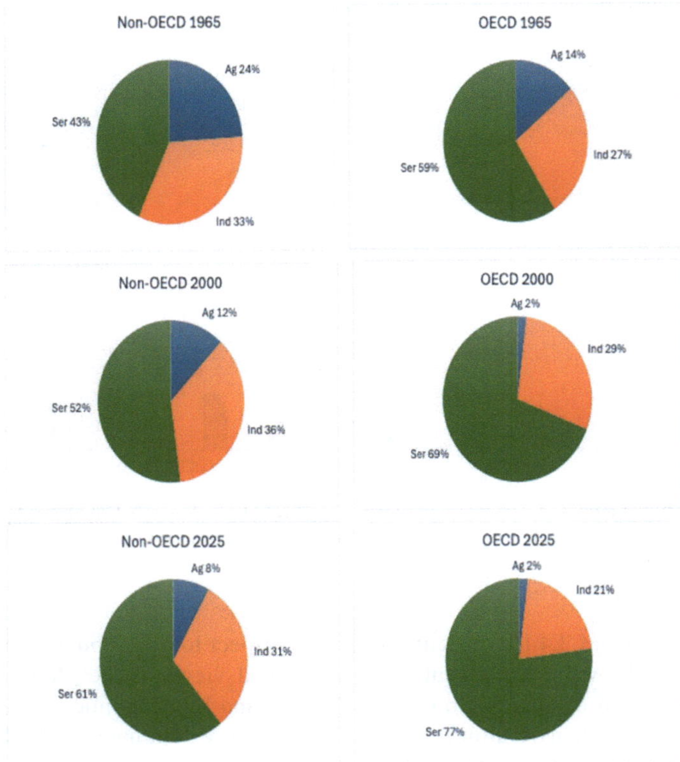

Fig. 2.5 Sectoral distribution of GDP by development level. *Source* Data from World Development Indicators. OECD member countries used as proxy for developed countries

yet confidently evaluate patterns in its historical size. Not only those who work in digital communications and data management, but teachers, scientists, and librarians could be placed in the information sector. Many other individuals (for instance, those in finance or insurance jobs) also prove difficult to classify (OECD Digital Economy Outlook, 2024). A sector as large and complicated as services really should be broken into at least 2–3 sectors, but often is not in looking at historical trends. We will do so in our forecasting.

There are many uncertainties in forecasting the global economy. Will the less industrialized countries of the world automatically follow the path of the first to industrialize and continue to move to services? To what degree has the transformation of many countries that industrialized early into service and information economies made the character of what workers do so different from traditional agricultural and industrial work that we can no longer adequately measure economic size, rate of growth, or productivity of workers? Such questions will challenge us later as we move from this initial historical exploration into forecasting and causal analysis.

2.5 Human Development Change

In close interaction with the demographic and economic patterns discussed above, the human condition has dramatically changed over the last few centuries and especially over the most recent decades. Average life expectancy has increased, a smaller percentage of people go hungry, more children survive until adulthood and more of those children will have access to clean water. Additionally, literacy has become almost universal among young people, people receive more years of education, and girls are increasingly educated at the same levels as their male peers or even beyond. A higher portion of the world lives in democracies than in earlier eras, and more of those democracies are led by women than would have been thought possible one hundred years ago when many women were first allowed to vote. There is more access to clean water, to electricity, and, thanks to the internet revolution, to information than at any point in human history. None of this is to say, however, that challenges, familiar and novel, do not remain. This section will explore the ways that the human condition has developed, where human development has lagged, and some of the challenges that remain.

2.5.1 Poverty

Poverty reduction is a core element of more general improvement in the human condition. It is no surprise that extreme poverty eradication is the first of the Sustainable Development Goals (SDGs).

Led by World Bank analysis, $1.90/day level was the global definition of extreme poverty in 2011 prices. With a focus on low-income countries it moved to $2.15 in 2017 prices and $3.00 in 2021 prices. That level of income (or more appropriately that level of daily consumption, even more difficult to measure) is judged as the absolute minimum to meet basic material needs of food, water, shelter, clothing, and so forth. In 2017 the World Bank began identifying higher poverty lines for lower-middle-income and upper-middle-income countries ($4.20 and $8.30 in 2021 prices).

We have seen drastic and impressive changes in poverty reduction over the last two centuries. Due to inflation, changing life styles, new technologies, and shifts in the global economy, it is, however, hard to meaningfully compare poverty levels over such a long period. For example, it is difficult to confidently say that someone earning $1.90/day in 1820 is in a similar situation to someone earning $1.90/day today, even with adjustments for inflation.

Nonetheless, focusing on purchasing power, analysts have estimated the percentages of global population living on such amounts over time, as well as additional percentages living at or below higher amounts (see Fig. 2.6). Today, less than 10% of the global population lives on less than $1.90/day (2011 prices), down from nearly 80% in 1820, with especially rapid progress since about the 1980s. This reduction is a remarkable human success story in spite of the need to continue such reduction across all of the lower income brackets of Fig. 2.6.

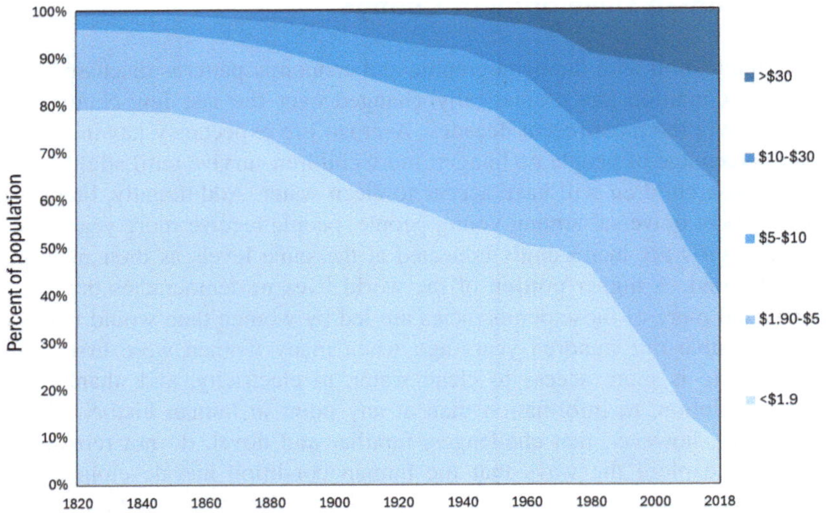

Fig. 2.6 Percentage of global population living on various daily incomes. *Note* Data are measured in international dollars at 2011 prices. *Source* Data from Our World in Data, using Key Charts on Poverty based on the Maddison Project Database and Moatsos (2020)

The global population has obviously grown dramatically over those two centuries. Fig. 2.7 helps us understand the patterns for the absolute numbers of those living in income brackets that are not identical, but similar to those of Fig. 2.6. It shows the relative stability over time of those living on less than $2 per day, in spite of the total population growth from about 1 billion to about 8 billion (reached in 2022). The growth in numbers of those living on $20 or more has accelerated in this century, as the numbers living on less than that have begun to decline.

Another important part of the story around global poverty has been the dramatic change in the global distribution of those living under the extreme global poverty line of $3.00 in 2017 prices. (Yes, the multiple and changing definitions of poverty over time and in our figures are confusing.) Fig. 2.8 shows the dramatic success in reducing the number of people in poverty since the 1990s in East Asia and since the 2000s in South Asia, demographically dominated respectively by China and India.

Overall, both the proportion and the number of the world's population living in extreme poverty have come down significantly since 1990. Yet, the absolute number of humans living in such poverty is still very high, approximately 750 million people.

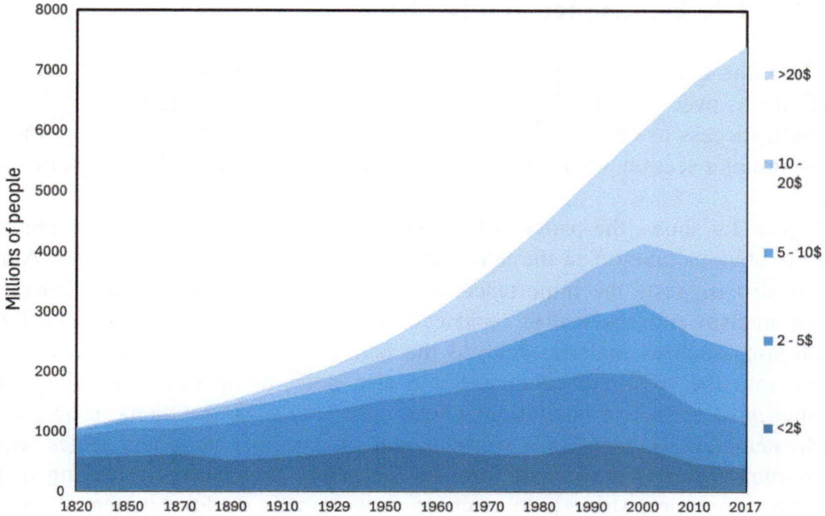

Fig. 2.7 Numbers of people globally living on various daily incomes. *Source* Data from Our World in Data, based on Lahoti et al. (2014) and the Global Consumption and Income Project, Maddison Project Database (Maddison, 2001), and van Zanden et al. (2014)

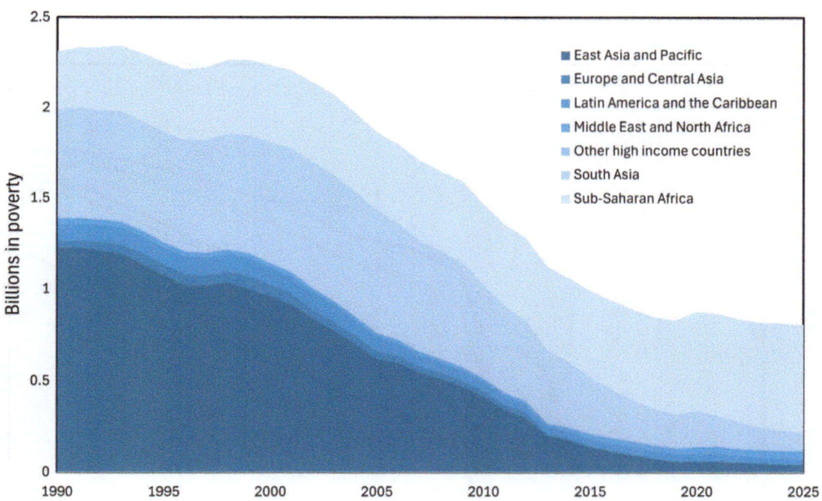

Fig. 2.8 Total population living in poverty by world region. *Note* Extreme poverty is defined as income below the International Poverty Line of $3 per day at PPP, expressed in international dollars at 2021 prices. *Source* Data from Our World in Data, using Key Charts on Poverty of the World Bank's Poverty and Inequality Platform for 2025, Hasell and Arriagada (2022) and Hasell et al. (2022)

2.5.2 Life Expectancy and Health

In 1900, the life expectancy of a child born in our world averaged 32 years. In 2025, it was over 74 years. That incredible extension is explained by many factors, including access to clean water and sanitation, to nutrition (to which we return in the next major section, focused on agriculture), and to ongoing advances in medical science.

Figure 2.9 shows the pattern of advance in life expectancy over 2.5 centuries and the obvious take-off in the late 1800s, starting with Europe and North America. It also suggests the importance of sociopolitical factors. The long histories of colonialism and absence of rapid economic advance, but also the prevalence of diseases such as malaria, delayed the take-off in rapid advance in Asia and Africa into the twentieth century. Some also point to institutional failings and the legacy of extractive institutions (Acemoglu et al., 2003). Others, prominently Wallerstein (2004), point to a world model of core and periphery states where economic surplus is drawn out from the periphery into the core, leading to the periphery countries' struggle to catch up economically or developmentally to the core nations.

Turning to specific examples, Mao's "Great Leap Forward" efforts in the 1950s caused a dip in life expectancy not only in China and Asia generally, but even in the global pattern of advance. And the impact of the COVID-19 pandemic in

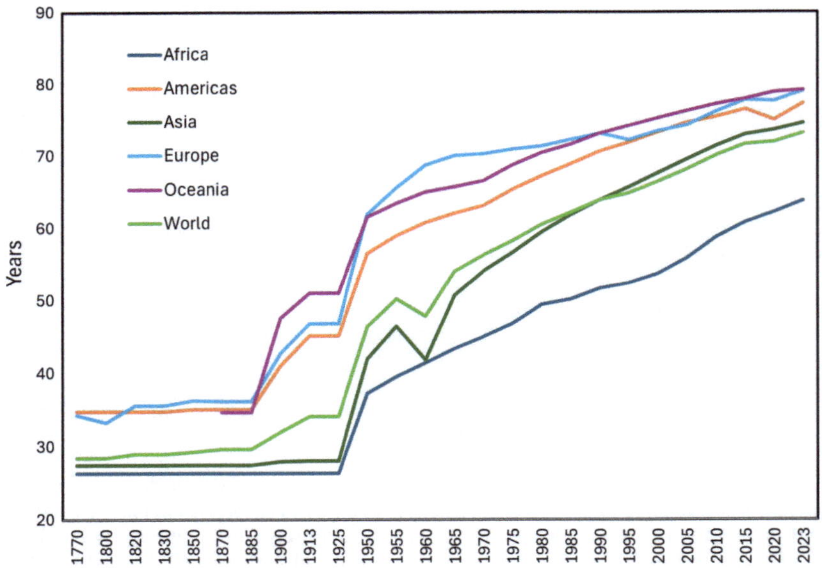

Fig. 2.9 Life expectancy by world region. *Source* Data from Our World in Data, based on Human Mortality Database, Dattani et al. (2023), Riley (2005), UN World Population Prospects, and Zijdeman and da Silva (2014)

the early 2020s is obvious. (The data points are inadequately detailed to show the costs of World Wars I and II, and of the Spanish flu during and after the first of those, but all had an impact.)

There is clearly further work to be done in improving life expectancy, and thus overall life quality, for people in Africa in particular. A very encouraging aspect of the patterns in Fig. 2.9 is the parallel increase in life expectancy around the world since the take-off in Africa. That suggests at least the possibility that, if advances in the high-income countries greatly slow as many expect, there could be a future pattern of global convergence at high levels. It is not impossible, of course, that medical advances could even accelerate globally (Hughes et al., 2015).

2.5.3 Education and Literacy for Women and Men

Strong links connect even basic primary school attainment by young women to lowered child mortality rates for their children. Successful democracies build on educated populations. Education thus improves the quality of life for families, communities, and countries.

It is estimated that in 1820 only 17% of adults had ever attended school and just 12 percent were literate (van Zanden et al., 2014). Education was the exclusive realm of the wealthy and aristocratic. In the twentieth century, a big push for education spread around much of the world and into Africa, the Middle East, and South Asia with decolonization after World War II. Although it takes much time to develop school systems and for societies to have basic levels of educational attainment across the entire age structure, nonetheless by 2020 global literacy had climbed to 82 percent, a remarkable achievement.

Figure 2.10 shows the portion of populations aged 15 or older who have completed primary education. Using a slightly longer data series, that rose from 41 percent globally in 1950 to 81 percent in 2025, but in low-income countries has reached only 46 percent. The better news on primary education is that the completion rate for current primary-age populations even in low-income countries reached 76 percent in 2025 and in lower-middle-income countries reached 93 percent. The push in recent decades for globally-universal primary education is not finished, but tremendous progress has been made.

Literacy rates are quite closely tied to the completion of primary education. For example, it is estimated that in 1970 approximately 55 percent of the world's population was literate, very close to the portion of the adult population shown in Fig. 2.10 to have completed primary education. The same is true of the approximately 80 percent of the global population in 2025 that completed primary education and the portion of the world that was literate. With the SDGs, the global community has now extended the target for universal education to include the secondary level.

Moreover, the focus has broadened in the SDGs to give more attention to gender equity in education than has been true historically. In developing countries,

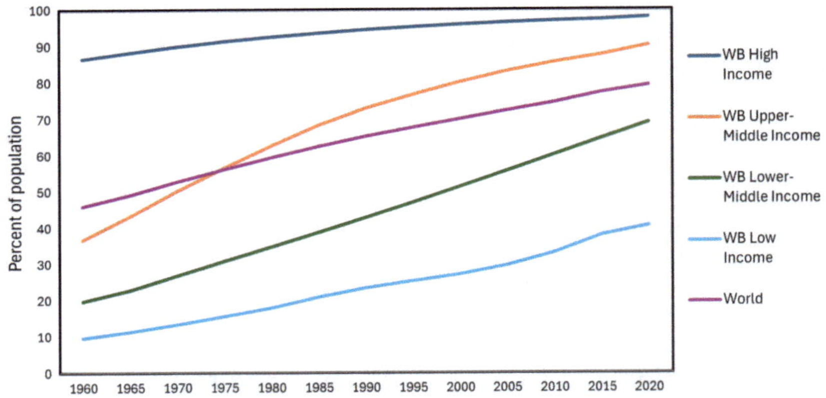

Fig. 2.10 Percentage of adult population having completed primary education. *Note* Countries grouped by World Bank income categories. *Source* Data from Wittgenstein Centre, Vienna, Austria

poor families have historically identified their sons as the most promising potential bread winners and therefore have used financial resources for their education. Also, due to traditional and patriarchal gender roles, daughters have more often been expected to help in child rearing and home making rather than to obtain education. Schools may not even have had proper restrooms for girls.

Still, Fig. 2.11 shows a great deal of progress with respect to gender equality in education and therefore in literacy. Global primary education completion and female literacy rates are converging with male literacy rates—and promisingly they are converging toward 100 percent.

With universality of primary education in sight, the struggle increasingly is to reach universality of secondary education. Both men and women's secondary educational attainment is growing in the developed and developing worlds, promising further progress toward a world of greater gender equality.

Figure 2.12 shows that even in high-income countries, fewer than 1/3 of adults had completed secondary education as recently as 1960. The rate for completion by adult females was about 10 percentage points below that for males (not shown). By 2025, however, the gender gap had been erased.

Significant educational challenges remain even in high-income countries. As of 2025, only about 85 percent of school-age children in those countries were graduating from the upper secondary level and only about 70 percent of adults had completed it. Given that such rates are far above the roughly 25 percent secondary completion rates in non-OECD countries, the global push for educational advance is very far from complete. Chapter 8 turns our attention to the future of global education.

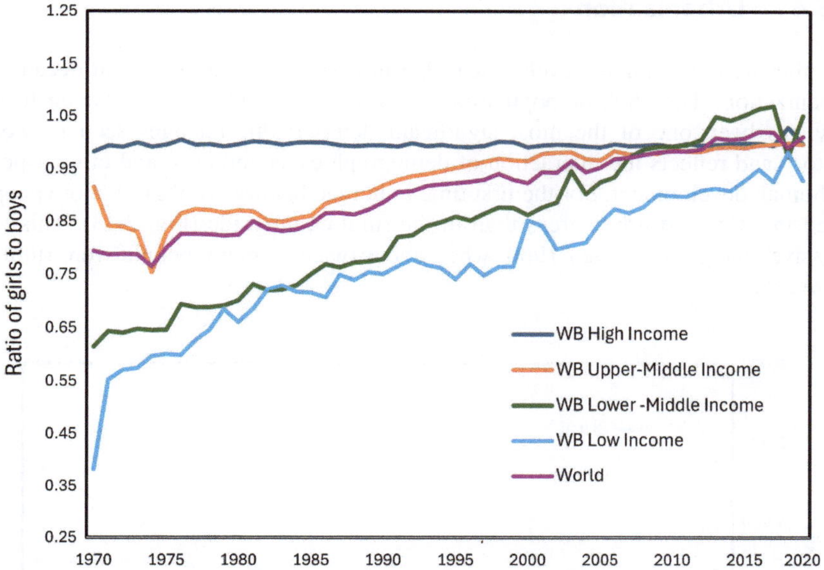

Fig. 2.11 Ratio of females to males in primary gross enrollment. *Source* Data from Wittgenstein Centre, Vienna, Austria

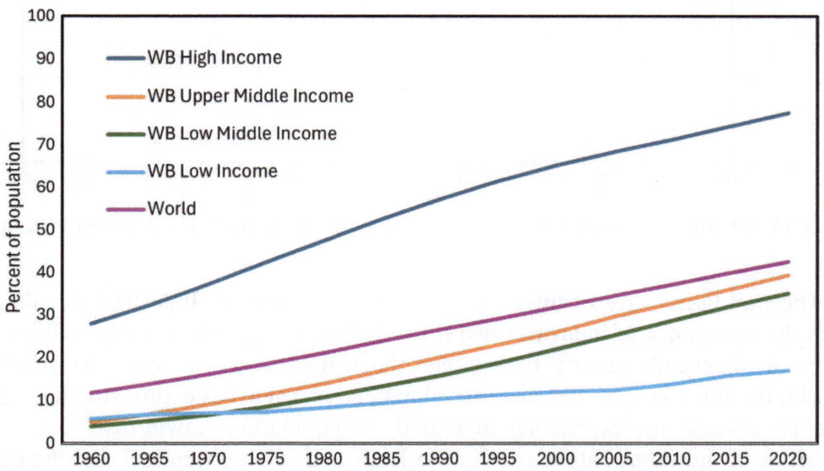

Fig. 2.12 Percentage of adults (males and females) having completed secondary education. *Source* Data from Wittgenstein Centre, Vienna, Austria

2.5.4 Urbanization

Another trend in human development that has been dramatic in recent decades is urbanization. The shift of populations from villages and the countryside to the city has been one of the most significant demographic changes seen in world history and reflects the interaction of demographics, economics, and other aspects of human development. For the first time in human history, as of 2008 more people lived in cities and urban areas than in the rural countryside (Fig. 2.13). This is a massive change from even 1960, when about twice as many people were rural as were urban.

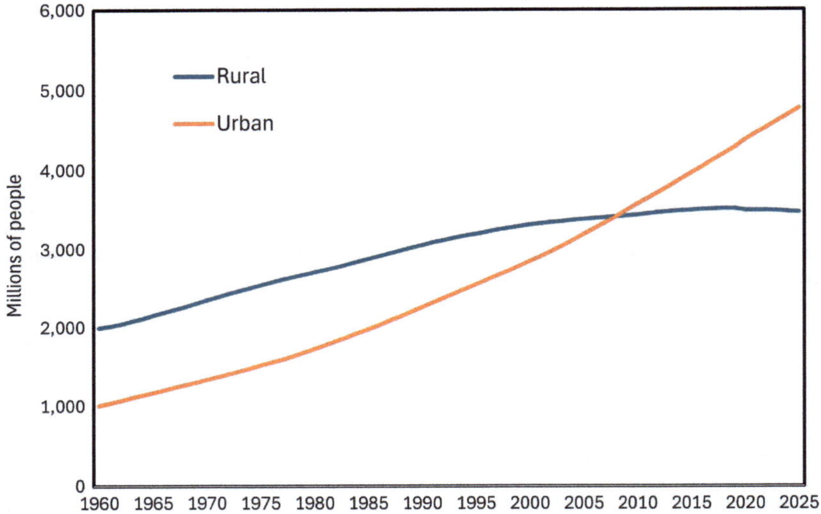

Fig. 2.13 Global rural and urban populations. *Source* Data from World Development Indicators

The first large wave of urbanization occurred during the Industrial Revolution with the emergence of industries and firms that brought people into the ballooning cities of nineteenth century Europe and the United States. Despite the grinding conditions for laborers, the underdevelopment of even basic infrastructure, and rampant disease, life expectancy increased and population growth rose.

Much of the urbanization seen in the latter half of the twentieth and the early twenty-first century has occurred in Asia and Africa. China, South East Asia, India, and sub-Saharan Africa have seen movements of people among the largest in human history. Their newly urban populations often face many of the human development issues that new urbanites experienced in the Industrial Revolution. There often is not the capacity to house or provide adequate physical infrastructure and social services for the influx of new people. Modern urban areas face new problems as well, such as urban air pollution. Outdoor air pollution may be responsible for up to 4 million deaths per year, with the majority of those exposed to bad air being the urban poor and working classes in the rapidly growing cities of the Global South (Rentschler & Leonova, 2023).

Nonetheless, these challenges of urban life are often outweighed by the opportunities presented by the dynamic economies and societies of urban areas. As we move into a world that is more and more urbanized, the range and dynamics of effects that it will have on the human condition remain uncertain.

2.6 Agriculture and Food Change

Repeated famines in Africa and regular expressions of concern about long-term food availability elsewhere might lead one to believe that world food production has not kept up with population growth. Figure 2.14 shows, however, an impressive record of per capita achievement, focusing on cereals production in the developing regions of the world. All developing regions except Africa have benefited significantly from such progress across the horizon shown. And although African per capita cereal production remained largely flat between 1965 and 2005, it rose 12 percent between 2005 and 2022.

Figure 2.15 explores the possibility that food production per capita is peaking by looking at the change over time in global production of three different food measures (rescaled so that they are easily compared on the same graph). The most basic measure is the global production of crops. On a global basis, crop production is the single largest portion of diets. Its growth has averaged about 2.3 percent per

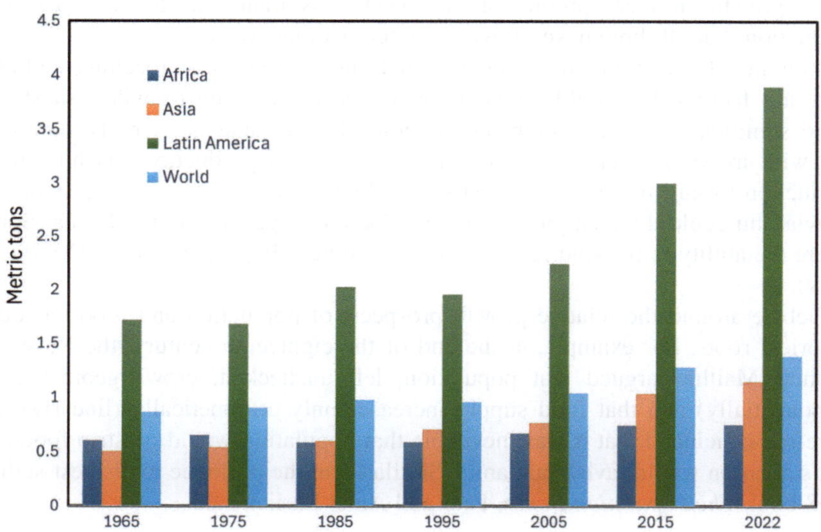

Fig. 2.14 Regional cereals production per capita. *Sources* Data from United Nations Food and Agriculture Organization and World Development Indicators

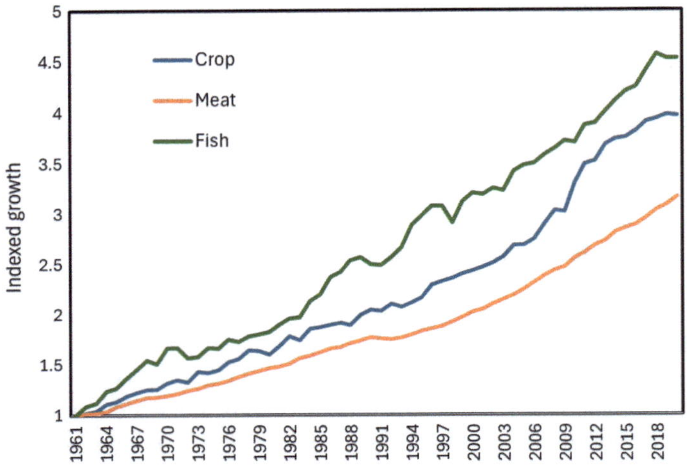

Fig. 2.15 Change over time in global food production by food type. *Note* Values are indexed to 1 in 1961. *Source* Data from the UN Food and Agriculture Organization

year since 1961. Growth in crop yields per capita has even accelerated in the first decade of the century. As always, there would be dangers in simple extrapolation and assuming progress will continue. A ceiling for agricultural production has, however, yet to reveal itself with any clarity.

Figure 2.15 also shows the growth in global meat and fish production. Since 1961, growth in meat production has been less than that of crop and fish production, but all three have grown in largely similar ways.

Although food production growth is not showing signs of reaching ceilings, there are factors that could arrest these trends. Overfishing of wild fish stocks could significantly reduce future production. Grazing land has largely stabilized and with irrigation there is often conversion to crop production, pushing meat production toward more costly livestock feeding. Novel viruses such as swine flu or avian flu could threaten global meat production, especially in the Global South where the ability to respond to such crises is limited (Espinosa et al., 2020; IPCC, 2019).

Debate around the relative growth prospects of population and food has deep historical roots. For example, at the end of the eighteenth century, the Reverend Thomas Malthus argued that population, left unchecked, grows geometrically (exponentially) and that food supply increases only arithmetically (linearly). He therefore concluded that it was inevitable that population would outstrip food and that starvation would ravish humanity. Similarly, in the prologue to his best-selling 1968 book *The Population Bomb*, Paul and Anna Ehrlich said

> The battle to feed all of humanity is over. In the 1970s and 1980s hundreds of millions of people will starve to death in spite of any crash programs embarked upon now. At this late date nothing can prevent a substantial increase in the world death rate. (1968: xi)

Yet despite the more localized famine deaths that did and continue to occur, mostly in conflict areas, the late twentieth century and early twenty-first century have been notable for huge improvements in health and longevity.

Most food analysts point to conflicts or to a shortage of income and therefore consumption power, not to a shortfall of absolute food production potential, as the culprit behind malnutrition. There are major differences in the food consumed by humans in different world regions. Figure 2.16 illustrates these differences by showing per capita caloric consumption for the world, for the non-OECD or developing countries, and for the countries of sub-Saharan Africa. Whereas citizens of the United States (not shown) consumed an average of 3,600 calories per capita daily already in the mid-1990s, calorie consumption in the poorest countries of the world ranged roughly between 2100 and 2600. In general, developing countries have been closing the gap with developed countries. The major exception is sub-Saharan Africa. Even there, consumption has climbed with some upward growth of production per capita and with the help of increased imports.

The differences in consumption across regions would be even greater were it not for the global trade of food. Substantial volumes of food, especially cereals, now flow from regions of surplus (primarily the Americas and Western Europe) to the rest of the world. Figure 2.17 shows that both sub-Saharan Africa and China have become more dependent on those flows, particularly in years of bad harvests. In the past, some analysts raised the possibility that China's growing income levels and imports might create difficulties for the world market and might take food from other regions such as Africa (Brown 1995). The reality has proven more complex than that. Counter to that prediction, Chinese food imports appear to have stabilized. Further, China and sub-Saharan Africa have both seen growth in their agricultural exports (not shown) as well as imports rise since the late 1990s.

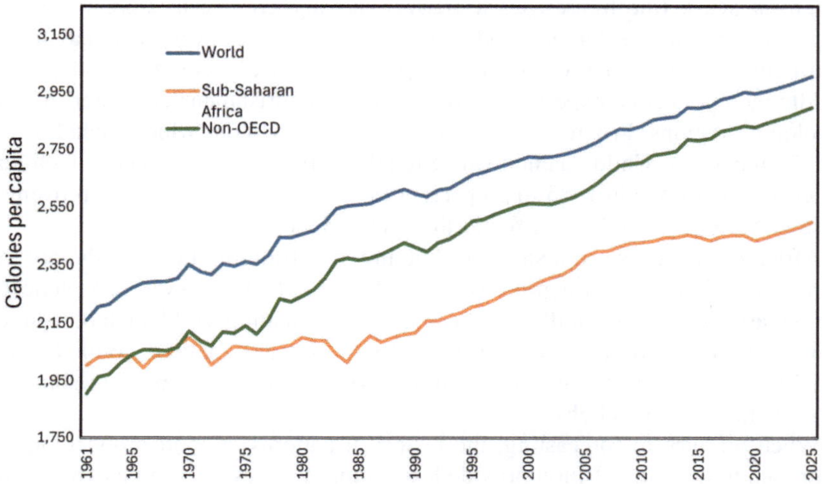

Fig. 2.16 Per capita daily calorie consumption. *Source* Data from World Development Indicators

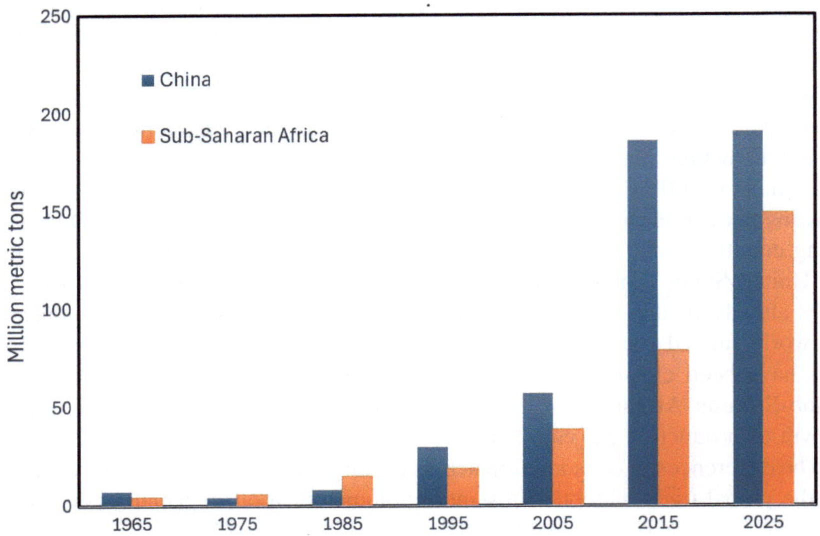

Fig. 2.17 Volume of crop imports of China and sub-Saharan Africa. *Source* Data from the UN Food and Agriculture Organization

Many developing countries have agricultural industries centered around the export of cash crops, such as cocoa and tobacco.

The bottom line with respect to food production, trade, and average calorie consumption is nutrition, which significantly reflects income and social support systems. Figure 2.18 shows the portion of children in Africa, South Asia, and Latin America that suffer from undernutrition's form of malnutrition as based on weight at age 5 (the figure uses a 10-year moving average to smooth reporting issues and reveal the larger trends). Even with its often very unequal income distributions, Latin America, with high per capita food production and rise into middle-income status, experiences low child malnourishment compared to other developing regions. Progress has been made in South Asia (which includes India and Bangladesh). Child malnourishment fell there from 50 percent of children under 5 years of age in 1985 to 28 percent in 2022. While this is an improvement, there is obviously *much* room for further improvement.

Africa has experienced a small decline in the portion of children who are malnourished. While this is a promising trend, it would be risky to conclude that sub-Saharan Africa is solidly on the path to eliminating child malnourishment. Many children who suffer malnutrition do so because of civil conflict, which will likely remain a barrier to the elimination of childhood malnourishment in sub-Saharan Africa and elsewhere.

When we turn to forecasting, the historical patterns of food production, consumption, trade, and nutrition provide less-tempting bases for simple extrapolation than did population size or even economic performance. In fact, most long-term

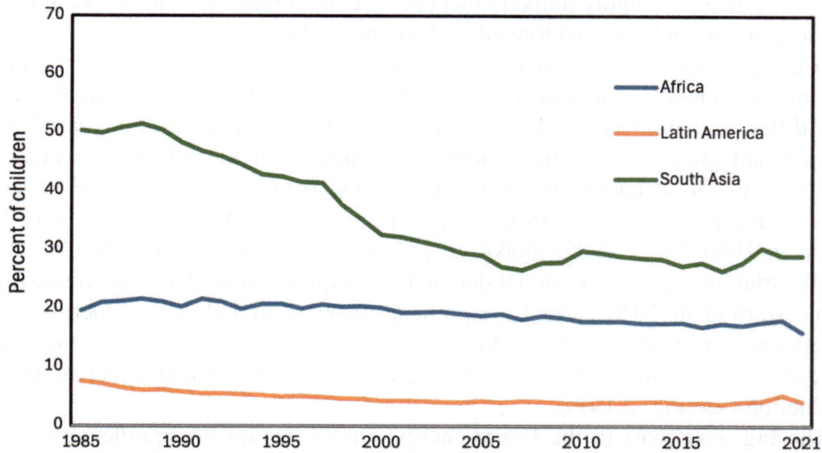

Fig. 2.18 Percent of children (under 5), malnourished by weight in selected regions. *Note* Showing a 10-year moving average to reduce fluctuation caused by reporting issues. *Sources*: Data from United Nations Food and Agriculture Organization and World Development Indicators

and even mid-range forecasters quickly move beyond the historical patterns of nutrition to causal analysis. Some are pessimistic. They point to the unavailability or shortage of land for the expansion of production relative to historical eras and to environmental degradation (for example, deforestation and desertification of land already under production). Some are optimistic. They direct our attention to new technologies for manipulating plant and animal life as well as to slowing population growth and rising incomes. We return to these issues of causal analysis in Chap. 9.

2.7 Energy Change

As food is energy for the human body, physical energy feeds the human economy. The amount of physical energy available to the average human has increased dramatically in the last 200 years. Watt's steam engine generated 40 horsepower in 1800, a dramatic advance over earlier models and over water mills and windmills. Very small modern car motors generate more.

Annual modern energy consumption per person in a low-income country today is little more than is available in one barrel of oil (about 42 gallons). Average consumption per person in a high-income country is about 25 times as much. Perhaps surprisingly, but related to increasing efficiency of use, the amount per person in high-income countries fell by about five percent between 1990 and 2025 while that per person in an upper-middle-income country rose by about a factor of three (but is still less than one-half of the value in high-income countries). Given these great global disparities and the different patterns of change in country groupings, it seems highly unlikely that purely extrapolative techniques would help us forecast demand changes toward midcentury and beyond.

Adding to both extrapolative and causal forecasting challenges is the almost certain movement of the world into a future with energy supply sources being very different than in recent decades. Annual global oil production increased very sharply and quite steadily from negligible amounts in the 1920s to a peak of nearly 22.6 billion barrels in 1979, then falling back significantly due to large price increases with the oil shock beginning in 1973 and the Iran-Iraq war that began in 1980. Since 1985, global oil production has grown quite steadily with small variations due to occasional demand and supply issues. The COVID-19 pandemic starting in 2020 caused a significant drop in global oil production from which it recovered only in 2023. While all these changes were occurring, the world quite steadily increased natural gas production and more irregularly increased coal production (see Fig. 2.19).

The big unknowns in the future energy supply picture lie significantly at the bottom of Fig. 2.19. Major growth is likely in the relatively smooth and historically slowly growing renewables sector, which includes solar, wind, hydroelectric, biomass, and geothermal energy production. This is due to the falling price of renewable energy production due to technological advance, rising prices of oil and other fossil fuels, and the threats associated with climate change (IPCC, 2023).

Yet, significant uncertainties remain even around fossil fuels. For many years it was believed that oil production futures would be related to ultimately recoverable oil resources as then understood. One at-the-time seemingly very successful long-term projection was made by M. King Hubbert in the 1950s for future U.S. oil production. Hubbert's bell-curve-shaped projection, known as Hubbert's Curve, predicted that U.S. oil production would peak in 1970 and slowly decline in the

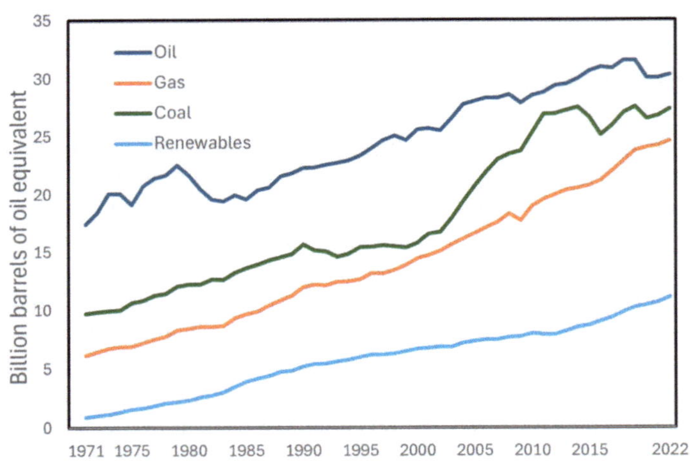

Fig. 2.19 Global energy production by type. *Sources* Data from the International Energy Agency and British Petroleum's Statistical Review of World Energy

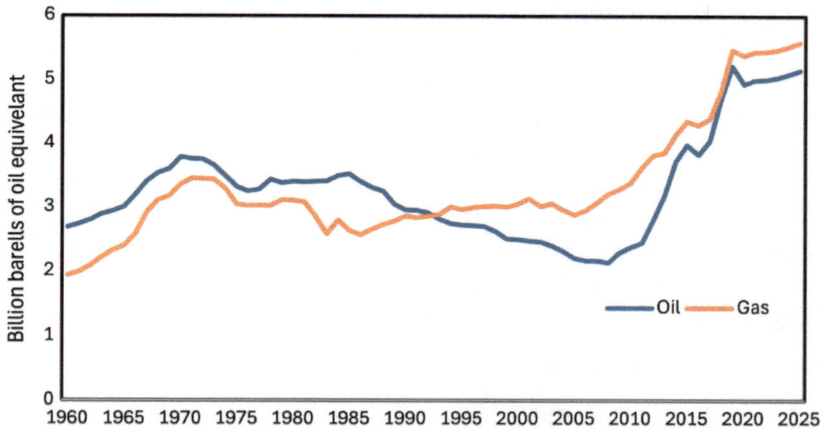

Fig. 2.20 U.S. production of oil and gas: Hubbert's curve and beyond. *Sources* Data from the International Energy Agency and British Petroleum's Statistical Review of World Energy

years that followed (and that the same pattern would characterize natural gas, often then associated with oil production). Despite some irregularities, Hubbert's projections held rather closely to what unfolded in American oil and gas production for several decades, past peaks in production in the early 1970s. However, as is often the case with long-term projections, unforeseen developments then challenged them (Fig. 2.20).

The challenges to Hubbert's projections came from the development of fracking technology and use of large deposits of previously unextractable oil and natural gas, both of which caused a boom in U.S. production. By the late 1980s, U.S. natural gas extraction was on an upswing and oil production surged after 2010.

Turning back to the global pattern and recognizing the uncertainties of both demand and supply, Table 2.2 shows data though 2020 and provides a forecast from IFs of the world energy supply by type for the year 2030. It is the forecasts of new renewable forms such as wind, photovoltaic, and geothermal power (encompassed in "Other Renewables" in the table) that are particularly murky.

This discussion has touched on some of the debates around energy to which we return in our discussion of causal logic in Chap. 10. These include the resource base of fossil fuels and the promise (and cost) of alternative energy technologies. They also include the prospects for economic growth and the relationship between that growth and energy demand. And they very much include the implications of the environmental impacts of greenhouse gas emissions and the national and global responses to those. Once again, we need to consider global trends but find it necessary to be thinking causally as well.

Table 2.2 World energy supply by energy type

		Percentage contribution to global supply			
Energy type	1960	1980	2000	2020	2030
Oil	34	46	39	31	26
Coal	47	27	25	27	31
Gas	14	18	22	25	26
Hydroelectric	6	7	7	7	3
Nuclear	0	3	7	4	4
Other Renewables	0	0	1	6	8

Source Data from the International Energy Agency

2.8 Environmental Change

We humans have long had an impact on our biological and physical environments, cutting down local forests, overhunting animals, depleting regional supplies of minerals, and destroying the productivity of plots of soil. We have also sometimes drained swamps (for better or worse), planted trees, and improved soil quality. It is only more recently, however, that our understanding of the scope of our impact has broadened from particular watersheds, valleys, and coastal regions to the globe.

It is almost impossible to touch on all the environmental concerns that have emerged in the last five decades. This discussion singles out three issues of wide-scale global importance: the release of greenhouse gases into the atmosphere; the increasingly heavy use and overuse of both renewable and nonrenewable water resources; and the destruction of forests. It is very difficult even to present data on environmental trends without some comment on issues of causal linkage.

Figure 2.21 shows atmospheric measurements of the primary greenhouse gas, carbon dioxide (CO_2), over the last several decades. Note the clear exponential trend surrounded by annual cycles. The cycles correspond to seasons and the ability of vegetation to remove some carbon dioxide from the atmosphere. Carbon dioxide allows sunlight to pass through relatively easily but reflects a substantial amount of heat radiation back toward the earth. Like the glass in a greenhouse, it raises air temperatures. The main driver of the increase in carbon dioxide is oxidization (burning) of carbon-based fuels.

We have noted the danger of utilizing too short a baseline for forecasting. Data from ice cores back 800,000 years show values that have never exceeded 300 parts per million by volume (ppm). Few forecasters anticipate levels less than 500–600 ppm by 2100. Many studies have also reconstructed global temperatures over hundreds of thousands and sometimes even hundreds of millions of years. The temperature reconstructions covary closely with those for carbon dioxide (Barnola et al., 2002). The globe is therefore moving into territory not seen in perhaps the last 15 million years (Krajick, 2023). Chapter 11 explores the issues around causality and climate change forecasting.

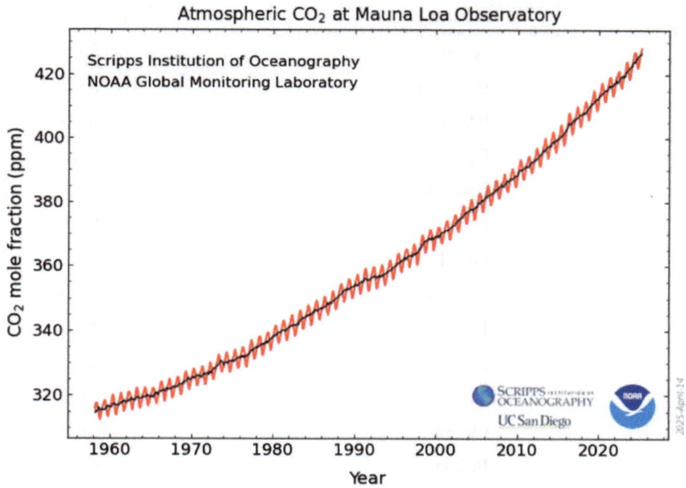

Fig. 2.21 Monthly mean carbon dioxide measurements at Mauna Loa. *Sources* Data prior to 1974 are from the Scripps Institution of Oceanography. Data from 1974 through August 2025 are from the National Oceanic and Atmospheric Administration

Turning to water use, Table 2.3 shows the ratio of water use relative to renewable resources in countries with the highest ratios of use to resources and therefore the most likely to have future water problems, including overuse of aquifers. Measurement is difficult and attention to the issue globally is relatively recent. This is an issue that strongly merits increasing attention and is also addressed in Chap. 11.

Concerning deforestation, the 2010s saw a net global loss of 47 million hectares of forest area. This is an improvement from the 1990s and 2000s, which saw net deforestation of 78 and 52 million hectares, respectively. Table 2.4 shows selected countries in which the annual rate of deforestation is greater than the average or that are especially important because of the extent of forest in the country, such as Brazil and the Congo. At current rates, Nicaragua, Angola, and Bangladesh are at fairly near-term risk of losing much or most of their forests. The table also shows that some countries, like Afghanistan, Pakistan, and Haiti, have already lost nearly all their original forest.

The condition of forests, especially tropical and rainforests, is a global problem, just as is atmospheric CO_2 from emissions. Reductions in forest extent contribute to global warming by releasing the carbon that forests "bank." Forests also serve as habitat for large numbers of species that exist nowhere else. It has been estimated that due to human impacts (such as climate change, hunting, and environmental degradation) species are going extinct at a rate that is 80 times higher than what would be expected without human interference (Cooke et al., 2023).

One uncertainty concerning environmental forecasts is the degree to which human-induced effects will be gradual or whether the equilibrium state of the global physical and biological systems might "tip" quite dramatically into another

Table 2.3 Water use as a percent of internal renewable resources in 2021

Country	Percent
Algeria	137.9
Bahrain	133.7
Barbados	87.5
Egypt	141.2
Eswatini	77.6
Iran	81.3
Israel	132.2
Jordan	102.5
Kuwait	3,850.5
Libya	817.1
Malta	78.3
Oman	116.7
Pakistan	162.1
Qatar	431.0
Saudi Arabia	974.2
Singapore	83.1
South Korea	85.2
Sri Lanka	90.8
Sudan	118.7
Syria	124.4
Tunisia	98.1
Turkmenistan	135.2
United Arab Emirates	1,533.3
Uzbekistan	121.8

Note All countries using more than 75 percent of their total renewable resources in 2021
Source: Data from Our World in Data, using data from the yearbook of the United Nations Food and Agriculture Organization (2024)

state. Our earlier discussion of forecasting mentioned threshold changes—a sudden change in the level of one or more variables. It is possible that the earth's living systems have a capacity to absorb and dampen much insult to them, but that at some point the damage becomes too great and fundamental patterns of temperature change, ocean water circulation, or the biosphere will change.

Table 2.4 Deforestation

Regions and countries	Total 2020 forest area	Annual change in forestation (2010–2020)	Extent of forest as percent of original
Africa			
Angola	66,590	−0.8	1.4
Congo, Dem. Rep	126,100	−0.8	5.6
Ethiopia	17,030	−0.4	6.4
Mali	13,290	0.0	NA
Nigeria	21,560	−0.8	NA
South Africa	17,010	−0.2	0.0
Sudan	18,330	−3.5	NA
Uganda	2,324	−1.7	0.2
Western Hemisphere			
Argentina	28,430	−0.6	4.9
Bolivia	50,820	−0.4	7.4
Brazil	496,200	−0.3	6.1
Colombia	59,100	−0.3	6.4
Costa Rica	3,015	0.5	5.3
El Salvador	573	−0.9	4.7
Guatemala	3,507	−0.6	5.7
Haiti	339	−1.1	0.3
Mexico	65,510	−0.2	7.5
Nicaragua	3,399	−2.1	4.6
Venezuela	46,200	−0.3	7.8
Asia			
Afghanistan	1,203	0.0	0.6
Bangladesh	1,817	−0.4	1.1
China	219,000	0.9	2.9
India	71,830	0.3	2.3
Indonesia	91,840	−0.8	5.7
Pakistan	3,706	−1.0	0.9

Source Data from World Resources Institute.

2.9 Technological Change

Technological change often exhibits exponential patterns in its early phases, commonly giving way to S-shaped patterns over time. Most of us probably have greatest familiarity with this *potential* pattern with respect to computing power. In 1965, Gordon Moore made his famous observation, to become known as Moore's Law (Roser et al., 2023), that the number of transistors per integrated circuit was doubling about every two years. Figure 2.22 traces the growth in transistors per chip, beginning with the invention of the transistor. The figure uses a semi-logarithmic scale; that is, each unit on the y-axis is a multiple of 10 relative to the unit below it. Were that graph to use traditional arithmetic scaling, the curve would show an exceptionally steep rate of exponential increase. Despite immense physical challenges, Moore's prediction has held for more than 60 years as processing power has continued to increase at a similar rate, confounding expectations of S-curve behavior. Relatively near-term predictions of plateauing could still again be proven false as new technologies, such as quantum computing, increase our computational power far into the future.

Not as well-known as Moore's Law is (Robert) Metcalfe's Law, describing the utility of a network. The law states that the value of a network increases with the number of potential connections, approximately the square of the number of nodes (like people) in the network (because no one can connect with themselves it is $n(n-1)$). With growth of the World Wide Web, this law has an obvious application and an intuitive appeal.

Figure 2.23 shows the growth of connections for fixed telephone, fixed broadband, and mobile broadband webs since 1975. The fixed telephone was, until recently, nothing short of revolutionary technology. Being able to communicate near-instantly, by voice, over hundreds and thousands of miles revolutionized communication in science, the business world, individuals' lives, and among political elites. Global fixed telephone connections peaked at 20 connections per 100 people. That innovation is now being dwarfed by the revolutions in fixed and mobile broadband connections.

In a little under two decades, the world went from not even 1 cellular subscription per 100 people to there being more cellular subscriptions than people on earth.

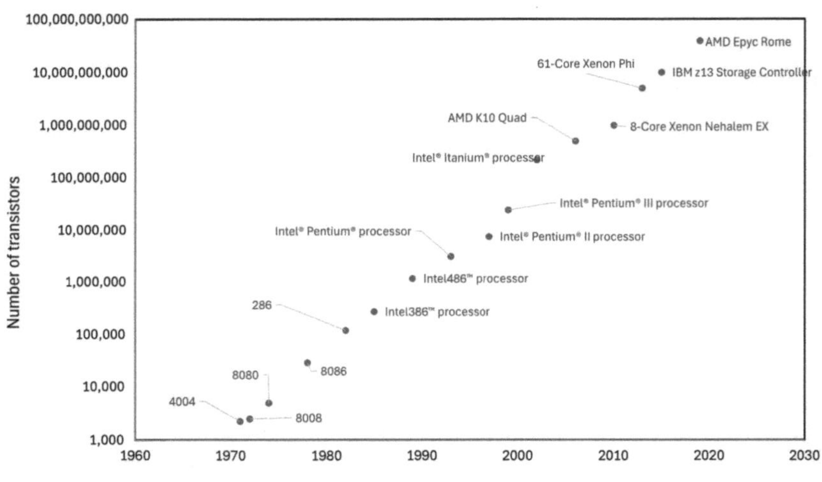

Fig. 2.22 Moore's Law. *Sources* Denver Post, July 10, 2005: K-8 and Our World in Data

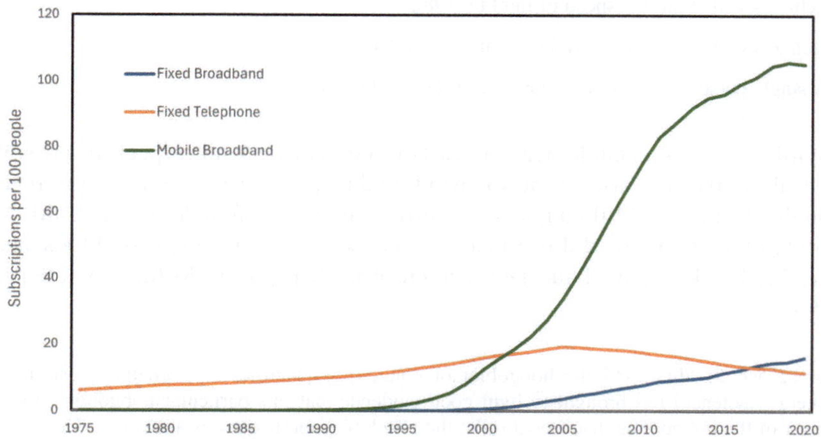

Fig. 2.23 Technology communication usage. *Source* Data from World Development Indicators

Even people near or in poverty often have mobile cell phones. Cellular technology has revolutionized the economies of developing nations, opening up pathways for formalization of their economies, facilitating more women's participation in the economy, and easing the process of starting a business and accessing credit. Mobile phones have also been tools that increase government accountability (also oversight of populations) and help communities fight corruption. They can help bolster social movements by spreading information about elections, demonstrations, and protests to citizens far faster than ever before. The integration of them with artificial intelligence access is now unfolding. At the same time, however, the change in communications patterns has contributed to destruction of traditional media aimed at entire populations, and its replacement by targeted media aimed at subpopulations with the effect of reinforcing informational and belief divisions among them.

At some point, process-specific technologies do reach saturation levels. Consider the growth of transportation speeds. These have advanced from fast horses to trains, to planes, to rockets escaping not only the earth's atmosphere but our solar system. A figure tracing that advance over time would be well served by using a representation called an **envelope curve** (enveloping a series of discrete technologies). Within technologies like transportation, such envelope curves can take an S-shape. In the case of human transportation, such a curve is generally assumed to be constrained by the speed of light.

Some might be tempted to extrapolate transportation speeds even further. Those familiar with Star Trek episodes (and other science fiction) know that the "warp speeds" of the Enterprise are vastly greater than the "impulse speeds" that it obtains from more traditional rocket engines. In a note of caution, however, one author pointed out that exponential extrapolation of earlier trends would have led to the following technological forecasts (Ayres, 1969: 20):

Vehicles will attain the speed of light by 1982

Humans will achieve immortality by the year 2000

A single person will control the power of the sun by 1981

While we want technological forecasts to be realistic with respect to (possible) physical limits, we also want to avoid undue pessimism about major breakthroughs. Robert A. Millikan, a Nobel Prize winner and founder of the California Institute of Technology, did just that in response to concerns expressed by science writer Frederick Soddy about the potential of nuclear power. Millikan wrote prior to 1945:

> Since Mr. Soddy raised the hobgoblin of dangerous quantities of available subatomic energy, [science] has brought to light good evidence that this particular hobgoblin—like most of the hobgoblins that crowd in on the mind of ignorance—was a myth … The new evidence born of further scientific study is to the effect that it is highly improbable that there is any appreciable amount of available subatomic energy to tap. (Sinsheimer, 1980: 148)

Technology obviously brings tremendous benefits to humanity. In addition to the provision of computing power, transportation and communication capabilities, and energy for the satisfaction of a wide range of needs and wants, technology has extended life expectancy, helped food producers more than keep pace with the growth in our numbers, and assisted in some protection of our environment (for instance, via new renewable energy forms). Technology also can have a dark side. While it allows us to control disease and produce food, it also helps us destroy forests and threaten the global oceans and atmosphere. We can see the dark side most clearly in human development of advanced military technology, which proceeds very rapidly. In 1500, new gunpowder "bombs" had the equivalent power of 0.001 tons of TNT. By 1961, the largest hydrogen bomb ever tested (by the Soviet Union) had power equivalent to 50 million tons of TNT.

This discussion has hopefully made clear the difficulties in extrapolating technological advance or its implications. Progress sometimes appears exponential and at other times appears bounded. There is no clear understanding of the degree to which (or if) technology advances regularly over time. In fact, a common argument is that such advance occurs in long cycles (Freeman & Louçã, 2001, nicely tied this to economic change). We return later to some of these uncertainties.

2.10 Sociopolitical Change

The threads of social and political change run through the fabric of other human system changes. Summarizing trends is challenging. To simplify the task somewhat, we divide it into two: changes within countries and changes in relationships among them.

The quality of domestic sociopolitical systems interacts strongly with most aspects of change that this chapter has discussed. Capable and efficient government is supported by human development and, in turn, normally advances it. Poverty reduction, a healthy population, and an educated citizenry all most often contribute directly to good governance, as does the economic advance that such governance in turn facilitates.

Democracy is generally responsive to the needs and desires of its citizenry, and long data series on its prevalence are more readily available than are series on variables that assess the effectiveness or efficiency of government. Setting aside very early manifestations of democracy such as those in Greece, modern democracy's origins followed the Industrial Revolution. Figure 2.24 shows three regions, Europe, Africa, and Latin America and their corresponding average score on the Varieties of Democracy (V-DEM) Electoral Democracy Index. Although subject to setbacks, the long trend toward more democratic societies has a history influenced in the twentieth century by two World Wars, the end of most colonialism in the 1960s, and the Cold War. In recent years, the trend has been set back as liberal democracies face challenges from populists on the left and right, augmented by the explosion of competing social media outlets and accelerated global immigration. Setbacks have also come from competing sectarian movements and sometimes military coups. It is unclear whether democracy's advance will continue or whether recent setbacks will significantly alter the pattern.

Turning to sociopolitical connections across countries, technological change and trade have increasingly tied peoples together globally. We can see the effect of these changes on the types and structures of social organization. With respect to the types of organization, one of the most dramatic trends has been the spread of European-style states (following the early emergence of England, France, and Spain) around the globe. In the contemporary era, the entire world is characterized

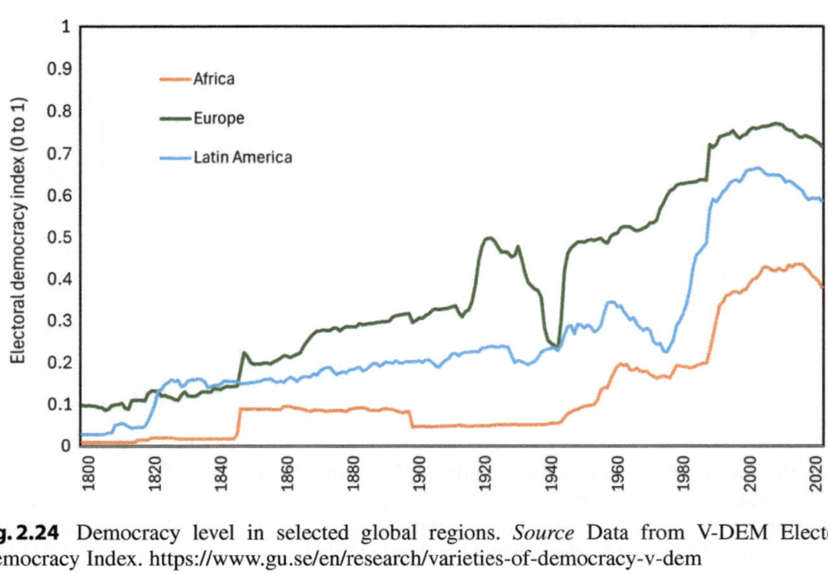

Fig. 2.24 Democracy level in selected global regions. *Source* Data from V-DEM Electoral Democracy Index. https://www.gu.se/en/research/varieties-of-democracy-v-dem

by such states and almost all the multi-ethnic empires that ushered in the twentieth century (or, like the former Soviet Union, were built during it) have given way to states built more closely around ethnic-cultural units.

A variety of forces, again including technology, have increased and changed interstate interactions. Figure 2.25 shows the growth of intergovernmental organizations (IGOs) based on country membership and shows international nongovernmental organizations (INGOs) during the post-World War II period with an explosion in international organization formation following the Cold War. IGOs like the United Nations, the North Atlantic Treaty Organization, and the European Union have proliferated considerably more rapidly than states. INGOs, like Greenpeace or Amnesty International, have proliferated still more quickly and numbered more than 36,000 as of 2022 (Yearbook of International Organizations, 2022). There has, however, been a slowdown in proliferation of organizations in recent years, perhaps reflecting some saturation of potential and/or need. While the number of international organizations (IOs) is still growing, a trend toward nativism and isolationism has also slowed the expansion of international organizations (Copelovitch & Pevehouse, 2019).

Sociopolitical change is obviously not just a cooperative phenomenon within countries (or other basic units). It always also involves potential or actual conflict among them. One element of that conflict is military spending and a related shifting of power balances across countries. Figure 2.26 shows such spending for five of the militarily most powerful entities in the world. The units are shown at

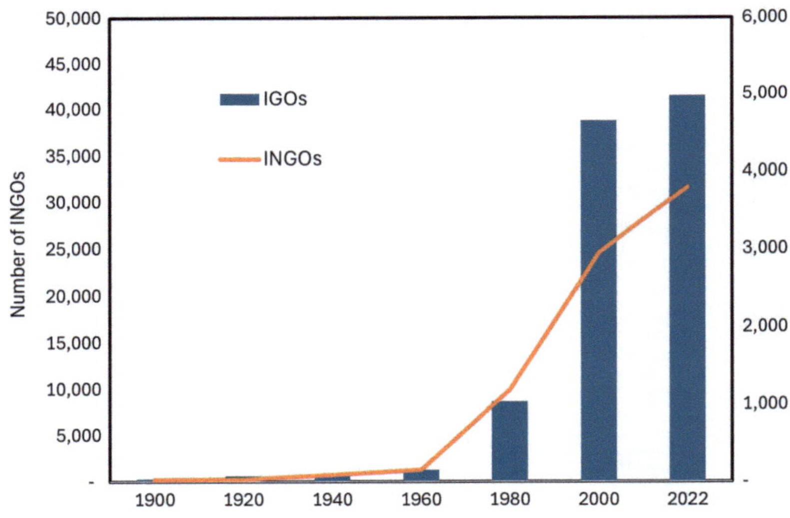

Fig. 2.25 Growth of international organizations. *Note* Criteria for counting IOs have changed over time, generally becoming more inclusive. Criteria for counting INGOs were tightened in 1989 to exclude less active or less international groups. *Sources* Data from Yearbook of International Organizations (2022) and Copelovitch and Pevehouse (2019)

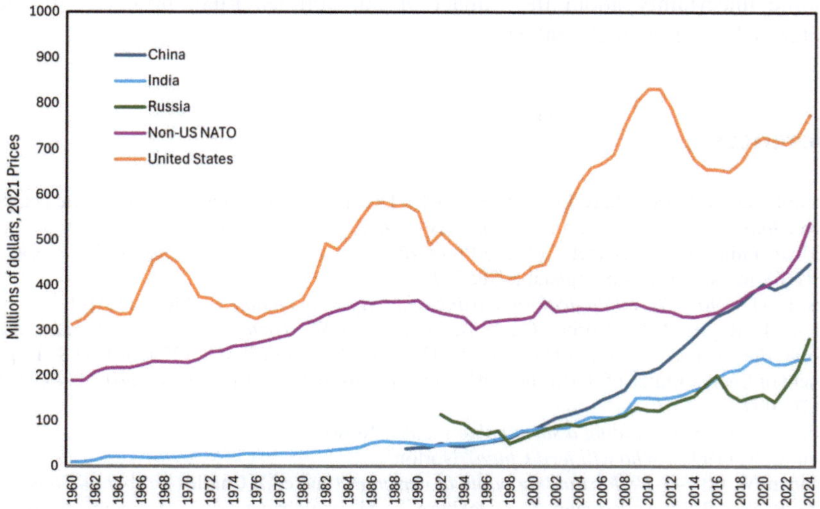

Fig. 2.26 Military spending of great power countries. *Note* Monetary values are at purchasing power parity. *Source* Data from Stockholm International Peace Research Institute (2024)

purchasing power parity (PPP) rather than the frequently used market exchange rates (MER) because much spending on equipment and personnel by great powers is domestic-based and domestic-priced. The figure shows that while the U.S. remains dominant, China has been rising quickly since the turn of the century. Also of note is the steady rise of Indian military spending. Non-U.S. NATO spending has risen in the 2020s and Russian spending in 2024 approximately doubles that of earlier years because of its war on Ukraine. Chapter 13 picks up forecasting of such change.

2.11 Conclusion

Throughout this chapter we have simultaneously visualized trends and possible extrapolations, while inveighing against uncritical trend extension. It is nearly impossible to know whether a trend will continue or reach a sudden limit, pass a directional turning point, or even end abruptly with a threshold change.

Nonetheless, trends and historical patterns are of great use in understanding the present and anticipating the possible future. Critical examination of them moves us quickly to a consideration of the causal dynamics that might underlie change in patterns. This chapter has suggested many causal linkages that we investigate further in subsequent chapters. That causal investigation will force us to move beyond looking in isolation at each chapter's focal issue area.

Despite the limitations of all forecasting, we must keep in mind the motivation that we face: pressures to act (including foregoing action), even in the face of some uncertainty about the impact of our actions. First, however, the next chapter elaborates casual analysis.

References

Acemoglu, D., Robinson, J., & Johnson, S. (2003). Disease and development in historical perspective. *Journal of the European Economic Association, 1*(2/3), 397–405.

American Philosophical Society. (1967).*Proceedings of the American Philosophical Society, 111*(6). https://www.jstor.org/stable/985727

Ayres, R. U. (1969). *Technological forecasting and long-range planning.* McGraw-Hill.

Barnola, J., Raynaud, D., Lorius, C., & Barkov, N. (2002). *Historical carbon dioxide record from the Vostok ice core (417,160—2,342 years BP).* (OSTI ID: 1394139). U.S. Department of Energy Office of Scientific and Technical Information. https://doi.org/10.3334/CDIAC/ATG.009

Brown, L. R. (1981). *Building a sustainable society.* Norton.

Brown, L. R. (1995). *Who will feed China?* Norton.

Buchanan, W. (1974). *Understanding political variables* (2nd ed.). Charles Scribner's Sons.

Cooke, R., Sayol, F., Andermann, T., Blackburn, T. M., Steinbauer, M. J., Antonelli, A., & Faurby, S. (2023). Undiscovered bird extinctions obscure the true magnitude of human-driven extinction waves. *Nature Communications, 14*(1), 8116. https://doi.org/10.1038/s41467-023-43445-2

Copelovitch, M., & Pevehouse, J. C. W. (2019). International organizations in a new era of populist nationalism. *The Review of International Organizations, 14*(2), 169–186. https://doi.org/10.1007/s11558-019-09353-1

Dattani, S., Rodés-Guirao, L., Ritchie, H., Ortiz-Ospina, E., & Roser, M. (2023). Life expectancy. *Our World in Data.* https://ourworldindata.org/life-expectancy

Diamond, J. (2004). *Collapse: How societies choose to fail or succeed.* Viking.

Ehrlich, P. R. (1968). *The population bomb.* Ballantine Books.

Ehrlich, P. R., & Ehrlich, A. H. (1972). *Population, resources, environment.* Freeman.

Espinosa, R., Tago, D., & Treich, N. (2020). Infectious diseases and meat production. *Environmental & Resource Economics, 76*(4), 1019–1044. https://doi.org/10.1007/s10640-020-00484-3

Freeman, C., & Louçã, F. (2001). *As time goes by: From the industrial revolutions to the information revolution.* Oxford University Press.

Hasell, J., & Arriagada, P. (2022). From $1.90 to $2.15 a day: The updated international poverty line. *Our World in Data.* https://ourworldindata.org/from-1-90-to-2-15-a-day-the-updated-international-poverty-line

Hasell, J., Rohenkohl, B., Arriagada, P., Ortiz-Ospina, E., & Roser, M. (2022). Poverty. *Our World in Data.* https://ourworldindata.org/poverty

Hughes, B. B., Kuhn, R., Margolese-Malin, E. S., Rothman, D. S., & Solórzano, J. R. (2015). Opportunities and challenges of a world with negligible senescence. *Technological Forecasting and Social Change, 99*, 77–91. https://doi.org/10.1016/j.techfore.2015.06.031

Intergovernmental Panel on Climate Change (IPCC). (2019) *IPCC special report on the ocean and cryosphere in a changing climate* (H.-O. Pörtner, D. C. Roberts, V. Masson-Delmotte, P. Zhai, M. Tignor, E. Poloczanska, K. Mintenbeck, A. Alegría, M. Nicolai, A. Okem, J. Petzold, B. Rama, & N. M. Weyer, Eds.). Cambridge University Press. https://doi.org/10.1017/9781009157964

International Panel on Climate Change (IPCC). (2023). Summary for policymakers. In *Climate change 2023: Synthesis report* (H. Lee & J. Romero, Eds.). Contribution of working groups I, II and III to the sixth assessment report of the Intergovernmental Panel on Climate Change (pp. 1–34). IPCC https://www.ipcc.ch/report/ar6/syr/downloads/report/IPCC_AR6_SYR_SPM.pdf

Krajick, K. (2023). A new 66 million-year history of carbon dioxide offers little comfort for today. *State of the Planet*. https://news.climate.columbia.edu/2023/12/07/a-new-66-million-year-history-of-carbon-dioxide-offers-little-comfort-for-today/

Lahoti, R., Jayadev, A., & Reddy, S. G. (2014). *The global consumption and income project (GCIP): An introduction and preliminary findings*. United Nations. https://www.un.org/ht/desa/global-consumption-and-income-project-gcip-introduction-and-preliminary-findings

Maddison, A. (2001). *The world economy: A millennial perspective*. Organisation for Economic Co-operation and Development.

Malthus, T. (1798). *An essay on the principle of population*. St. Paul's Church Yard.

Meadows, D. H., Meadows, D. L., Randers, J., & Behrens, W. K., III. (1972). *Limits to growth*. Universe Books.

Mesarovic, M. D., & Pestel, E. (1974). *Mankind at the turning point*. E.P. Dutton & Co.

Milanović, B. (2023). The great convergence: Global equality and its discontents. *Foreign Affairs*. https://www.foreignaffairs.com/world/great-convergence-equality-branko-milanovic

Moatsos, M. (2020). *Global absolute poverty, present and past since 1820* [Dissertation]. Utrecht University. https://doi.org/10.33540/129

Organisation for Economic Co-operation and Development (OECD). (2024). *OECD Digital Economy Outlook 2024 (Volume 1)*. https://www.oecd.org/en/publications/oecd-digital-economy-outlook-2024-volume-1_a1689dc5-en.html

Rentschler, J., & Leonova, N. (2023). Global air pollution exposure and poverty. *Nature Communications, 14*(1), 4432. https://doi.org/10.1038/s41467-023-39797-4

Riley, J. C. (2005). Estimates of regional and global life expectancy, 1800–2001. *Population and Development Review, 31*(3), 537–543. https://doi.org/10.1111/j.1728-4457.2005.00083.x

Roser, M., Ritchie, H., & Mathieu, E. (2023). Technological change. *Our World in Data*. https://ourworldindata.org/technological-change

Sinsheimer, R. L. (1980). The presumptions of science. In H. E. Daly (Ed.), *Economics, ecology, ethics* (pp. 146–161). W.H. Freeman.

Stockholm International Peace Research Institute (SIPRI). (2024). *SIPRI military expenditure database*. https://www.sipri.org/databases/milex

Union of International Associations. (2022). *Yearbook of international organizations 2022–2023* (Vols. 1A & 1B. 2 Vols.). Koninklijke Brill NV.

United Nations Department of Economic and Social Affairs. (2006). *United Nations demographic yearbook 2003*. United Nations. https://doi.org/10.18356/707c3a6a-en-fr

United Nations, Department of Economic and Social Affairs. (2024). *World population prospects 2024, online edition*. Population Division. https://population.un.org/wpp/downloads?folder=Standard%20Projections&group=Most%20used

United Nations Food and Agriculture Organization (UNFAO). (2024). World food and agriculture statistical yearbook 2024. *FAO*. https://doi.org/10.4060/cd2971en

van Zanden, J. L., Baten, J., Mira d'Ercole, M., Rijpma, A., Smith, C., & Timmer, M. (Eds.). (2014). *How was life? Global well-being since 1820*. OECD. https://www.oecd.org/en/publications/how-was-life_9789264214262-en.html

Wallerstein, I. (2004). *World-systems analysis: An introduction* (5. print. ed.). Duke University Press. https://archive.org/details/worldsystemsanal0000wall

World Bank. (2024). Poverty, prosperity, and planet report 2024: Pathways out of the polycrisis. *World Bank*. https://doi.org/10.1596/978-1-4648-2123-3

Zijdeman, R., & Ribeiro da Silva, F. (2014). Life expectancy since 1820. In J. L. van Zanden, J. Baten, M. Mira d'Ercole, A. Rijpma, C. Smith, & M. Timmer (Eds.), *How was life? Global well-being since 1820* (pp. 101–116). OECD. https://pure.knaw.nl/portal/en/publications/life-expectancy-since-1820-2/

Causal Analysis and World Views 3

Abstract

Data like those discussed in Chap. 2 greatly help us in structuring mental and computer-based models of the global system. The thousands of data series and the statistical analysis capabilities embedded in the International Futures (IFs) system (discussed in Chap. 4) provide a strong basis for understanding possible global futures. By itself, however, a massive data collection is an inadequate foundation for that understanding, no matter how sophisticated our statistical analysis might be. In addition, we need structural models to organize our thinking. Causal analysis and world views in combination provide those. Extended causal analysis generally involves identification and exploration of feedback loops. As we study and struggle to organize those into broader understandings of global dynamics, our minds and models rely upon and elaborate world views. This chapter surveys basic elements of dynamics involving feedback loops and their elaboration in world views. It also suggests how attachment to particular world views can endanger broad-scale, open, and adaptive thinking if we fail to recognize and consider alternative understandings and perspectives. The IFs system relies heavily on both statistical and structural analysis. And although this volume's usage of IFs and discussions of global change are not organized by competing world views, it is important that we recognize their importance and consider our own.

3.1 Introduction

Chapter 1 identified an important challenge: *We must act in the face of an uncertain future.* It suggested three questions to make such action a bit easier:

1. What kind of future would we prefer?
2. Where does the current path appear to be taking us?

B. B. Hughes, *Exploring and Shaping International Futures*, World-Systems Evolution and Global Futures, https://doi.org/10.1007/978-3-032-22084-4_3

3. How much leverage do we have to bring about a future we want?

That chapter identified the Sustainable Development Goals (SDGs) as helping us address the first question, although such goals and associated targets (and our own) change across time. Chap. 2 began to explore the second question by considering historical trends across a wide range of global issue areas. Most chapters of this book will further help us address that question by undertaking issue-focused analysis with attention to forward and backward linkages across issues.

This chapter lays some foundation for consideration of the third question. To explore leverage, we need to dig into causal analysis, a foundation of the International Futures (IFs) model system and of the world views that often frame causal understandings and future thinking.

3.2 Causal Analysis

As we looked at trends and inevitably thought about extrapolations in Chap. 2, we could see that simple linear and exponential extrapolations would very often be inadequate. Some growth processes are clearly limited. For instance, literacy cannot surpass 100 percent of the population and food production cannot grow indefinitely in the face of fixed resource bases. We therefore sometimes turn to alternative forms of extrapolation, such as S-shaped and bell-shaped curves, that *implicitly* recognize possible constraints.

Causal analysis helps us *explicitly* recognize such constraints and more generally assists in understanding how complex processes might evolve in the future. At the root of causal analysis is the distinction between independent and dependent variables (cause and effect). Looking at population growth, we all know that many factors affect it. The availability of contraception changes birth rates. So, too, does the level of education and availability of opportunities for employment of (and freedom of choice by) prospective mothers and fathers. Similarly, the quality of medical care and the survival rate of infants and children not only affect death rates but also birth rates.

We can divide causal relationships into two categories. **Positive relationships** exist when an increase in the independent variable (such as malaria or COVID-19) leads to an increase in the dependent variable (such as death rate). In such relationships a decrease in the independent variable most often leads also to a decrease in the dependent variable. In contrast, **negative relationships** exist when an increase or decrease in the independent variable (such as contraception availability) causes the reverse direction of change in the dependent variable (such as birth rate).

It is, however, sometimes difficult to distinguish independent and dependent variables clearly. For instance, a high rate of population growth in an Asian country might lead (cause) the government to institute a family planning program and increase the availability of contraceptives, which we have already suggested might cause population growth to decline. In such a situation we have a system of variables that *feedback* on each other in complex cause and effect dynamics.

If population growth gives rise to family planning programs that control that growth, the process is reminiscent of a home thermostat that controls temperature by turning heat on and off as necessary (when temperature increases, the thermostat ensures that heating ceases, which in turn leads temperature to decline). Figure 3.1 portrays the two causal linkages in both the family planning and home-heating examples. Note that one linkage in each case is positive and one is negative and that increases in a variable like population eventually feed back in a loop of linkages to decreases in the same variable. We call such a combination of linkages a **negative feedback loop**, and, as these two examples suggest, such loops often tend to produce relative stability in the processes they represent, especially when, as in the case of a thermostat, an explicit target helps control the dynamics. A multivariate feedback loop becomes a negative loop whenever it includes an odd number of individual negative linkages (for instance, one, three, or five negative linkages out of six or seven total linkages).

In contrast, consider what might happen if the death rate from malaria, AIDS, or some other disease in an African country overwhelmed the medical establishment in that country, making treatment less effective, which in turn made it impossible to cope medically with the disease and led its incidence to rise further. This would resemble the situation of a thermostat with its "wires crossed": sensing too high a temperature, this rogue thermostat turns up the heat. Figure 3.2 portrays the two causal linkages in these examples. We call the dynamics in these examples **positive feedback loops**. A feedback loop will be positive or self-reinforcing whenever there is an even number of negative linkages (such as zero, two, or four). Thus, the disease incidence example with two negative relationships is also a positive

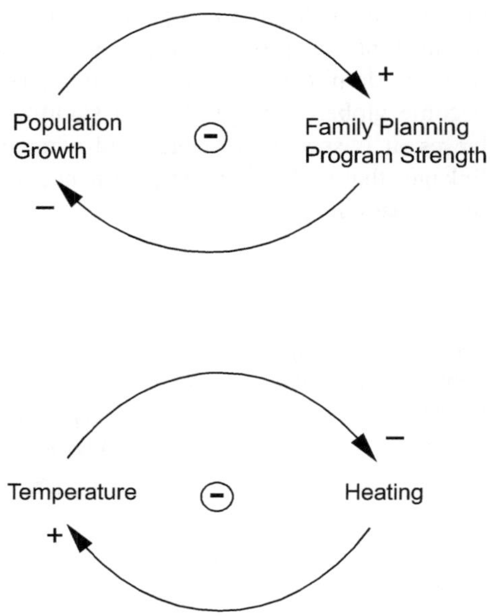

Fig. 3.1 Negative feedback loops. *Source* Author

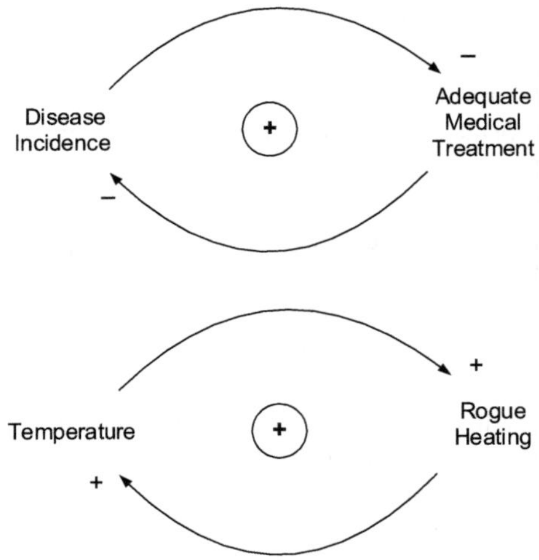

Fig. 3.2 Positive feedback loops. *Source* Author

feedback loop. Processes in positive feedback loops tend to either collapse precipitously (like population in the disease-afflicted country could) or to grow without bound (like the temperature in the heat-afflicted home).

Many growth processes are combinations of positive and negative feedback loops. Consider, for instance, the growth of world oil production. On the one hand, increases in production can facilitate further increases in production (for instance, by leading to improved technology or by fueling deeper drilling). On the other hand, increases in production deplete resources, which can lead to restraints on production. If the first and positive loop is dominant, as it has been globally for a long time, oil production grows steadily. When the second and negative loop becomes dominant, as it has become in Trinidad and Tobago and could eventually become globally, oil production stabilizes and eventually declines or even collapses. It is specific variables and parameters in the equations representing loop linkages that will determine dominance. Figure 3.3 represents this interacting pair of feedback loops.

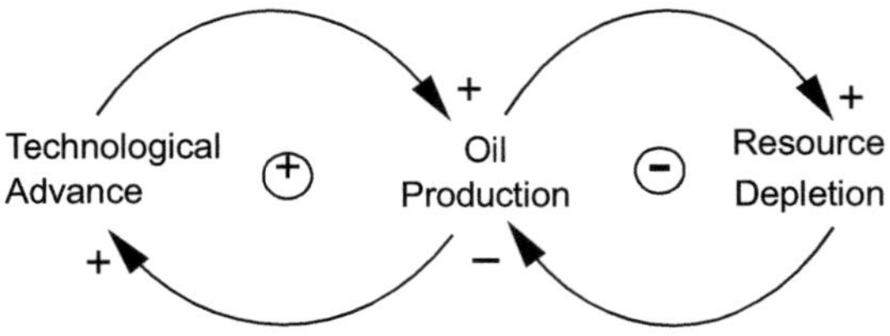

Fig. 3.3 Interaction of feedback loops. *Source* Author

3.3 Causal Understandings and World Views

Building on the simple causal linkages discussed above, more complex causal models greatly help us understand change. All of us already have at least rudimentary causal models concerning the world development system in our minds and those models help us make sense of the world. Consider, for instance, a very simple, but useful model from the field of world politics. That model suggests that the attempt by countries to balance power with power often sets up an escalating feedback loop. What one state views as a defensive buildup in reaction to the challenging power of another, the second will likely view as a potentially offensive threat. Thus, it will, in turn, undertake a buildup. Scholars and practitioners of international politics know this logic as that of the **security dilemma**. They understand that the logic sets up arms races that may increase the probability of war and point to the arms race before World War I and the U.S.-Soviet Cold War as examples. Figure 3.4 represents such a pattern of interaction.

If it is dangerous to rely on rational judgment of states in power balances, and the competitive logic of the security dilemma regularly works to initiate and fuel dangerous arms races, can we conceive of an alternative model of world politics? Fig. 3.5 suggests one that portrays growing global community, accompanied by increasing levels of cooperative interaction. Mutual strengthening of community and cooperative interaction creates a positive feedback loop. It also makes war increasingly expensive because it would disrupt valuable trade and other connections. Other linkages, including positive feedback loops with democracy, strengthen the dynamic.

The very simple diagrams of Figs. 3.4 and 3.5 tend to be embedded in much larger world views of global politics, often called **realism** and **liberalism** (earlier sometimes labeled “idealism”). Those who focus strongly on security values often look to the traditional state (country) as the most reliable guarantor of that security. States use police to maintain order internally and rely upon military forces to pursue it externally. Realists recognize that the system is imperfect but argue that

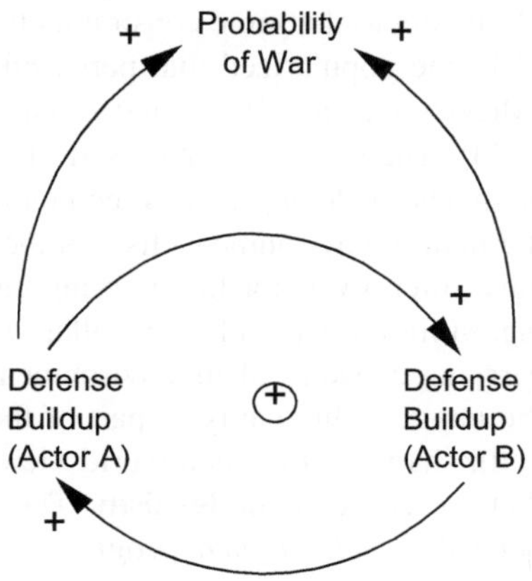

Fig. 3.4 The security dilemma. *Source* Author

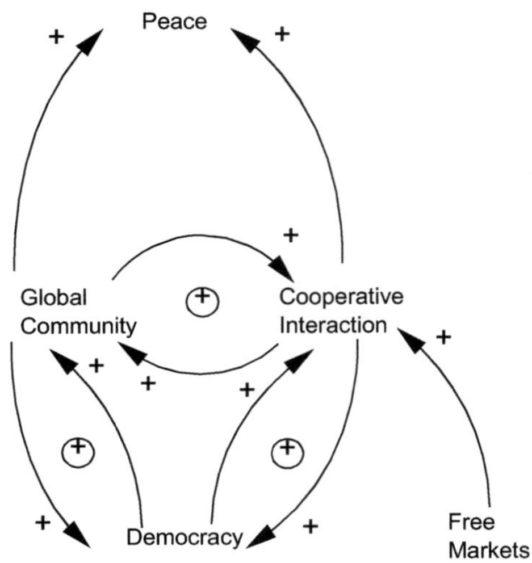

Fig. 3.5 Cooperative interaction and global community. *Source* Author

a world without central government is fundamentally a world of anarchy. In such an environment, protecting the power of states is seen as the best hope.

Critics of that understanding, especially political liberals, point out that the security of the state system has always come at the cost of intermittent warfare. The weapons of the modern era, especially nuclear weapons, have raised that cost far too high. Moreover, the pressures for interstate cooperation on environmental, economic, and security issues have grown simultaneously with communications and transportation technology that make such global cooperation increasingly possible. Liberals and realists can therefore emphasize quite different aspects of causality.

Similar to the analysis of world politics, key world views surrounding environmental sustainability are integrated, multi-systemic visions. We can portray two such world views. Neither has names that are widely accepted, but the first can be called **eco-wholism** (Hughes, 1985; see variations in Pirages (1983) and Haas (1990)). It portrays humanity as very much part of a larger eco-system, subject ultimately to the same constraints that such systems place on all species. The Club of Rome popularized that perspective via volumes such as *The Limits to Growth* (Meadows et al., 1972) and *Mankind at the Turning Point* (Mesarovic & Pestel, 1974). The Club of Rome used the term *problematique* to capture the interrelated elements of demographics, economics, energy systems, agricultural systems, and the natural environment. Its members frequently warned that just as other species have often overshot the carrying capacity of their environment and, after a time lag, suffered a population collapse, so too humanity has been overshooting the carrying capacity of its environment. Mental models like that of Fig. 3.6 convey the image of humanity as part of the environment and subject to its constraints.

In sharp contrast is a world view that again lacks a widely accepted name, but that we can call **modernism.** The core of modernism is not the *problematique*, but rather *technological progress.* Figure 3.7 represents that model in very simple,

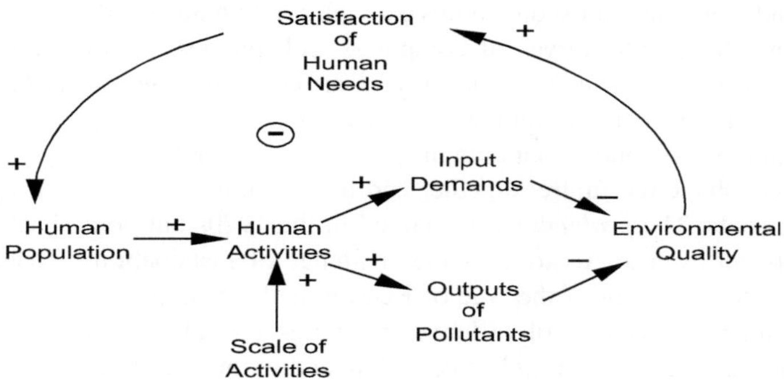

Fig. 3.6 A simplified model of Eco-Wholism. *Source* Author

almost caricatured form. The more important feedback loop in that model is the one that links human technological knowledge to itself. This perspective argues that the stronger our existing base of understanding, the faster and easier humans add still more to that base. Kahn et al. (1976) built a vision of the long-term future largely on that loop. Simon (1981) did so even more explicitly, portraying population growth not as a burden on the environment, but in human knowledge terms as *The Ultimate Resource*.

These competing world views are not historical perspectives, but rather something we might call "enduring understandings." Today, eco-wholists point to the growth curves of atmospheric carbon dioxide and freshwater usage as two of the

Fig. 3.7 A simplified model of Modernism. *Source* Author

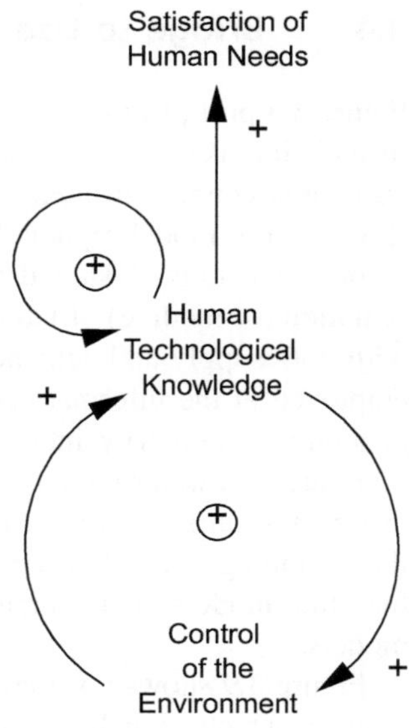

clearest and potentially most dangerous examples of human-based overshoot. Modernists point to growth curves of computers, cell phones, internet linkages (the global networked knowledge society), and artificial intelligence as the clearest examples of an explosion in human knowledge.

The reality is, of course, that both the *problematique* and *progress* are real. They will, indeed, shape the future: together and in interaction. And it is an interaction with deep roots. The *problematique,* named in the 1970s but going back not only to Malthus but even to Aristotle, is the challenge of sustainability. Technological *progress,* a core concept of the Age of Reason, is the promise.

To summarize, because of different emphases on values and varying causal understandings, we face a complicated task in addressing our three questions: what future we would prefer; where global dynamics are taking us and what leverage we have. Value orientations and understandings of the world tend to shape and reinforce one another. In fact, values tend to shape discussions even of where current changes appear to be taking us. Different value-and-understanding orientations (we can call them **world views**) emphasize different trends or can interpret the same trend in quite different ways.

In this volume we cannot lose touch with the recognition that causal understandings do differ. Once again, we cannot predict the future, but we can think more explicitly and systematically about it to expand our understanding, better communicate with one another, and hopefully enhance decision-making.

The *Help* system of IFs contains causal diagrams on IFs models, including a large-scale network diagram of the entire system, interacting with elaboration of these and other world views.

3.4 Bridge to Use of IFs and Scenario Analysis

Figure 3.8 portrays the IFs system's collection of models and a very small fraction of their interactions. Such portrayal is somewhat arbitrary because boundaries of issue-area connected models are not definitive. For instance, the figure does not show a labor model separately from the population and economic models, but the labor connection illustrated between them is actually in IFs an elaborated representation (or module) of labor by sex, household type, and sectors of employment, with both supply and demand sides. IFs also recognizes that some of that labor is employed in the informal economy, sometimes called the shadow economy, with generally less security, income, and productivity. Similarly, inside the environmental model is a water model with different sides of a balancing process representing different sources of water and types of demand. In contrast, what is shown as the technology model is representations of technological advance scattered across the other models of IFs, including in the health, economy, energy, and agriculture models.

Figure 3.9 shows a somewhat different representation of IFs models, including finance, which arguably could be considered a submodule within the economic model. The economic model represents financial flows among firms, households,

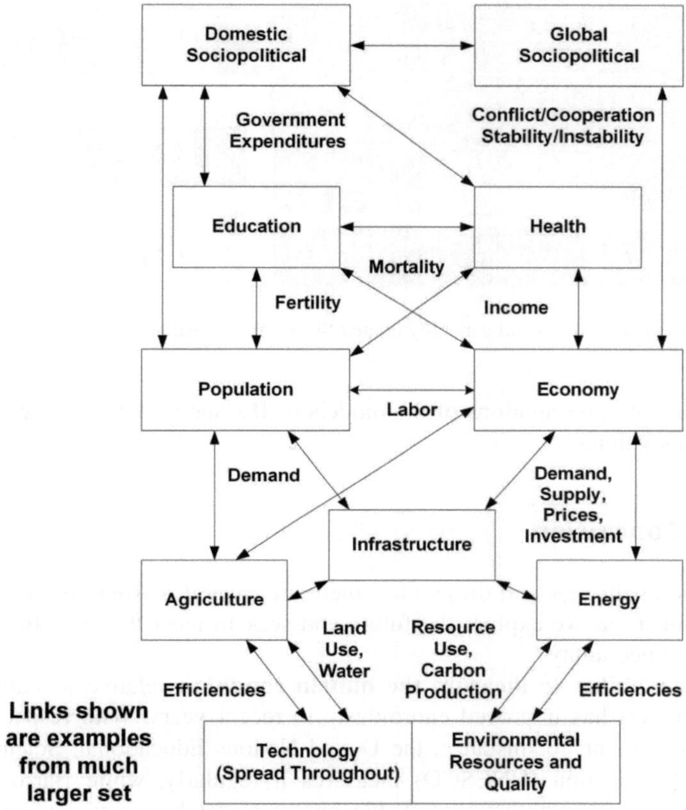

Fig. 3.8 Basic models of the IFs system and illustrative linkages. *Source* Author

and the government in what Chap. 7 on the economic model will describe as a "social accounting matrix," the dynamics of which have the complexity of a separate model.

The identification of IFs models in Fig. 3.9 and their colors allow us to portray very roughly how those models connect to the subject matter of the SDGs. Demographics, economics, health, and education connect closely to the human development goals. Agriculture, energy, and the environment connect to the SDGs that put a heavy emphasis on environmental sustainability. The models in the middle area have strong relationships to both human development and environmental sustainability, but they also might be considered instrumental socioeconomic systems that are important in helping pursue those two SDG categories. The obvious linkages of the IFs portrayal in Fig. 3.9 to the three levels of "spheres of action and values" in Table 1.1 are no accident. Although the boundaries of these categorizations and linkages between IFs models and the SDGs are fuzzy, they will help

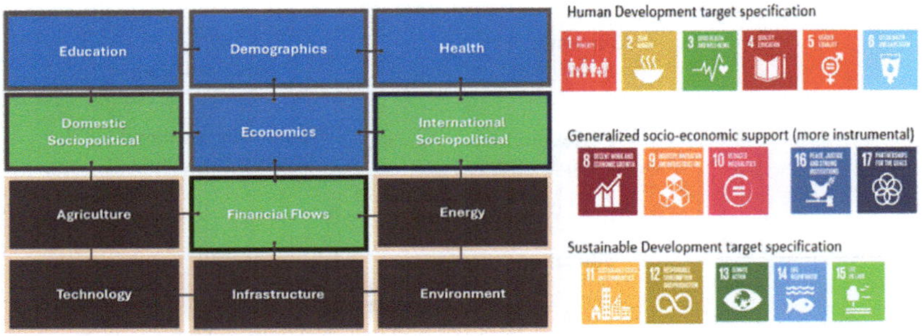

Fig. 3.9 The models of IFs and generally related SDGs. *Source* Author

us in organizing presentations of the models of IFs and how to use them in Parts III-V of this volume.

3.5 Conclusion

Despite the challenges and difficulties, there are several reasons for at least measured optimism as we explore the future and seek to meet the need for acting in the face of uncertainty.

First, our **ability to measure the human condition** relative to national and global goal sets has improved enormously in recent years. With respect to educational attainment, for instance, the United Nations Educational, Scientific, and Cultural Organization (UNESCO) measures it regularly, while Barro and Lee (2001) put together another remarkable historical database upon which the IFs project has drawn heavily. Infant mortality measures are widely available (see, for example, the World Bank's annual *World Development Indicators*, which also provide information on access to safe water and malnutrition levels among children). The United Nations, the World Health Organization, and the Global Burden of Disease project track mortality patterns by cause. The International Monetary Fund provides information on gross domestic product and its growth twice a year. The United Nations Development Programme has created a summary measure called the Human Development Index (HDI) that builds upon infant mortality rates, education levels, and economic well-being. Similarly, the National Oceanic and Atmospheric Administration and others have given us precise information on atmospheric carbon concentrations from the top of Mauna Loa in Hawaii and on the extent of global warming, while the Food and Agriculture Organization tracks our use of the oceans and croplands for food. Windows into the world's condition have opened.

In summary, the second half of the twentieth century was a period of incredible explosion in our assessment of all the elements of human sustainable development. It is remarkable to recall that in the middle of the twentieth century even the Gross National Product was a relatively new measure and that the human database concerning worldwide individual life conditions, economic well-being, and social capacity was skimpy at best.

Second, as discussed in Chap. 1, our **values and goals** have become much clearer. Goals and targets are policy levers in and of themselves. It is true in nearly all domains of social activity that you "get what you measure." Thus, the clear statement of goals and the regular release of measurements with respect to those goals are among the strongest indications of emerging human commitment to action on issues around sustainability.

A third foundational element, as also discussed in this chapter and much elaborated in the later chapters on specific issues, is **increased human understanding** of the dynamics of systems (demographic, economic, social, and environmental) in which we function. It is very encouraging to consider the extent to which our ability to represent the operations of individual systems has grown, and even our capacity to trace the interconnections across complex interacting systems has expanded. Integrated assessment models (IAMs), of which IFs is one, have proliferated around the world, greatly enhancing our ability to explore global futures (Hughes, 2019). Despite inevitable differences around understandings, insights from model systems are critical if we are to comprehend the trajectories on which we find ourselves and the opportunities for leverage that we might exert to attain our goals.

Fourth, human development has increasingly given us **new levers for action**, should we choose to use them. These include the vast benefits of human development to date: the advance in the life conditions and individual capacity of so many; the growing wealth of humanity; the growth of our social capacity; and the expansion of a broad knowledge base. The emergence of new information and communication technologies (ICT), and now of artificial intelligence, expand quite dramatically our ability to access existing knowledge and to use networks for its application and, perhaps even more exciting, for the accelerated creation of still more knowledge. They also, of course, create some risk for human futures. Subsequent chapters will focus on many of these levers within and across issue areas.

The first three chapters in this volume have provided context for our explorations and potential contributions to shaping international futures. Some readers may be ready to move into use of the IFs system and can jump to Chap. 6, which begins the substantive focus portions of the book with attention first in Part III (Chaps. 6–8) to human development and well-being. For those not familiar with the mechanics of IFs use, however, Part II (Chaps. 4 and 5) will help you gain that familiarity.

References

Barro, R. J. & Lee, J.-W. (2001). International data on educational attainment: Updates and implications. *Oxford Economic Papers*. https://barro.scholars.harvard.edu/publications/international-data-educational-attainment-updates-and-implications

Haas, E. B. (1990). *When knowledge is power*. University of California Press.
Hughes, B. B. (1985). *World futures: A critical analysis of alternatives*. Johns Hopkins University Press.
Hughes, B. B. (2019). *International futures: Building and using global models*. Academic Press.
Kahn, H., Brown, W., & Martel, L. (1976). *The next 200 years*. William Morrow.
Meadows, D. H., Meadows, D. L., Randers, J., & Behrens, W. K., III. (1972). *Limits to growth*. Universe Books.
Mesarovic, M. D., & Pestel, E. (1974). *Mankind at the turning point*. E.P. Dutton & Co.
Pirages, D. (1983). The ecological perspective and the social sciences. *International Studies Quarterly, 27*(3), 243–255. https://doi.org/10.2307/2600681
Simon, J. L. (1981). *The ultimate resource*. Princeton University Press. https://muse.jhu.edu/pub/267/monograph/book/118538

Part II

The International Futures (IFs) System

Understanding IFs and Its Basic Use

4

Abstract

This chapter is the first of two in the volume's Part II that discuss the structure and use of the International Futures (IFs) toolkit. Understanding three features of IFs facilitates its use. First, a huge database and associated tools within the IFs system make possible both statistical analysis and the placing of forecasts (also called projections within the integrated assessment modeling community) in historical context. Second, a significant set of display tools, including tables, graphics, maps, and multiple specialized displays, supports the analysis of both data and forecasts. Third, a scenario creation and storage system helps users frame both simple and complex interventions in their exploration of alternative futures and possible strategic leverage points. This chapter discusses the use of the first two features described above and Chap. 5 turns to the third. This chapter also identifies the multiple interacting models that make up the IFs system, briefly summarizing their key structural elements (to be elaborated in the usage-oriented chapters of this volume's Parts III–IV).

4.1 Introduction

Chapter 1 emphasized the reality that as we move into the global future and hope to shape its unfolding in directions we prefer, we do so with great uncertainty. Examination of global trends, like we undertook in Chap. 2, contributes considerably to helping us understand where change may be taking us. Thinking about the future often involves extrapolation from recent or long-term trends or considering historical analogs. We know, however, the importance of the old saying that "a trend is a trend is a trend – until it bends." Understanding the causal dynamics underlying trends is fundamentally important. Chapter 3 emphasized attention to values as well as to such understandings.

B. B. Hughes, *Exploring and Shaping International Futures*, World-Systems Evolution and Global Futures, https://doi.org/10.1007/978-3-032-22084-4_4

This chapter introduces use of the computerized tool named International Futures (IFs). It can be an aid to our mental models and to understanding human leverage for consciously shaping our futures. Computer models formalize mental models mathematically. Computer models are *not* automatically superior to mental models because of that formalization. In fact, computer models can often be less sophisticated than a good mental model.

Nonetheless, the careful use of computer models has some significant advantages. The first is that a computer model is highly explicit. One can examine the model very carefully. In contrast, we often have difficulty explaining our own mental models and may find ourselves clueless with respect to the mental models of others. Second, a computer model sometimes can attain a level of complexity and sophistication that enhances our mental models. Such sophistication might allow the computer system to probe the secondary consequences of change and action in ways that we cannot. (It is important to remember, however, that complexity can also introduce error.) Third, we can use a computer model to investigate experimentally a variety of different assumptions related both to uncertainties and to value-based desired futures. For instance, we might examine the possible consequences of several courses of action, just as chess programs can now investigate far more possible moves than even expert human players. Or we might investigate the consequences of a single plan under differing assumptions about the workings of some key element in the environment of our action.

We need to keep both these benefits of computer models and also the cautions about their limitations in mind as we turn our attention to IFs. IFs is a large-scale integrated global modeling system whose broad purpose is to serve as a thinking tool for the analysis of longer-term country-specific, regional, and global futures across multiple, interacting issue areas. It began as an educational tool but is increasingly being used in support of policy analysis.

IFs is heavily data-based and deeply rooted in theory. With respect to data, it uses, and provides access to, an extensive database for longitudinal and cross-sectional analysis. In so far as possible, thousands of data series represent 188 countries since 1960. Then, using theoretical guidance, IFs represents major agent-classes (households, governments, firms) interacting in a variety of global structures (demographic, economic, social, and environmental). Flexible forecast horizons can extend to 2100 or even beyond.

The system draws upon standard approaches to modeling specific issue areas whenever possible, extending and integrating them across issue areas. The most extensive documentation on IFs is available in its Help system (accessible via the Main Menu screen of the model or by direct link to its Wiki; see also the system's network diagram). Those include full documentation through causal diagrams, equations, and computer code. See also Hughes (2019) which evaluates IFs in the context of other integrated assessment models (IAMs). (Those interested in the philosophical approach to modeling of IFs and the history and plans of the project are invited to read Appendices 1 and 2.)

The rest of this chapter introduces you to IFs by:

1. Providing guidance for basic use of the model. (Chapter 5 provides additional detail on undertaking scenario analysis with it.)
2. Surveying briefly the core elements of the model in preparation for their elaboration in subsequent chapters.

As you move into the next sections of this chapter, it will be very useful if you sit down with the computer and iterate between "playing" and reading.

4.2 Use of IFs: Display and Data Analysis

One important goal of the IFs project is that the model be easy to use. Its size and scope complicate that, but its interface helps. Basic use has three core functionalities (*Display*, *Scenario Analysis*, and *Data Analysis*), corresponding to menu options from the Main Menu of IFs shown in Fig. 4.1, to be discussed.

The menus and options and the modeling itself change gradually over time, so please be flexible if you see some discrepancies as you read through the following. Also there will be some differences in the version of IFs available for use on the web and the one you can download onto your own computer. The downloaded version gives you some extra flexibility, notably in terms of saving your own scenarios and analyzing or elaborating them over time, an issue we address in the next chapter. The two versions do not differ significantly with respect to *Display* and *Data Analysis*.

The first basic option on the IFs menu is *Display* of results because most users begin with considerable exploration of the model's Base Case (the "this is where global dynamics seem to be taking us" scenario) or other pre-packaged scenario

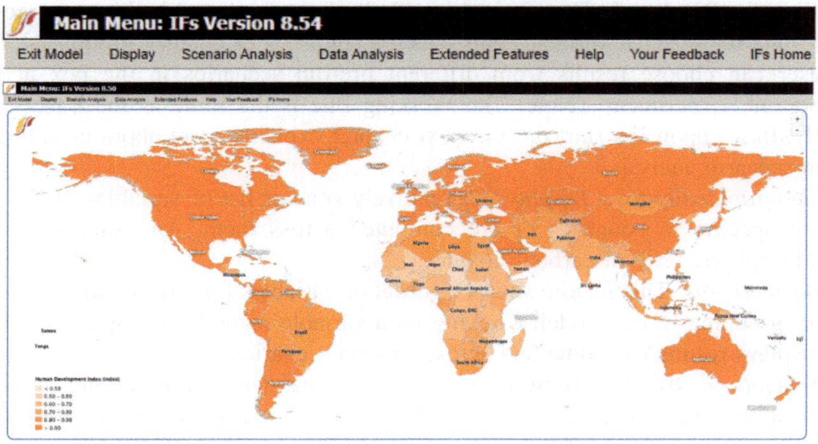

Fig. 4.1 The main menu of IFs (top) and the main screen (below). *Source* Author

forecasts before building new scenarios. The second is *Scenario Analysis*, the feature that supports exploration of alternative futures that we may wish to avoid or to build. Because of the relative complexity of *Scenario Analysis*, we give it special attention in the next chapter. The third option is *Data Analysis*, which facilitates exploration of a huge amount of historical data. Many users of IFs around the world do so primarily to explore those data for countries or globally and to study trends and relationships.

In this chapter we begin with *Display* and then turn to *Data Analysis*. This volume leaves support for learning by more experienced users about the fourth menu option, *Extended Features,* to the Help system. One of those features, however, merits your recognition. IFs includes a "data preprocessor" that is activated with a menu sub-option to process the constantly growing and temporally updated database to fill holes, reconcile data, and otherwise create the initial conditions that anchor the first or base year of IFs forecasting to data.

4.2.1 Display Results

Turning to the forecasts produced by IFs, the *Display* capabilities from the Main Menu of the IFs platform contain most standard formats such as tables, line graphs, bar charts, and pie diagrams, as well as maps. In addition, computational capabilities exist to combine and/or transform existing variables into ones newly defined by the user.

The wealth of variables and parameters in IFs (see Table 4.1 for important terminology) can make it difficult for beginning users to identify important focal points. Facilitating such identification are five sub-options under *Display* on the Main Menu:

1. *Flexible Displays.*[1] These "packaged" displays facilitate looking at model results for selected variables and selected countries or country groupings. The user can choose results from different pre-run scenarios in the box at the bottom. Experiment! (Note: The Working File is the same as the Base Case (IFsBase) upon IFs start up; the next chapter explains more about changing it in scenario analysis.)
2. *Geographically-fixed Displays.* A relatively smaller list of variables in IFs are not specific to countries or are computed across them. An example is the atmospheric level of carbon dioxide.
3. *Basic Report.* This feature shows a subset of variables for any country or country grouping in the model. Clicking on a variable name brings up a table that displays results over time and across pre-run scenarios.
4. *Specialized Displays.* These forms provide a great deal of focused information, most often on a single issue area of the model, such as population or education

[1] Abdelrahman Ibrahim led development of a major reworking of the IFs *Flexible Display* form.

Table 4.1 Important terminology in use of IFs

Parameters	Numbers in the model equations that specify the strength and character of causal relationships among variables. (IFs represents the names of parameters in lower case, and this volume shows them also in bold, e.g., **tfrm,** a multiplier on the total fertility rate.)
Variables	Concepts (such as population) computed over time dependent on initial conditions, the behavior of causally related variables, and parameters that link them to such causally related variables. (IFs shows variable names in upper case and the volume shows them also in bold, e.g., **POP**.)
Endogenous	Internal to the model (computed)
Exogenous	External to the model (user-specified). Most exogenous specifications are only for data-based initial conditions of variable series in the base year of the model or for parameters. Some data series such as GDP and population can be specified exogenously based on other forecasting work. This capability is useful in scenario analysis
Scenario	A story. In the model, a scenario is a forecast based on specified initial conditions and parameters and possibly also on exogenous specifications. A scenario helps us consider the possible implications of its specifications upon the evolution of variables of interest

by age and sex; values (using the dimensions from the World Value Survey); economic growth accounting showing contributions to it; social accounting matrix flows and "stocks;" human development and sustainability (many targets and indicators associated with the Sustainable Development Goals); poverty analysis at flexible levels of income; or global distribution patterns for any variable in IFs (using maps, Lorenz curves, or Gini coefficients). In some of these, such as the social accounting matrices, clicking on cells drills deeper and/or provides information over time. Please explore and play with these extensive display screens.

5. *Self-Managed Display*. Although this is the most complicated display form to use, it is also the most flexible and complete. It allows the user to display any variable or parameter (see Table 4.2 for a list of important variables) in the model in any display format. Computational options further allow combinations and transformations of variables.

Learning by doing is fundamental to understanding and using a complex system such as IFs. This volume therefore includes exercises throughout to help the active reader. The → symbol and indentation set them off from other text.

→ Explore the various Display options of the Main Menu. Pay special attention to Flexible Displays. It facilitates display of variables across the entire IFs system, often with historical trends for them, and it does not require knowledge of the variable names in the model. Spend at least a few minutes exploring, perhaps with the goal of learning something new about an issue area of specific

Table 4.2 Most-used variables across IFs issue areas and models

Demographics	Economy	Development	Infrastructure	Education	Agriculture	Health	Sociopolitical
Mortality	**Production**	**General**	**Cookstoves**	**Attainment**	**Production**	**Communicable Disease**	**Domestic**
CDR	GDPPC	GDI	COOKSTOVES	EDYRSAG15	AGP	HIVCASES	
CHLMOR	GDPPPP	HDI, HDIMF		LIT	LD, YL	HLHIVMORT	**Capacity**
INFMOR	GDPR	LIFEXP	**Electricity**	EDYRSSLE	IALK	HLMALARIAMORT	GOVINDCAPAC
	PRI		INFRAELECACC		LIVHERD		GOVCORRUPT
Fertility	WGPD	**Poverty**	INFRAELEC	**Basic Education**	LOSS	**Finance**	GOVEFFECT
Births		POVEXTREME	INFRAELECPROD	EDSECLOWRGRADS		HLCOST	
CBR	**Productivity**	POVMDLOW		EDBASENRG	**Consumption**	HLCOSTMORT	**Finance**
CONTRUSE	MPFTOT	POVMIDHIGH	**Finance**		FDEM		FIRMTAXR
TFR			INFRACOSTADJFAC	**Finance**	AGDEM	**Illicit Drugs**	GS
	Investment	**Values**		EDTOTCOST	CLPC	DRUGUSEGRAMOPIATE	HHTAXR, FIRMTAXR
Migration	I	MATPOSTR	**ICT**		FEDDEM		
MIGRANTS	ST	TRADSRATLD	ICTCOMPUTERS	**Primary**	INDEM	**Mortality**	**Inclusion**
MIGRATE			INFRANET	EDPRIGRADS		DEATHS	GEM
MIGRATEIN	**Labor**	**Environment**	ICTMOBIL	POPEDPRI	**Markets**	HLYLD	GII
MIGRATEOUT	LAB, LABS	**Carbon Cycle**	ICTSPEND		FPRI	WLIFE	DEMOCPOLITY
MIGRANTSIN		CARANN	ICTBROADMOBIL	**Secondary**	WAP		
MIGRANTSOUT	**Income**	TES		EDSECPER	AGX, AGM	**Risk Factor**	**Security**
	GDI		**Transport**	EDSECLOWRGRATE		HLSMOKING	SVDTHSCONFLICT
General	HHINCEARN	**Climate Change**	CARSTOT		**Energy**	HLOBESITY	
POP		ENVTEMP	INFRAROADPAVEDPCNT	**Tertiary**	**Demand**	HLBMI	**International**
HHSIZE	**Consumption**	CO2PPM		EDTERPER	ENDEM		**Civil Society**
LABORPR	C, CS	CO2PER	**Water and Sanitation**	EDTERGRADS		**Undernutrition**	DIPLOINGOADVOCACY
POPR	FIRMEARN		SANITATION		**Price**	MALNCHIL	DIPLOINGOPROGRAM
POPLE15	CSHHS, CSHHS	**Land**	SANITHH	**Quality**	ENPRI	MALNCHP	
POPGT65		LANDAREA	WATERHH	EDQUALSECALL	WEP	MALNCHPSAM	**Diplomacy**
POPWORKING	**Distribution**	LD			ENX, ENM		DIPLOEMBASSY
	GINIDOM, GINIFULL						DIPLOTREATY
		Pollution			**Production**		DIPLOIGO
	Global	ENVPM2PT5			ENP		
	X, XS				WENP		**Foreign Aid**
	M, MS	**Water**					AID
	IMFCREDIT	WATUSE					AIDDONATE, AIDREC
	XFDIIN, XFDIOUT	WATSAFE					
	WTRADE	WATUSE					**Power**
		WATSAFE					CPOW, NPOW
	Savings						
	SAVINGS						**Security**
							GOVNUCWP

interest to you. The Help system also has step-by-step lessons that may be useful in your exploration.

→ The most difficult of the Display options to use is Self-Managed Display. It is, however, also the most flexible, so a basic understanding is important. Test it by preparing a table and a graph of a single variable, population (**POP**), for a single country. Selecting Self-Managed Display under Display on the Main Menu will take you to the Full Variable/Parameter Selection form. Before you can use the Go to Display button on that form's bottom to take you to options, including Table or Graph, you must identify variable(s). As a test, select **POP** to display. To do so, either type **POP** in the input box on top of the long list of variables (upper case) and parameters (lower case) on the form (followed by touching your Enter key) or use the slider bar on the right-hand side of the large list to scroll down to **POP** and then click on it. In either case, a new box will appear with a list of the countries in IFs and you should select one. You will then see the variable name **POP** and your country selection in the Status Box at the bottom of the form. The Go to Display button will take you to the Display Menu form and allow you to display your table or graph.

→ Note that tables have a Save option that allows the user to save them to a file or to the clipboard. For tables that involve groupings of countries, options on the top of the Full Variable/Parameter Selection form include allowing you to switch between displaying individual countries and groups of them. Another option allows the automatic decomposition of a group into the country members. Many other options are available in the menus that accompany tables and graphics.

It is sometimes difficult to learn about the variables that will be of interest to you even with the help of *Flexible Displays* and *Self-Managed Display*. Table 4.2 should help, but you will not always know from the variable name if it is exactly what you want. In the last exercise, you saw that short definitions are available to you in the list of variables and parameters on the *Full Variable/Parameter Selection* form. It is also easy to learn much more about the variables. After you have selected a variable, the name can be seen in a Status Box at the bottom of the *Full Variable/Parameter Selection* form. If you click on a name in the Status Box, options at the top provide several ways of getting more information. For instance, *Explain* will show a flow chart from the area of the model in which the variable is computed; *View Equations* shows equations; *Drivers* shows forward and backward linkages.

→ Experiment with a small number of variables of interest to you from Table 4.2. Select them in *Self-Managed Display* and then click on them in the Status Box to learn more about them. You can find more integrated information about the entire model in the *Help* system, but this procedure often adds targeted information.

4.2.2 Data Analysis

The Main Menu of IFs is also the port of entry into *Data Analysis* options and capabilities. The database of IFs includes nearly 7,000 variables from more than 80 sources. It draws widely from standard sources, which are credited as you explore them, including the United Nations (population data, food and agricultural data, and substantial social data), the World Bank (very considerable economic and social data), the International Monetary Fund (international financial data), and the Organisation for Economic Co-operation and Development (social expenditures). In addition, many specialized sources such as the University of Maryland's Polity project, the World Values Survey, and the World Resources Institute (environmental data) have been used in preparation of the database over time.

A subset of the database is used by a data preprocessor to initialize model variables for a specified base year from which forecasting begins. An *Extended Feature* in the Main Menu of the interface calls up the preprocessor, which is a relatively unique feature among IAMs and helps the Pardee Institute regularly update that base year.

IFs includes a variety of tools for analyzing the associated data, including a basic Geographic Information System (GIS) or mapping capability and longitudinal (over time) display of individual variables. The tools also include cross-sectional (relationship) analysis for two or more variables, with graphical display as well as many alternative statistical computations.

→ At this point explore the *Data Analysis* options.

- Look at a *WorldMap* for some variables of interest to you and explore switching across variables, zooming in on regions in countries, and using other options available.
- *Analyze Across Countries* facilitates cross-sectional analysis. Look at the relationship between two variables, such as total fertility rate (**TFR**) as the dependent variable and GDP per capita at PPP in 2021 U.S dollars at purchasing power parity (**GDP2017PCPPP**) as the independent variable for data years of your choice. Generate a plot of the two variables against each other in which each point will represent a different country. Can you fit a line or curve to the relationship and see a statistical relationship?
- *Analyze Across Time* allows longitudinal analysis. Look at the change over time for individual or all African countries in variables like life expectancy (**LIFEXPECT**) at birth and at democracy levels for Latin America & the Caribbean (either **FREEDOM** or **POLITYDEMOC**). Can you do trend analysis and create a simple extrapolation (extension) of the time pattern of life expectancy in Latin America and determine what functional form (e.g., linear, logarithmic, or polynomial) has the best visual and/or statistical fit? Sections in the *Help* system from the Main page will take you deeper into the analysis.

Understanding at least the basics concerning Display and Data Analysis capabilities in IFs is foundational to any use of the system. The above survey should facilitate your use of the system to explore the Base Case forecasts and alternative pre-run scenarios with the version of the model you are using. It also should help you explore how those forecasts extend historical data series. Even further, it should facilitate your digging further into the incredibly rich database that comes with IFs, both for longitudinal and cross-sectional analysis.

What the information conveyed in this chapter **does not help you** with are (1) a deeper understanding of the models within the IFs system and how they interact with each other; and (2) the use of IFs to build and explore your own scenarios, whether that be to better understand where global dynamics may be taking us or to consider a variety of possible changes in those dynamics, some of which human intervention can bring about. The rest of this chapter provides an introductory overview of the structures that make up the IFs system, an important foundation for that deeper understanding and for building your own scenarios in the coming chapters.

4.3 Elements of the IFs Model System

This section provides a quick survey of the structure of IFs, one issue area/model at a time. The bullet listings of elements in the key models of IFs include terminology that will be very familiar to those who have worked in the respective issue areas but may not always be known by others. Subsequent chapters will help you understand that which is unfamiliar and to expand your mental models as you proceed.

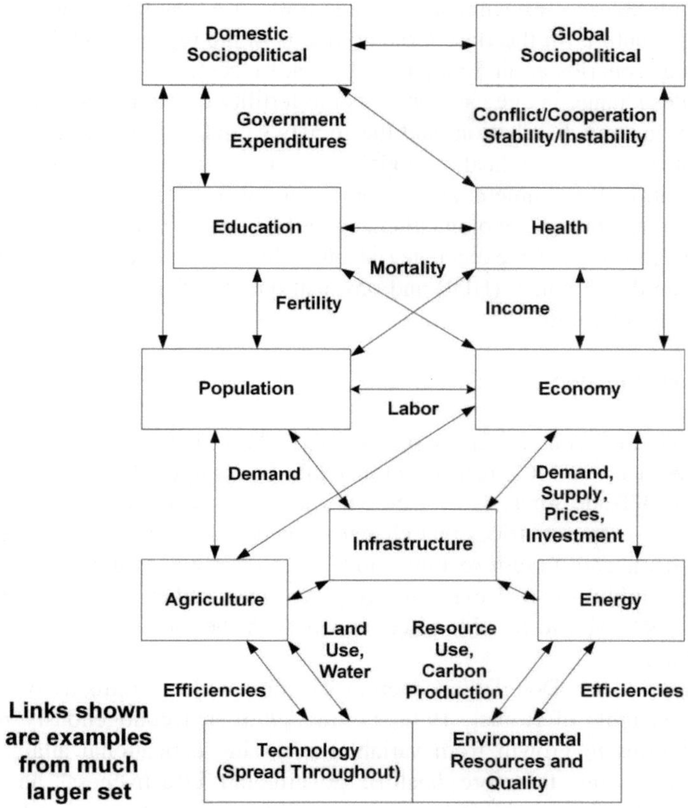

Fig. 4.2 An overview of IFs. *Source* Author

Figure 4.2 is the same as Fig. 3.8 and shows the major conceptual blocks of the International Futures system. The elements of the technology block and parts of the environment blocks are, in fact, scattered throughout the model. The named linkages between blocks and the linkages themselves are illustrative and not at all exhaustive.[2]

Chapters 6–13 will each present one or more of the models of IFs in the order shown below. The listing below identifies key structural elements.

The population model:

[2] The model is a dynamic, recursive (stepping forward one year at a time with results from the last year) system that can run without intervention from its initial year through time horizons to 2100 or beyond.

- represents 32 age-sex categories to age 150+ in a standard cohort-component system structure (in the Base Case no one is in the highest categories, but they facilitate scenarios around major medical advances);
- calculates change in age category-specific fertility of households in response to income, income distribution, and the society's total fertility rate;
- calculates (using the health model) change in mortality rates in response to educational attainment and income, mortality-specific cause drivers, and assumptions about technological change affecting mortality;
- computes average life expectancy at birth, literacy rate, and overall measures of human development (HDI) and physical quality of life;
- represents migration.

The economic model:

- represents the economy in six sectors: agriculture, materials, energy, manufactures, services, and ICT (other sectors could be configured, using raw data from the Global Trade Analysis Project, but only with substantial model revisions);
- uses input–output matrices that change dynamically with development level;
- is a general equilibrium-seeking model that does not assume exact demand–supply equilibrium will exist in any given year; rather it uses inventories as buffer stocks and to provide price signals with the model chasing equilibrium over time;
- contains a Cobb–Douglas production function with a productivity term (following insights of Romer, 1994; Solow, 1956) that endogenously represents contributions to growth from variables including educational attainment, life expectancy (and therefore health), government effectiveness, an index of domestic conflict, infrastructure extent, R&D expenditures, and extent of connection to the global economy via trade;
- uses a Linear Expenditure System to represent changing consumption patterns;
- utilizes a "pooled" rather than bilateral trade approach for international trade;
- is embedded in a social accounting matrix (SAM) envelope that ties economic production and consumption to inter-actor financial flows.

The health model:

- represents 15 different causes of mortality and morbidity across the communicable disease, non-communicable disease, and injuries/accidents categories;
- differentiates mortality and morbidity patterns by age and sex across causes;
- has underlying dynamics that vary by type of mortality;
- drives mortality in the population model.

The education model:

- traces schooling by year through primary, lower secondary, upper secondary, and tertiary levels;

- feeds completed years of schooling to a structure that maintains information on adult educational attainment by age;
- includes a rudimentary representation of educational quality;
- reconciles demand for education spending with supply of funds from governments.

The infrastructure model:

- represents development of paved roads, unpaved roads, electricity generation capacity, rural and urban access rates to electricity, irrigation, safe water and sanitation access across 3-step and 5-step ladders of access, wastewater treatment capacity, and information and communication technology adoption rates for land-line telephone, mobile telecommunications, broadband cable, and broadband mobile categories;
- tracks construction and maintenance costs and reconciles those with spending available including governmental.

The agriculture model:

- represents production, consumption, and trade of crops and meat; it also carries ocean fish catch and aquaculture in less detail;
- maintains land use in crop, grazing, forest, urban, and "other" categories;
- represents crop demand for food, for livestock feed, and for industrial use of agricultural products;
- is a partial equilibrium model in which food stocks buffer imbalances between production and consumption and determine price changes;
- overrides the agricultural sector in the economic module unless the user chooses otherwise.

The energy model:

- portrays production of multiple energy types: oil, gas, coal, nuclear, hydroelectric, solar, wind, geothermal, and other renewable energy forms;
- represents consumption and trade of energy in the aggregate;
- represents known reserves and ultimate resources of fossil fuels;
- portrays changing capital costs of each energy type with technological change as well as with drawdowns of resources;
- is a partial equilibrium model in which energy stocks buffer imbalances between production and consumption and determine price changes;
- overrides the energy sector in the economic module unless the user chooses otherwise.

The environmental model:

- allows tracking of remaining resources of fossil fuels, of the area of forested land, of water usage, and of atmospheric carbon dioxide emissions and concentration (with impact on temperature and precipitation by country and therefore on agricultural production and economic growth);
- represents water demand in agriculture, industrial, and municipal categories and represents water supply from multiple sources, thus effectively being a separate water model;
- provides a display interface for the user that builds upon the Advanced Sustainability Analysis system of the Finland Futures Research Centre (Kaivo-oja et al., 2002).

The sociopolitical model, in association with the economic model:

- represents fiscal policy through taxing and spending decisions;
- shows six categories of domestic government spending: military, health, education, R&D, infrastructure categories elaborated in IFs, other infrastructure, and a residual category; also represents net foreign aid;
- represents changes in social conditions of individuals (like fertility rates or literacy levels), attitudes of individuals (such as the level of materialism/postmaterialism of a society from the World Values Survey), and the social organization of people (such as the status of women);
- represents the evolution of democracy;
- represents the prospects for state instability or failure.

The international political model:

- traces changes in power balances across states and regions;
- allows exploration of changes in the level of interstate threat;
- has an option for switching these international variables plus trade, migration, and other international flow variables from monadic (country total) to dyadic (country-to-country) representations.

The implicit technology model:

- is distributed throughout the overall model system;
- allows changes in assumptions about rates of technological advance in agriculture, energy, and the broader economy;
- represents the extent of electronic networking of individuals in societies;
- is tied to governmental spending on R&D.

Note that the models interact in a multitude of ways. Very illustratively: Economic growth affects population growth and vice versa; changes in assumptions about energy resources affect prices and investment decisions and, again, economic growth. Rising education levels and wealth affect fertility and mortality, values and opinions, quality and character of political institutions, political stability, and even war and peace. Some of these interconnections are well-specified, based on extensive research. Others are more speculative, more of a starting point than an end point in anyone's work with the system.

4.4 Conclusion

As indicated earlier, basic use of the IFs system requires at least an understanding of the display-related features of the interface, so that you can explore the Base Case scenario and other scenarios pre-run with your version of IFs. This chapter has provided a survey of aspects of the display system, including Flexible Displays, Specialized Displays, and Self-Managed Displays. It has also introduced you to the huge database that underlies the models of IFs and that facilitates longitudinal and cross-sectional display.

In addition, this chapter has provided a very basic overview of the models within IFs. Both use and understanding of the overall system will require considerably more experience and elaboration before any sense of mastery is likely. It is very common for new IFs users to feel somewhat tentative (and perhaps even somewhat overwhelmed) at this point. Yet, learning about and using IFs can reward your efforts.

The next steps toward understanding the model and learning to use it are (1) an introduction to scenario analysis with IFs (the next chapter), and then (2) issue-by-issue, model-by-model, and scenario elaboration (the following chapters).

References

Hughes, B. B. (2019). *International futures: Building and using global models.* Academic Press.

Kaivo-oja, J., Luukhanen, J., & Malaska, P. (2002). *Methodology for the analysis of critical industrial ecology trends: An advanced sustainability analysis of the Finnish economy.* Finland Futures Research Centre.

Romer, P. M. (1994). The origins of endogenous growth. *Journal of Economic Perspectives, 1*(Winter), 3–22. https://www.jstor.org/stable/2138148

Solow, R. M. (1956). A contribution to the theory of economic growth. *Quarterly Journal of Economics, 70*(1), 65–94. https://doi.org/10.2307/1884513

Scenario Analysis with IFs

5

Abstract

Chapter 4 began presentation of the structure, features, and use of the International Futures (IFs) system. It focused on the display and data analysis capabilities of IFs. This chapter turns attention to undertaking scenario analysis with IFs. To support the use of IFs in the volume, it defines scenario analysis not only in terms of broad and integrated stories but also in terms of more focused interventions that can support causal thinking and potential policy analysis. The Base Case scenario of IFs is not a set of individual variable extrapolations; it is instead the product of complex dynamics across the many interacting issue areas represented by the multiple models within IFs. The interactive interface of IFs facilitates not only the development of scenarios based on simple or elaborate interventions controlled by parameter changes but also the saving of such scenarios for future exploration, elaboration, and comparison. A seven-step exercise and a discussion of typical elements in more elaborate intervention sets support learning those processes.

5.1 Introduction

The previous chapter summarized how to use the many *Display* options in International Futures (IFs) to explore the IFs Base Case and other pre-run scenarios. It also outlined at least initial use of the *Data Analysis* feature to explore historical patterns of countries and the world across the wide range of issues that IFs represents and to do statistical analysis across data series. Although those features provide everything that many users of IFs want, much of the system's value lies in supporting analysis of the leverage we have in improving global futures. That analysis lies substantially in building and exploring alternative scenarios.

B. B. Hughes, *Exploring and Shaping International Futures*, World-Systems Evolution and Global Futures, https://doi.org/10.1007/978-3-032-22084-4_5

5.2 Understanding Interventions and Scenario Analysis

A scenario is a story or story outline. Thinking about the future normally involves creating alternative scenarios or stories about its possible evolution. Some such scenarios are exploratory and consider the possible unfolding of different futures around key uncertainties, such as the rate of some aspect of technological advance or the fragility of some element in the global environment. Other scenarios are normative and develop stories about preferred futures, such as a global transformation to sustainability, requiring social expenditures or other action. There is a scenario analysis guide scenario analysis guide on the IFs Wiki that can be very useful; it has more information than we can provide here, but this chapter provides the foundation for undertaking scenario analysis in subsequent ones.

Scenarios in a computer model typically are built from multiple interventions that collectively help build a coherent story about the future. Often however, and in this volume, the word "scenario" is used more loosely to refer to any intervention (such as the change of a fertility rate for a country or an alternative assumption about changing costs of a renewable energy form). The interventions almost always involve changes to parameters in the IFs system, much less frequently to initial conditions of selected variables.

When IFs or other computer simulations are "run" without making any changes to parameters and initial conditions specified as the default values, they generate a forecast that in IFs is called the Base Case (or sometimes "reference run"). The IFs Base Case, always available when a model session is initiated, is itself a scenario. Sometimes the Base Case is incorrectly referred to as a trend extrapolation or a "business as usual" scenario. That is incorrect because the dynamic formulations, and therefore the forecasts of IFs, often generate changes from historical or contemporary patterns (for instance, the emergence of dominance by renewables in energy production or substantial changes in fertility and mortality patterns with economic development). Sometimes also in the IFs project it is called the Current Path scenario, which too can be somewhat misleading. In summary, the Base Case of IFs is a computation that involves the full dynamics of the model and therefore typically displays non-linear behavior, often resulting in paths that are quite different from past patterns or current trends.

The Base Case is a good starting point for scenario analysis for two reasons. First, it is built from data-based initial conditions of all variables and on parameters that have been given reasonable values from data or other analysis. These initial conditions and parameters make up the package of interventions that, in interaction with the model's structure, generate the Base Case scenario. Second, the Base Case is periodically analyzed relative to the forecasts of many other projects across the range of issue areas covered by IFs and is to a degree "tuned" to reproduce the general behavior expected by respected forecasters and our team.

In addition to the Base Case, most versions of IFs include several other pre-run scenarios (see Table 5.1 for additional important terminology). The other pre-run scenarios are those that the Pardee Institute has developed in projects with partner organizations or in its own analysis.

Table 5.1 Scenario terminology and management

Scenario files	
Base case	The "standard" forecast that comes with IFs without any parameter change interventions. It is not a prediction. Clearly some of the multitude of assumptions in IFs must be wrong, but we need to start somewhere. The Base Case gives us a reference point before we begin changing assumptions. It is always scenario number "1"
Working file	Opening IFs copies the Base Case into the Working File. When you make any parameter changes, the Working File becomes distinct. Unless you save a Working File, thereby providing it a name, it will disappear when you exit IFs. It is scenario number "0"
Pre-run scenarios	Any scenario that is available when you start working with IFs is a pre-run. Pre-run scenarios have names (such as SDG Push.run.db) with numbers of "2" or higher
File management	
Creation of a scenario intervention package	To create a scenario intervention package, use the Scenario Analysis "tree" to specify new parameter or initial condition values. In the tree you can Name and Save the scenario intervention package. You can reopen such a file in a later session
Run a scenario intervention Package and Create New Pre-run	When you create a scenario in the tree, it consists only of a set of intervention parameters. To see the implications of those assumptions for the future, Run Scenario from the tree, calculating all variable values into the Working File. Then use Scenario Analysis/ File Management/Save Working File As … to make the Working File a pre-run scenario for later use
Editing of a scenario intervention file	If you Named and Saved a scenario intervention package and wish to work on it later, open or add it into the scenario tree. Then you can edit it and re-run the model. You may want to Name and Save the intervention file again and after the Run again Save Working File As …

If you look at the *Flexible Displays* form, you will see a list of pre-run scenarios in the box at the bottom of the screen. Because those have already been run, based on a set of interventions constituting their foundations, the user can immediately display their results.

→ If you have not already done so, use *Display* options to explore the differences across the pre-run scenarios available to you with IFs in some area of interest to you. For instance, in *Flexible Displays* you can hold down the Ctrl (control) key and click on/highlight more than one scenario at a time, then create tables or graphs across multiple scenarios. With focus on selected variables, explore the range of variations offered by the scenarios in your model version and ask yourself if it seems reasonable and what your own preconceptions about likely futures might be. World population and world GDP are two useful variables that often frame variations in global future analysis.

5.3 Simple Intervention Scenario Analysis

To go beyond the Base Case and other pre-run scenarios, and to create your own scenario, use one of the key elements of the IFs system, the scenario tree. To call up the tree, look on the Main Menu under *Scenario Analysis* and pick the *Quick Scenario Analysis with Tree* option. Figure 5.1 shows the tree with intervention subcategories elaborated.

The scenario tree allows you to call up an extensive number of your own interventions and/or a set of stored scenario intervention files. It is important to understand that the world as represented by IFs consists substantially of *stocks* and *flows* (see also Appendix 1 on The Philosophical Approach in IFs to Modeling). Stocks carry over across time and can be incremented or decremented. Examples are population, financial and physical capital, energy resources, atmospheric carbon level, and knowledge. Flows are what increment or decrement those stocks across time. Examples are births, deaths, economic investment, capital depreciation, and carbon emissions. A great many of the intervention points in IFs for analysis target changes in flows directly and stocks only indirectly. For instance, the total fertility rate (number of births per woman across a lifetime) affects births and the share of the energy production system that is based on fossil fuels affects carbon emissions.

Before we discuss scenario analysis involving multiple interventions, however, let's take a short, guided tour through the scenario process with a single intervention so that you understand the conceptual steps in creating a scenario. To make effective use of the rest of this book, you *must be able to create your own scenario,* so take a little time with this extensive exercise.

→ **Step 1: Driver/Parameter Identification.** Parameters control relationships in IFs and drive scenarios. Much less frequently, a change in the base year's initial value of a variable will drive a scenario. So, to make an intervention into the model you almost always choose a parameter and change it. Although all the parameters exist like fruit at the end of the branches of the scenario tree, your first challenge is to find the one you want to change. The remaining chapters of this book will help you do that by identifying the names of some of the key parameters related to issues and interventions of interest to you. For instance, the next chapter will identify a multiplier on total fertility rates as an important parameter in creating scenarios involving population change. That multiplier

Fig. 5.1 Scenario management system. *Source* Screen capture from IFs 8.57 by author

has the parameter name **tfrm**, but you have no reason to know that. What you will know is that you want to affect fertility, mortality, productivity, energy, or whatever. The tree offers you two ways to find and select the specific parameter you want to change. Steps 2a and 2b sketch those alternatives. Table 5.2 lists some of the most frequently used parameters in the model. You might want to bookmark, print or otherwise simplify access to Table 4.2 (frequently used variables) from the preceding chapter and Table 5.2 here, because you may refer to them often throughout this book as you undertake scenario analysis. For more elaboration of parameters and their use, see the Wiki's Guide to Scenario Analysis.

→ **Step 2a. Searching by Name or Key Word.** Use the *Parameter Search* menu option. If you know the parameter name, you could for example type in **tfrm** to directly call up the parameter and the option of jumping to it in the tree and manipulating it. Typing a word of interest (like "fertility" or "tfr") can also

Table 5.2 Most frequently used parameters across IFs issue areas and models

Demographics	Economy	Health	Education	Infrastructure	Agriculture	Environmental	Sociopolitical	
Fertility	**Capital**	**General**	**Enrollment**	**Roads**	**Production**	**CO_2**	Domestic	International
tfrm	invm	hlmortm	edpriintngr	infraroadm	tgrld	carbtax	**Security**	**Power**
contrusm			edtergradgr	infraroadpavedpcntm	ylm, ylmax	carfuel (1, 2, 3)	sfintlwaradd	cpowf
	Productivity	**Disease**	edprisurtrgtyr		landirareaactual	envclimsens	wpextinterv	drcpow
Mortality	mpfadd	***Communicable***	edtergradtrgtyr	**Electricity**	fishcatchm	envylchgm	govriskm	
mortm	mfleadr	watsafem		infraelecaccm	aglosstransm			**Threat Levels**
		sanitationm	**Finance**	ensolfuelm	aglossprodm	**Land Use**	**Capacity**	wpgreat1
Migration	**Social Accounting**		gdsm (education)		ldcropm	forestm	govcorruptm	wpcon
migrateinm	govrevm	***Non-Communicable***	edbudgon	**Water & Sanitation**		carforst	goveffectm	glocom
migrateoutm	gdsm	hlsmokingm	edqtqltrm	landirareaequipm	Energy		govregqualm	ally
	firmtaxm	hlobesitym		sanitationm	**Demand**	**Water**	econfreem	
Labor	hhtaxm	malnchpsamm	**Gender Parity**	watsafem	endemm	watres		**War**
workageentry	ginidomm		edprigndreqintn		enrgdpgr	desald	**Inclusiveness**	waron
workageretires	gdprext	***Injuries & Accidents***	edprigndreqsur	**ICT**			democm	cdmf
labparm		deathtrpvm		infratelem	**Supply**	**Air Pollution**	freedomm	
labfemshrm	**Trade/Aid**		**Quality**	ictbroadm	enpm, eprodr	envpm2pt5m	gemm	**Diplomacy**
femshrgr	xshift	**Technology**	edqualag15m	ictbroadmobilm	resorm	ensolfuelm		diploembassymaxm
	termtrm	hltechshift	mfpedqual		rdm	hlsolfuelsw		diploigomaxm
	aiddon, aidrecm			**Finance**	qem			diplotreatymaxm
			Grade-Level	infraroadpavedcostm	eninvm	**Economic Impact**		
	Informality		edpridrpgrdm	watsafecostm	enml, enxl	climeconimpsw		
	labinformshrm		edprienrgrdm	ictmobilcostm		climeconimpstk		
	gdpinformshrm			landircostm				

often help you find the appropriate driver in the scenario tree. Try searching on "fertility" to find **tfrm** and then load it.

→ **Step 2b. Searching by the Logic of the Tree.** You could alternatively open up the fertility multiplier parameter by exploring the structure of the tree. Looking at the Scenario Tree, click on the *Households/Individuals* trunk (this is a banyan-type tree with multiple trunks). That certainly would seem a more likely place to find a way to change fertility than would *Environmental Uncertainties* or *Firms/Businesses*. Among the branches you will see the obvious place to look next, namely *Demographic/Population*. Click on it and a box of drivers will open. You will find the parameter name and a brief definition.

→ **Step 3: Driver Change.** Whichever route (2a or 2b) you followed to find the parameter, click on the TFR (total fertility rate) multiplier. That will open a box providing a list of country names. Pick the geography of interest to you. If you want to use a grouping of countries (like sub-Saharan Africa) rather than a single country, you click on (toggle) the *Set Group or Country* option on the menu of the form before you select the driver. After you select a parameter and geography, a graphic will appear showing the current value(s) of your parameter over time. Because with **tfrm** you selected a **multiplier**, the Base-Case value is always "1". The model uses multipliers to easily raise or lower base-case values of many variables and the multipliers are always "1" in the Base Case so that values are unchanged by them. You could use the *Fully Customize* option to create any pattern of intervention over time you desire, and at some point you should explore its use. For now, select the *High* option above the graph and note the change in the figure. This option will phase in a higher fertility for your chosen country or group.

→ **Step 4: Run the Model.** Now you have changed a parameter, but it will affect nothing else until you run the model and recalculate all the variables in it for all the countries. Pick the *Run Scenario* option from the menu. Your parameter will first be loaded into the *Working File* (which until now has been identical to the Base Case as discussed in Table 5.1). You will then be shown the Running IFs form. You can change the end year or leave it as is and *Start Run*. The first year of model runs takes much more time than subsequent years, because it sets up many variables for future use. When the message appears saying "Run Successful – Click to Continue," click on that button.

→ **Step 5: Analyze your Results.** When all variables in the model have been recomputed and you have been told that the run was successful, the Working File will, in essence, be your new scenario. Use the techniques you learned in Chap. 4 to compare the Working File with the Base Case. Either use a pre-prepared display from *Flexible Displays* or use *Self-Managed Display* to compare **POP** from it with **POP** for your chosen country or grouping in the Base Case—it should be higher in the Working File, because you increased fertility. But your intervention will also have changed much else. Look, for instance, at Gross Domestic Product (**GDP**) of your country before and after the intervention and at GDP per capita (**GDPPC**). If your country or grouping was fairly large, you may also see changes in variables elsewhere around the world.

→ **Step 6 (optional): Save your new Working File with a Name of Your Choice.** To do this, look under *Scenario Analysis* for *File Management/Save working file as ...* If you do that, your scenario run will appear with the name you then provide when you open other forms like *Display/Flexible Displays* and when you open IFs again in the future. **Important note: Saving your working file as a new pre-run scenario is only available in the downloaded version of IFs. When using the web version, you will need to retrieve the scenario intervention file and re-run the model to produce a new Working File for your analysis.**

→ **Step 7 (optional): Save your parameter change(s) as a scenario instruction file.** To do this, look under *Scenario Analysis* for *Quick Scenario Analysis with Tree.* The menu of that form contains a *Scenario Files/Name and Save* option that allows you to name a folder for saving the scenario and the instruction file itself. If you do that, the scenario instructions will be saved for retrieval later (in the Scenario folder under IFs on your machine when using the stand-alone IFs and on the web for one month if you use that version of IFs). As the information that comes up when you *Name and Save* says, there will be two ways you can later retrieve the saved scenario instruction file. If you are simply opening the scenario instruction file in the tree, touch the *Scenario Files/Open* option, move your mouse to *Other* at the bottom of the list that pops up, move your mouse to *User Defined Scenarios*, and keep moving to your defined folder name and your scenario name. Clicking on that scenario name will repopulate the tree with your parameter change(s). Alternatively, use the *Add Scenario Component* option and move to *a User Defined Scenarios*, etc. With either approach, once your parameter changes are in the scenario tree you can make additional ones and then save the revised scenario instruction file and/or run the scenario.

5.4 Building More Extensive Scenario (Story) Analyses

Now that you have mastered (to some degree) the scenario of the above exercise with a single intervention, we can step back and discuss scenarios more generally. The tree structure implicitly (although not fully in its category names) distinguishes among key categories of interventions. There is no perfect typology of the blocks for building scenarios, but the conceptual approach found useful in the IFs project includes the following elements:

- **Framing scenario foundations.** Much scenario analysis implicitly or explicitly identifies one or two key dimensions of uncertainty that are not fully under direct human control. In global scenarios, for example, a common dimension is the pace of technological change and of economic growth driven by it (for example, because of AI advances). A second important dimension of uncertainty that is not under direct human control is whether the environment is fundamentally robust or fragile with respect to both demands for inputs to human systems and reactions to outputs.

- **Elaborated agent interventions**. Governments and households are the primary agents in IFs, allowing the movement of scenario elaboration closer to the level of specific levers of action. Scenarios adding elements of agent behavior can specify changes in the levels and patterns of foreign aid, different fiscal priorities by government (such as more R&D, military, or education spending), an OPEC oil production restriction in a given year, economic liberalization within India, and so on.
- **Shocks, surprises, wild cards.** These can refer back to either the technological or environmental dimensions of framing scenarios or to surprises in human agency. Breakthroughs in human life expectancy or, conversely, a new plague illustrate this category.
- **Changes in causal understandings**. This is not a typical element in scenario analysis, but mental or computer models clearly can have specifications that are wrong. IFs allows the user to have access to initial conditions and all relationship parameters of the model from the Scenario Tree form. For instance, one can change an initial condition like the government debt level of a country as a percentage of GDP (**GOVDEBT**) or the HIV infection rate (**HIVRATE**).

Most users will focus their attention on key framing assumptions and agent/policy interventions. Again, see Table 5.2 for the most frequently used parameters and see Table 5.3 for some others that will sometimes be of interest. Remember that each intervention in the tree changes a single parameter of the model.

Although users have total control over scenario development with the tree, many will want to draw also on stored scenario intervention files. The *Add Scenario Component* menu option of the tree allows retrieval into the tree of pre-existing intervention packages. Loading and running scenario parameter files that can be found in the category named *World Integrated Scenario Sets* allows IFs to produce scenario runs that exhibit behavior related to Pardee Institute projects, including for the Global Trends reports (2020, 2025, and 2030) of the U.S. National Intelligence Council; the Global Environmental Outlook (GEO-4) work of the United Nations Environmental Program; and projects with the United Nations Development Programme (UNDP). As of this writing, the SDG Push scenario from work with the UNDP is a pre-run scenario that was loaded and run with these tools. So are the pre-run Shared Socioeconomic Pathway (SSP) scenarios, replicating and extending that set, which is used heavily by integrated assessment models in looking at possible futures across a wide range of alternative assumptions. Subsequent chapters will discuss these integrated scenarios further (Chapter 8 explores SDG Push and Chap. 14 looks across many scenario sets).

Table 5.3 Other frequently used parameters across IFs issue areas and models

Demographics	Economy	Health	Education	Infrastructure	Agriculture	Environmental	Sociopolitical	
Fertility	**Productivity**	**General**	**Intake & Survival**	**Roads**	**Demand**	**CO_2**	**Domestic**	**International**
eltfrcon	mfpedspn	hlmorbm	edterintgr	Infraroadraitrgtval	clpcm	enrgdpgr	**Security**	**Power**
tfrmin	mfpinfrindtrad	hlstddthsw	edseclowrtrangr	infraroadraisetar	elascd	carabr	sfusehist	cpowdf
	mfpinfrindict		edseclowrtrantrgtyr		indemm	envco2fert	sfconv	govnucwpadd
Mortality	mfpenpri	**Disease**	edtergradm	**Electricity**		carabrchng	govriskweight	wpconmax
mortconv		***Communicable***		enelecshrendemm	**Production**			
	Social Accounting	malnelimstartyr	**Finance**	infraelecacctrgtval	envylchgadd	**Water**	**Capacity**	**Threat Levels**
Migration	indirecttaxm	malnelimtargetyr	gdsedm	infraelecaccsetar	livhdpro	watdemandelas	govcorruptsetar	wpgreat2
migrateinr	ginidomr		edexppconv		elagind	landirareaequipelas	goveffectsetar	wppowtran1
migrateoutr	govhhtrnpenm	***Non-Communicable***	edexpsuconv	**Water & Sanitation**	aglosstransm	desali	govregqualsetar	wpconsq
	gdsedm	envpm2pt5m		watsafenewsw	aglossconsm	desald		wpthrconv
Labor			**Gender Parity**	landirareaequipm	ldcropm		**Inclusiveness**	
labunemprm	**Trade/Aid**	**HIV/AIDS**	edseclowrgndreqtran			**Air Pollution**	swseffects	**War**
labmarketsw	protecm	hivm	edtergndreqint	**ICT**	**Energy**	ensolfuelsetar	democimpus	wpthrwar
labdemsm	xfdistockm, xfdistoutm	hivtadvr		ictbroadtrgtval	**Demand**	hlsolfuelsw	democimpoth	cwarbase
	xfdiwgrm	aidsdrtadvr	**Quality**	ictbroadsetar	elasde	ensolfueltrgtyr	democpolitysetar	nwarsv
			edqualpriallm	ictbroadmobilsetar	carbtax		gemsetar	cdmf
	Informality	**SAMs**	edqualsecallm			**Economic Impact**		elndpow
	gdpshadowcoefflabinf	malnchpsamadj		**Finance**	**Supply**	climeconimplin		
	labinformcoefftax	malnchpsamm	**Grade-Level**	infrainvnewpubshrm	resorunconm	climeconimpsq		
	labinformcoeffbus	samadmissionadd	edmodelsw	Infrainvmaintpubshrm	etechadv			
	labinformcoeffed	samavertedm	edseclowrenrgrdm	infraelecaccruralcostm	etechadvuncon			
				ictmobilcostm				

→ Running a large-scale scenario drawing upon two scenarios developed for the UN Women organization. From the Main Menu choose *Scenario Analysis* and then *Quick Scenario Analysis with Tree.* Once inside the Scenario Tree two paths lead to the folder with those scenarios: (1) use the *Scenario Files/Open/Other* menu options on the Scenario Tree and search under World Integrated Scenario Sets for the UN Women scenario set; or (2) use the *Add Scenario Component* menu option on the Scenario Tree and search under *World Integrated Scenario Sets* for the UN Women scenario set. Put the specific scenario instruction sets dealing with equality of wages and education for women into the tree. The names of those sets are *unws1WageRat* and *unwS4EdFem.* After opening the first (which will delete the earlier **tfrm** intervention if you have not already done that), add the second. Run this new and quite complex scenario and again compare it with the Base Case. From the Flexible Displays form, use the Search Lists option to find variables related to "gender.“ Look at some of the displays such as the Gender Development Index for the World and World Bank Low Income groups to explore how the scenario compares with the Base Case.

In summary, the Scenario Tree can be considered a kind of scenario cockpit because it allows you great flexibility in creating and managing your own scenario, combining your own specific intervention assumptions and the stored scenario intervention files. Remember that the Scenario Tree form allows you to *Name and Save* your own scenarios and come back to them later. The pre-run scenarios and the saved scenario parameter files typically have annotations that explain their foundations to you (see in Fig. 5.1 the *Annotate Scenario* option that will call up any existing annotation concerning a scenario you have loaded and allow you to annotate any scenario you build). Lessons in the *Help* system provide further guidance about scenario analysis.

5.5 Taking Stock

When you have finished your investigation of IFs in Chaps. 4 and 5, you should have basic proficiency with respect to the six major steps of analysis with the model:

1. Exploring the initial forecast or scenario of the model called the "Base Case" using various display options. Again, the Base Case is not a prediction of the future. It is a statement of how the future might look, ***if*** the initial conditions, equations, and Base Case parameter values of the model were all correct (and they can never all be completely correct).
2. Changing assumptions of the Base Case. You will use the *Scenario Tree* form for this. But before you do it, think through the objectives of your scenario analysis. What is the question you wish to address? What is the geographical scope of importance for your analysis? What kinds of interventions will either

(a) represent specifications of uncertainty about key developments, or (b) appear to be areas where policy choices might be important?

3. Running the model again to obtain new forecasts or "scenarios." Your changes will often create what you consider to be a more accurate or more desirable representation of the world. You will use the *Run* option on the *Scenario Tree* form to do this.
4. Comparing the forecast(s) generated by the new assumptions with the Base Case. Make sure that you understand why the new scenario has produced different results. This will often require that you understand quite a bit about the model structure as outlined at the end of Chap. 4 and elaborated across the rest of this volume and in the IFs Wiki. Your search for understanding may require still more changes of assumptions and runs of the model.
5. Saving the new scenario you have made for future use (optional). You will often save the scenario intervention file from the scenario tree. In the stand-alone version of IFs, you will frequently save the Working file into a new pre-run file.
6. Evaluating what you have learned about the model and about the world it represents. To what degree are the results you have obtained artifacts of a model that you either do not understand or with which you disagree? To what extent have you stretched your own understanding of the world and therefore improved your own mental model of it? Your work with a computer model can productively continue until you feel that the model cannot take you further with your analysis or your mental model is superior to the computer model (and can continue still further, if you are then willing to extend the computer model itself).

These steps of scenario analysis are somewhat idealized, but you may periodically refer to them to help you make the most efficient use of IFs.

5.6 Conclusion

As indicated earlier, instruction on IFs use and the survey of model elements in this chapter and the preceding one are intended to be foundational for your use. That usage and understanding of IFs will require considerably more experience and elaboration before any sense of mastery is likely. It is very common to be feeling somewhat tentative at this point.

The next steps toward understanding and using IFs are more easily mastered on an issue-by-issue basis. Therefore, the following chapters introduce you individually to the various issue areas of IFs. They will help you undertake specific interventions and issue-specific scenario analysis. The final chapter will return to scenario analysis integrated across the issue areas.

Part III
Human Development

Population

6

Abstract

This chapter is the first of nine in Part III that focus on specific models and issue areas in the International Futures (IFs) system. All nine review forecasts and uncertainties concerning global futures in the issue area before discussing core dynamics within IFs and then moving to scenario analysis. This chapter begins with long-term forecasts of country-specific and global demographics from other sources. Other forecasts facilitate comparisons and show that demographic forecasts with IFs are often well placed within those from other institutions and models. IFs represents age and sex cohorts of population that progress through time, responsive to births, deaths, and migration. One unique feature of the IFs demographic model is calculation of deaths in an integrated health model (Chap. 8) that represents 15 categories of mortality and morbidity. Another is causal elaboration of fertility dynamics, including the use of a large-scale model of education (also elaborated in Chap. 8) that tracks children through years and levels of education and tracks attainment levels of adults as they age. Economic conditions (elaborated in Chap. 7) also affect fertility, mortality, and migration. Model users have many leverage points for intervention analysis.

6.1 Introduction

Chapter 2 discussed the accelerating growth of world population over centuries, including the surge following World War II that reached peak rates around 2.1% per year in the 1960s. The accelerating growth was due to many factors, including improvements in diets and public health. It is not surprising that many who were attentive to that pattern were greatly concerned about the growing pressures it put on resources, including food supplies.

B. B. Hughes, *Exploring and Shaping International Futures*, World-Systems Evolution and Global Futures, https://doi.org/10.1007/978-3-032-22084-4_6

Much has happened since that global peak in population growth rate. Even though the rate has quite steadily fallen since the early 1970s, dropping to 0.82% in 2025, global population has climbed from 3 billion in 1960 to 8.3 billion in 2025. We need to understand not only the possible patterns of future growth, but also the drivers of that growth as they affect both fertility and mortality patterns. It will be important to consider also how the age and sex structures of the population may change.

Future population patterns are fundamentally important to all the issues addressed in this volume and are of great interest to broad forecasting communities. Unlike the situation with respect to many of the issues explored in coming chapters, we can compare century-long forecasts of population and economics from IFs with those made by others. Demographic forecasts from two institutions stand out. (Note: Both institutions use the term "projections," common also across the integrated assessment model (IAM) community, but because that term can suggest an extrapolative methodology, the IFs project and this volume generally refer to "forecasts" and "scenarios.")

The United Nations Population Division (UNPD) collects data globally and typically provides data and new multiple-variant projections biennially. IFs uses UNPD historical data for model initialization and includes its projections in Flexible Displays facilitating comparison. A second source of forecasts, also attentive to UNPD data, is the Wittgenstein Centre (WIC) for Demography and Global Human Capital in Vienna, Austria, associated with the International Institute for Applied Systems Analysis (IIASA). WIC/IIASA work includes population projections, most recently updated in 2023, for countries of the world for the five Shared Socioeconomic Pathway (SSP) scenarios (O'Neill et al., 2014, 2017). IIASA maintains those in its broader SSP database, used widely across the IAM community. Flexible Displays in IFs provides access to the values of WIC/IIASA scenarios as well as those of the UNPD when the Complete List option is chosen from the bottom of the window.

Figure 6.1 shows forecasts from the IFs Base Case in comparison with those from the UNPD (low, medium, and high scenarios) and WIC/IIASA (the SSP2 scenario, often called "middle of the road"). Although there is some probability that almost any scenario will prove prescient, the High Variant of the UNPD (orange line) looks almost impossible in a world where fertility rates are declining substantially and more rapidly than life expectancy is rising. It is not surprising that the UNPD's high value for 2100 was reduced by nearly one billion people between its 2017 and 2024 releases. In 2017, its medium projection was about one billion people higher than the WIC/IIASA SSP2 and IFs Base Case scenarios, but all three are now almost identical despite quite different methodologies. The UNPD low projection, which would bring global population in 2100 back down below that of 2025, would require an exceptionally rapid decline in fertility and/or a significant rise in mortality.

UNPD, WIC/IIASA, and IFs forecasts of population all rely on underlying expectations for births and deaths, driven in turn by fertility rates and mortality

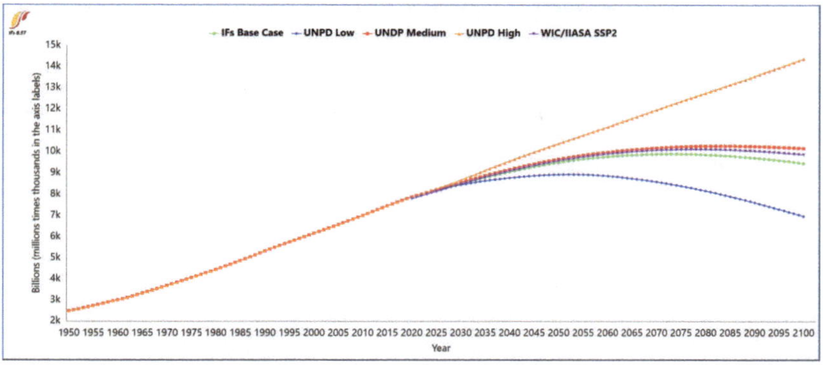

Fig. 6.1 Global population history and forecast. *Sources* IFs 8.57 Base Case; UNPD projections of 2024; and IIASA/ WIC SSP2 2023 Revision (for IIASA/WIC methodology, see Lutz et al., 2014, 2018)

patterns by age. These variables underlie the rise and then decline of global population growth rates over the long run and determine also the variation of those rates across countries.

Significantly driving the rise of global population rates over the long-term was a rise in life expectancy, not least from the survival of more newborns and children into adulthood and reductions in maternal mortality. Global life expectancy in the World Bank's low-income country category rose from about 45 years in 1960 to 52 years in 1980. It reached 67 years in 2025, an extraordinary achievement. In lower-middle-income countries, the climb was from 48 to 71 and in upper-middle-income countries it rose from 58 to 77. Even in high-income countries, life expectancy rose from 69 to 80 over that same 65-year period.

The forecasts in Fig. 6.2 come from the UNPD's medium projection. Those from IFs are similar globally, but in IFs the gap across income categories gradually closes across the century to seven years in 2100, rather than the stabilization at 14 years after 2030 in UNPD projections.

Features of interest in Fig. 6.2 include the large early-period dip in life expectancy of upper-middle-income countries. That is a result of the great rise of famine-associated mortality in China (currently upper-middle income and treated that way throughout) during the Great Leap Forward years (1958–1960). Also, the roughness in the line for low-income countries can be attributed significantly to famine years in Africa with inadequate foreign assistance to offset food shortages, including in the early 1980s. Also see the dip around the world in the early 2020s associated with COVID-19.

Figure 6.3 turns again to forecast comparisons and shows the UNPD medium projection, WIC/IIASA SSP2, and IFs Base Case forecast of global life expectancy across the rest of this century. There is remarkably little variation across them and all three pretty much continue the somewhat slowing rise of recent years.

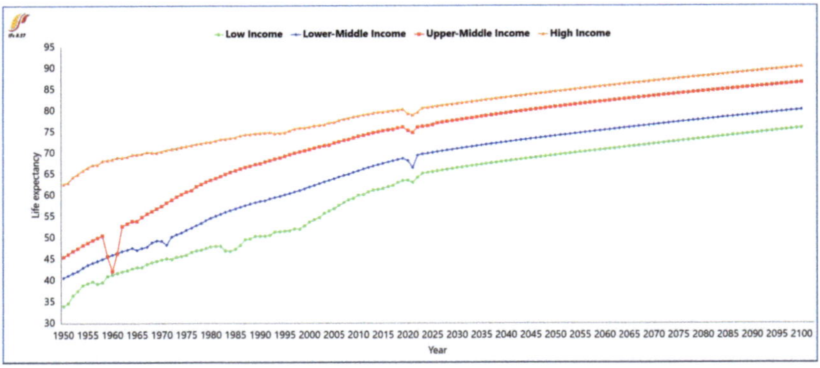

Fig. 6.2 UNPD life expectancy history and forecasts for global income groupings. *Source* IFs 8.57 showing UNPD medium projection (2024) aggregated by World Bank country-income groupings

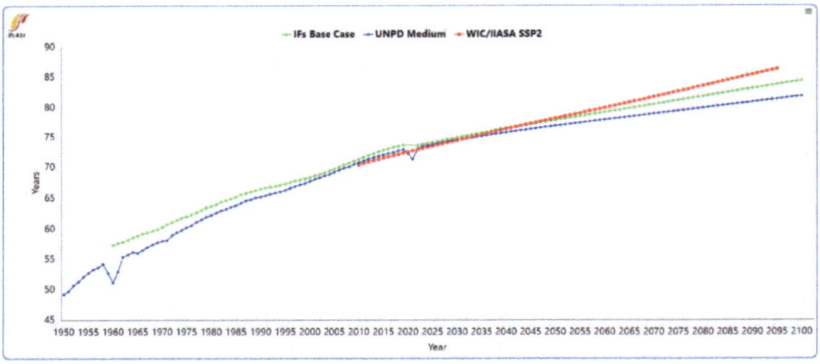

Fig. 6.3 Global life expectancy history and forecasts. *Sources* IFs 8.57 Base Case, UNPD Medium Projection (2024), and IIASA/WIC SSP2 (Lutz et al., 2018)

In forecasting of population, much attention in recent decades has been on fertility, where patterns have changed even more dramatically than those for mortality. Figure 6.4 shows historical data and UNPD medium projections of the total fertility rate (TFR) for the four World Bank country-income groups. The TFR is the average number of live births during lifetimes of women. Note that TFR dropped very substantially for all groups between 1955 and 2025. The decline was most dramatic in the upper-middle-income category, greatly influenced by China's implementation of two-child and then one-child policies. It was also very substantial for lower-middle-income countries. The decline has come later for low-income countries, beginning about 30 years after their rise in life expectancy. Significant continued low-income decline appears likely given the experience in other income groups and the likely rise in GDP per capita (discussed in the next chapter). Again, the IFs Base Case scenario anticipates more convergence across the World Bank country-income categories than does the UNPD.

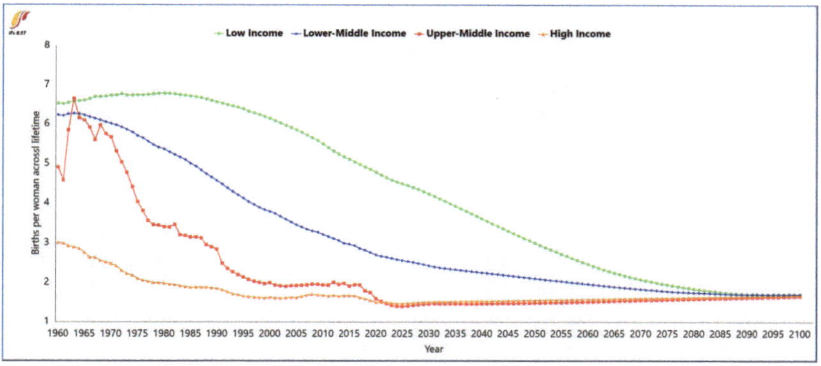

Fig. 6.4 TFR History and forecasts across global income groupings. *Source* IFs 8.57 showing UNPD medium projection (2024) aggregated by World Bank county income groupings

A TFR of 2.1 is often identified as the population "replacement rate," recognizing that not all children will reach adulthood and be fertile. Even though the TFR in lower-middle-income countries may fall to 2.1 by about 2050, and low-income countries may reach that by 2070 (IFs Base Case 8.57), population in the former set of countries may well continue to grow until near the end of the century and that in the latter set may be growing significantly into the next century. Population growth continues for decades in countries after TFR falls to replacement rate. That is because the age structure is then such that the elderly population facing nearer-term mortality is substantially outnumbered by the population in their child-bearing years. Figure 6.5 illustrates this point by showing the age-sex structure forecasted by IFs for sub-Saharan Africa (SSA) in 2025, 2050, and 2100, in comparison with that of the 27 European Union (EU) countries in 2025. Although the IFs Base Case suggests that SSA may reach replacement TFR in 2070, its population growth rate in 2100 could still be about 0.2% and its age structure will be younger than that of the EU countries.

If current patterns and such forecasts prevail, global population is likely to begin shrinking near 2090 and is already doing so in the EU and some of East Asia. The discussion in much of the world has thus shifted from concern about population growth to concern about its shrinkage. While shrinking population can have positive implications for the environment, it generates at least two interacting concerns for societies, namely (1) relative growth in the size of the aged population who often need various forms of social support; and (2) relative shrinkage in the number of those entering the workforce and able to generate incomes and financing for social well-being (Sciubba, 2022). Figure 6.5 illustrates the imminence of these issues for the EU where the two largest population cohorts in 2025 were 55–59 and 60–64, both about 25% larger than the 10–14 and 15–19 cohorts.

One aid to understanding some of these concerns is the concept of "demographic dividend," which is measured by the ratio of numbers of people of working

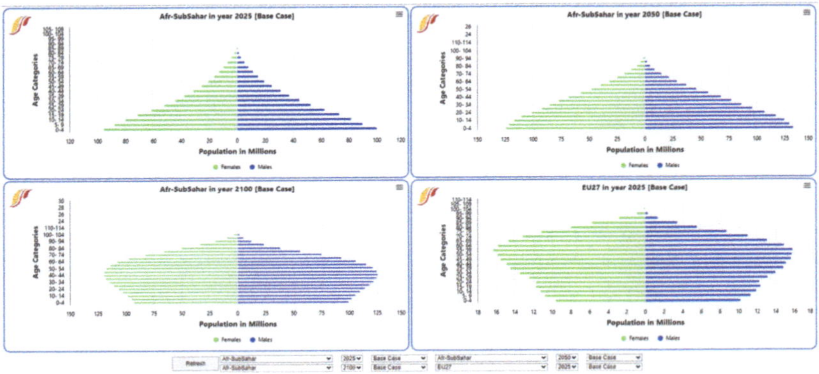

Fig. 6.5 Age-sex distribution in sub-Saharan Africa (2025, 2050, and 2100) and in the European Union (2025). *Source* IFs 8.57 Base Case

age (15–64) to the sum of elderly (65 and older) and school-aged children numbers. In 2025, those ratios were 1.26 and 1.81 for low-income and high-income countries, respectively, a real economic advantage for the latter category. The IFs Base Case suggests the ratios will be reversed at 1.85 and 1.24 in 2100, mostly because of the decline of youth dependence in low-income countries and the rise of elderly dependence in high-income ones. The ratio will fall for the two other world income categories, most dramatically in upper-middle-income countries from 2.2 to 1.2. Clearly, such a coming transition will pose significant issues for aging countries, not least growing government debt in many of them with large-scale need for programs of transfers to, and medical care for, the elderly.

Challenges from high fertility rates, including bulging populations of school-aged children and restricted choices for women leading to high birth rates, are not going to disappear quickly in much of the lower-income world. Migration, mostly of men, from the lower- to the higher-income countries will lower population growth rates in origin countries and increase them in recipient countries. In fact, such a flow, long existent, has accelerated since the 1960s as shown in Fig. 6.6.

The outflow patterns (not shown) are insightful. From low-income countries, there was great variation that can be attributed to periodic famines and conflicts in those countries that drove people abroad. In contrast, the outflow from lower-middle-income countries has been more stable, but also generally increasing. It is frequently suggested that many people in poor countries want to move to richer ones, but that it requires some income to finance migration. The increasing GDP per capita in lower-middle-income countries appears to be facilitating that.

With respect to the absolute value of net migration rates, they are clearly highest and mostly rising for the high-income recipient countries. Rates of three to four per thousand since 2000 means that migration has been contributing roughly 0.35% per year to population growth. While that has been offsetting low and even

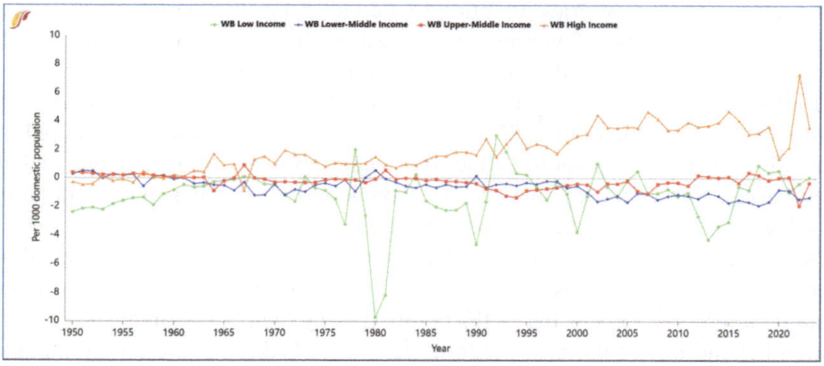

Fig. 6.6 Net migration flow rates into World Bank country-income groupings. *Source* UNPD data

negative domestic birth rates in many recipient countries, the accumulation of such inflows over even 2.5 decades can lead to foreign-born populations rising to close to 10% of the domestically-born population. The rise of domestic populist and anti-immigrant social movements in high-income migrant recipient countries has become substantial. At the same time, the issue is interacting with significant domestic conversations related to the implications of below-replacement fertility. Governments that have tried to increase fertility rates have generally had very limited success. Illustratively, China, which used considerable coercion to reduce TFR starting in the late 1970s, has not succeeded in reversing that reduction. The uncertainties about the unfolding of competing pressures with respect to immigration is one reason that forecasting of migration patterns by the UNPD, WIC/IIASA, and IFs must be treated with great caution.

There are, of course, other important demographic-related issues. Controlling infectious and parasitical diseases (IPDs) plays a critical role in the effort to increase longevity, especially in lower-income countries. More than 25% of all deaths globally are attributed to IPDs and their prevalence waxes and wanes with the evolution of bacteria, viruses, and parasites and with the effectiveness of sanitary practices, vaccines, and drugs to prevent or cure them. The identification in recent decades of diseases like AIDS, COVID, Ebola, and Bird Flu and the resurgence of long-known diseases like malaria have caused some observers to speculate that humanity faces resurgence in IPDs and the devastation they cause. The importance of health, not simply in terms of the demographic implications of age-sex-specific mortality and resultant life expectancy, was reason for the development within IFs of an extensive health model (see Chap. 8).

Another important demographic issue is urbanization. From 34% in 1960, global urban population passed 50% of total population in 2008 and has continued to climb, reaching 58% in 2025, with variation across countries from 14% in Papua New Guinea and Burundi to 100% in Singapore.

Social scientists also draw attention to a concept labeled "youth bulge," said to emerge when the ratio of young people—especially young men—comes to represent a larger than normal share of the population, making it hard for them to find jobs and often creating a breeding ground for resentment, revolution, and/or terrorism. In 2020, there were 26 countries in which the portion of the population aged 15–30 was 30% or more of the total under 65 years of age. Those included Afghanistan, Syria, Yemen, and the West Bank and Gaza Palestinian populations, all subject to various degrees of social unrest.

As indicated earlier, population dynamics are affected by and affect almost every issue of interest to us in global forecasting. It will be useful to know more about demographic forecasting in IFs before turning to scenario analysis. For information beyond that in the next section, see the IFs Wiki, the IFs network diagram , and the Pardee Institute's home page.

6.2 Population Dynamics and Leverage

The simplest forecasts of population futures, including some of the high variants in global population forecasts, are no more than extrapolations of current growth patterns. They may use a "simple growth model," an equation that computes population at time point (t) as population at a prior point ($t - 1$) times one plus a growth rate (r). Computer models can generalize this formula to predict population at any time point in the future, based on an initial point ($t = 0$).

Yet, not a single demographer believes that population growth rates will fail to change over a long period of time. Instead of looking at aggregate growth rates, modeling population growth normally focuses on forecasting births and deaths, often framing changes in those rates in terms of the demographic transition that has been going on for the last several hundred years—first in Western Europe and its offshoot-states, now in most of the rest of the world. The classical theory of demographic transition[1] states that death rates and birth rates start out high, but about equal, leading to small or no changes in population. Then, advances in knowledge and improvements in income result in a decline in death rates, causing rapid population growth. Later, socioeconomic development brings about sustained reductions via decline in fertility rates—at which point births and deaths may come back into balance.

While this description fits the historical data reasonably well, it does not account for the rapid spread of sub-replacement fertility rates in many rich countries in recent decades nor for the very rapid decline in fertility rates in almost all poor countries, whether they are experiencing much economic advance or not. Neither does it do a good job in explaining exactly which "socioeconomic" forces are

[1] This characterization of the "classical theory" is by Eberstadt (2005; see also 2000) who attributes this theory to Frank Notestein (see Notestein and 1967). For a failed forecast using a different demographic theory see Ehrlich and Ehrlich (1968), who predicted that population would outstrip food supplies, leading to hundreds of millions of deaths by starvation in the 1970s.

pushing world birth and death rates back toward equilibrium (or whether those forces could overshoot and result in sustained population decline). How much of the decline is due to economic forces, how much to improved communication, women's empowerment, or government population control policies? It is because demographers are unsure of the answers to those questions that forecasts by the UNPD, WIC/IIASA, IFs, and others can show a broad range of estimates.

In terms of modeling, professional demographers rely on an approach to forecasting population called cohort-component analysis. Cohort-component analysis represents the population of a geographic region by age category (called a "cohort" as its members age) and by sex. Demographers refer to these representations as age-sex distributions (see again Fig. 6.5). It is common to maintain the categories for five-year age groupings, such as for people who are 5–9 or 20–24 years of age. Demographers also collect and maintain data on the number of births that occur annually for every 1,000 women in each age category and the number of deaths that occur each year per 1,000 people in every age category. These data create fertility and mortality distributions, respectively. Like age-sex distributions, they vary by country. In summary, before delving into the complications around driving variables, it is mechanically simple to combine age-sex, fertility, and mortality distributions into a dynamic model of population growth.

In its displays IFs represents the population, fertility, and mortality of each country or region in 32 5-year age categories, with the oldest containing people 150 years of age or older (in case of major medical advances). By multiplying the fertility distribution by the age distribution, the model calculates the number of births and it adds those infants to the bottom cohort. Multiplying the mortality distribution by the age distribution provides the number of deaths in each cohort, and those diminish the respective cohort each year. In addition, *roughly* one-fifth of each cohort passes to the next age group in every year. However, "under the hood" in IFs, the model actually tracks 1-year rather than 5-year cohorts for more accuracy in that dynamic, appropriately for a model that advances time in 1-year steps.

This approach to forecasting population directs our attention to age-specific fertility and mortality distributions. It builds upon the core causal understanding of population dynamics shown in Fig. 6.7, which shows the interaction of two feedback loops. A positive feedback loop connects births to population; all else being equal (the implicit assumption of all causal analysis), as population rises, births rise, and as births increase, population grows. A negative feedback loop connects deaths to population. Whether population grows or not obviously depends on the relative strength of these two loops and our scenario analysis in the next section focuses heavily on the underlying strength in each. (The IFs Wiki has flow diagrams that elaborate the structural detail of this and other causal diagrams in this text; for discussion of this one look under Contents/Understanding the Model/Demographic Flow Charts. See also Chap. 3 for a general discussion of such feedback loops.)

The focal points in population forecasting therefore are the fertility and mortality rates (deaths by cause), which affect the numbers of births and deaths,

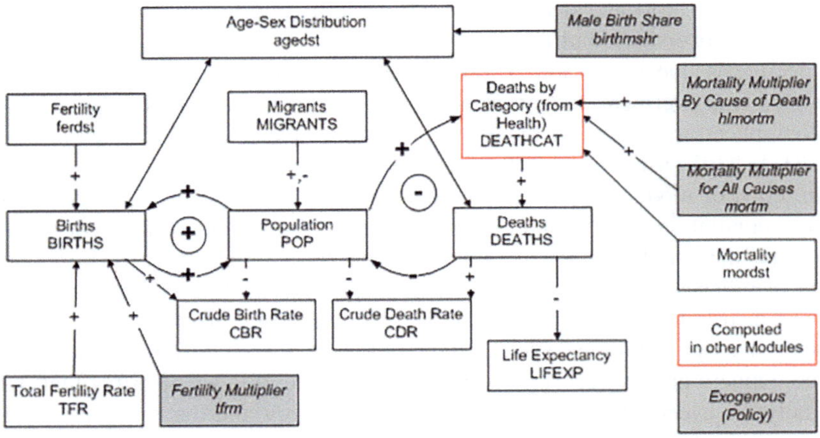

Fig. 6.7 Primary population dynamics in IFs. *Source* IFs Wiki https://pardeewiki.du.edu/index.php?title=Population

respectively. IFs computes those rates as functions of multiple driving variables, briefly discussed below.

Rich countries tend to have low fertility and mortality rates, while poorer countries tend to have higher rates. The explanation for the impact of income on mortality is fairly obvious. Individuals in societies with higher income are more likely to have access to good nutrition, clean water, safer working conditions, better health care, and other supports of a longer life.

Explanation of the depressing effect of higher income on fertility is somewhat more complicated. Whereas children in poor families often provide economic benefits through both early earnings and support of parents in old age, in rich families the costs of child care and education more commonly exceed any income that children add to the family. Another very important factor in reducing fertility is the status of women in a society. When women have more access to education and to job opportunities, the average number of children they bear in a lifetime decreases sharply. Governmental or private programs that make information about family planning available and that provide access to contraceptives also have an unmistakable impact in reduction of fertility.

While we face important sources of uncertainty, this review suggests that humans exercise much leverage on the growth of our population. The next section turns to development of alternative scenarios for global population in the face of uncertainty.

6.3 Interventions and Scenario Analysis

IFs facilitates investigation of how we might help shape a future consistent with our own understandings and values. We can do that through scenario analysis, as introduced in Chap. 5.

In this, the first of the chapters focused on specific models and issue areas, it is useful to reiterate the two basic steps of scenario analysis. Specifically, you will first select a set of indicator variables upon which you wish to focus. Then you will choose and manipulate parameters within the model that represent points of possible human leverage or other uncertainty to push those indicators in the directions you desire or that represent alternative assumptions about uncertainties. This will often be a process of trial and error, one that will help you think through causal linkages, constraints, and trade-offs.

6.3.1 Indicator Identification and Review

You will often select multiple sets of indicator variables upon which to focus as you develop scenarios. One set will be specific to the issue area in which you are exploring leverage (population, economics, energy, and so on). Other sets will be drivers of your focus variables and variables that they in turn influence. You will need to be attentive to causal dynamics across the IFs system.

Multiple sub-options of *Display* on the Main Menu (discussed in Chap. 4) will help you identify both specific issue and bigger picture variables of interest:

- *Flexible Displays* has multiple categories to population issues: *Demographics* is the macro category, and meso categories include *Mortality*, *Fertility*, and *Migration*. Categories and variables in them will change across IFs versions, so explore.
- *Basic Report* contains variables of interest in all issue areas.
- Options in *Specialized Displays* include Cohorts of Population for all countries and groupings.
- *Self-Managed Display* provides access to all variables and parameters. *Variable Search* on the resulting form, with entries you make like "population," "fertility," or "migrate," can help identify relevant variables. To assist your search, the IFs Wiki elaborates the model structure and names. From the Main Page of the Wiki go to Understand IFs and then to the model in IFs of interest to you.

You will most likely want in *Flexible Displays* or *Self-Managed Display* to look at **POP** for total population, **POPR** for population growth rate, **TFR** for total fertility rate, **LIFEXP** for life expectancy, **CBR** for crude birth rate (births per 1,000 people), **CDR** for crude death rate (deaths per 1,000 people), **INFMOR** for infant mortality, **MIGRATE** for the rate of net migration, and possibly **POPURBAN** for the size of the urban population.

In addition, you will want to keep an eye on indicators of the bigger picture that changes in population (or any other issue you are exploring) might affect. You might wish to look at variables such as world and region-specific gross domestic product per capita (**WGDPPC** and **GPPPC**), the human development index (**HDI**), calorie availability per capita (**WCLPC** and **CLPC**), and literacy rates (**WLIT** and **LIT**). Population changes in model simulations may also affect the global level of atmospheric carbon dioxide in parts per million or relative to pre-industrial levels (**CO2PPM** or **CO2PER**) and the world forest area (**WFORST**), as well as model variables that point to changes in domestic political stability and interstate rivalry (**SFINSTABALL**, for example, and **THREAT**).

6.3.2 Choosing and Using Points of Intervention

Once you have explored variables of interest in the Base Case scenario, you will dig into causal analysis. Scenario analysis with an IFs model always involves the interaction of two elements. The first is understanding the endogenous (or internal) working of the model. This is the "all else being equal" component of causal analysis, critical because in any analysis most of the IFs system is calculating numbers in any scenario with the same equations and parameters it uses in the Base Case scenario. The second element is exploration of the exogenous (or external) assumptions imposed on the model.

You will change parameter values—possibly only a single parameter and seldom more than a handful of parameters—to intervene in the model. Even a single intervention, however, will change a calculation that then normally provides a different number to another calculation, which in turn provides a new number to still another calculation, and so on, a process that uses the structures of the model to propagate a limited intervention through all model calculations.

To make an intervention, select the *Scenario Analysis* option from the Main Menu, then click *Quick Scenario Analysis with Tree.* You will see the main intervention categories for IFs and, if you click on any of them, numerous possible intervention points will be listed. The demographic intervention points are found in several different places. If you click on *Technological Change* and then click on *Demographic/Population,* you will see parameters you can change, such as a trend of mortality reduction with time and technology (**tmortr**) or a brute force mortality multiplier (**mortm**). In the *Households/Individuals* category, *Demographic/Population* subcategory you can see a multiplier on contraception use (**contrusm**) and a direct multiplier on fertility rate (**tfrm**). There are also demographic intervention points under *Governments/Sociopolitical Systems*, *Selected Initial Conditions*, and *Relationship Parameters.*

BUT: rather than searching the scenario tree**, most users of IFs** take a different approach to identifying parameters. The scenario tree listing shows a total of about 350 intervention points. This is an intimidating number of issues to deal with, and depending on your interests you probably will want to use only a small proportion of that number. If you know the name of a specific parameter, like "**tfrm,**" you can use the *Parameter Search* option to search the tree for it. There are two ways of identifying specific parameter names. You can again use the IFs Wiki, which both identifies the parameters and shows their logic of use in the model flow diagrams and equations. Or you can in *Parameter Search* provide a keyword like "fertility," "mortality," or "population." Table 5.2 provided a listing of frequently used parameters across the IFs system so may also be helpful here. Table 5.3 shows less frequently used parameters.

Where might you intervene in IFs to create new and interesting population scenarios? In our discussion of population dynamics, we saw that two key drivers were fertility and mortality rates and the effect they have on births and deaths. IFs has multipliers available for those rates (**tfrm** and **mortm**, respectively). Multipliers allow you to effectively increase or decrease variables that are otherwise calculated endogenously. That is, the multiplier is used to increase or decrease the effective value of the variable it multiplies in calculations where the variable is used. When the value of a multiplier is "1," it has no effect. There are multipliers for a great many variables throughout IFs.

Important Note 1: Multipliers. When you introduce a multiplier value other than "1," the combined value of the multiplier and the variable you are multiplying sets up changes in the model that can propagate widely, returning through feedbacks to alter the variable being multiplied in future years. Although sometimes the model feedback dynamics will amplify the effect of the multiplier, more often the dynamics will dampen the effect of your scenario. This is because the model, like the real world, has many negative or equilibrium-seeking loops that resist movement. This tendency for the system to "fight" against an intervention is familiar to, and frustrating for, policymakers. When you find it happening in IFs, it suggests that real-world leverage may be hard to achieve.

→ From the Scenario Tree, click on *Households/Individuals*, then click on *Demographic/Population*, and then click on the total fertility rate multiplier driver/parameter (**tfrm**). A pop-up box appears with five entries. *Define* gives a concise description of what this variable—the leverage point—does in the model. *Drivers*, *Explain*, and *View Equation* give more detailed information about the parameter and its use in the model. Click on *Select* and pick a country. Or click on the menu option *Set Group or Country*, pick *Groups*, and select a country grouping. Whichever you do, frames appear that make it easy to increase or decrease the fertility rate now embedded in the model for this country or grouping by some multiple. Use the slider bar or hit the *Low* button to decrease the value of the multiplier to 0.8. This change asks the model to assume there will only be 80% of the births that would otherwise be calculated. Note that the model gradually introduces your new value over time, which is usually the most sensible assumption—policy or technology changes rarely have their full impact immediately. If you want to have even more control of the process, click on *Fully Customize*. This button opens another window that allows you to make bigger or smaller changes in the multiplier and to change the timing of the change.

→ The *Run Scenario* menu option will run the model with your change. Then examine the size of population and its growth rate in comparison with the Base Case. How much effect does that assumption have by 2050? You may have to run the model longer to see really significant change.

You should have in mind a basis or rationale for changes that you make in multipliers or other parameters. Do you believe that technological breakthroughs may prolong life substantially? If so, you might wish to consider changing the values of the mortality multiplier (**mortm**).

We often use multipliers to force into the model changes that we understand might come about in multiple ways. For instance, a reduction of **tfrm** from 1.0 to 0.8 could represent a postulated breakthrough in birth control technology, an intensification of family planning programs, increases in income, or improvements in the health and education of prospective mothers. IFs contains many "general" multipliers because it cannot always represent all the more specific policy levers that you might want to consider. In the population area, however, it does include some such specific levers. For instance, **contrusm** allows you to increase or decrease the portion of the population using contraception (read about this and other parameters in the Wiki by selecting *Help*). Or you can increase government spending on health or education (**gdsm**). Several parameters affect migration patterns, including **migrateinm** and **migrateoutm** (most countries have both inward and outward migration, so the model provides multipliers on both). Looking beyond demographic variables to those in other models of IFs that will more indirectly affect demographics, you might accelerate or decelerate economic growth via changed assumptions about technological advance (**mfpadd** or **mfpleadr**). (The next chapter will explain how these work in the economic model.)

Important Note 2: Saving and Retrieving Scenarios. See again Table 5.1 for information about scenario files and file management. When you specify interventions in the *Scenario Tree,* they persist only as long as your current session with the model. To save them for a subsequent session, use the *Scenario Files, Name and Save* option of the tree. In the web version of IFs, you are not allowed to save the actual file of full results created when you run the model—those files are too large and would clutter the server for other uses. Thus, when you want to recreate a scenario in a future session, you should reload the saved intervention file into the tree (using *Add Scenario Component*) and re-run the model. But in the stand-alone version of IFs, you can save complete run files as well as intervention files. To save the run files, look in the Main Menu under *Scenario Analysis, File Management* for the *Save working file as ...* option. Use of it will copy the working file to the name you specify and you can later *Open* that file, copying it back into the working file of a subsequent session—or simply use the file with the given name in comparative analysis across scenarios. It is suggested that you save the intervention file as well in the stand-alone version, making it possible to retrieve and/or change later (use the *Scenario Files/Name and Save* option when your intervention is in the Tree).

Changing intervention variables from other models/sections of IFs can have a big impact on population. Try increasing health expenditures to see what happens to births, deaths, and population. Change the government health expenditure multiplier **gdsm** for the World grouping from 1.0 to 2.0 and run the model to 2100.

When you introduce a scenario intervention into the tree, your change stays there until you explicitly reverse it or exit from the model. If you wish to introduce another parameter change, it is important that you be able to distinguish its impact from that of the earlier one, and you will normally remove the earlier one so you can compare your new scenario with the Base Case. To do so, use the options *Scenario Files/Clear Tree.* You may want to save your old scenario first (see Important Note 2).

→ If you do the intervention with health expenditures, what happens to WPOP and to the distribution of global population between young, working age, and old people? Do the differences in population change between the OECD countries and the non-OECD countries (the rich-developed countries vs. the less-developed countries) seem reasonable to you? What happens to the other categories of government expenditure for any country or grouping of special interest to you?

6.4 Conclusion

This chapter has been the first on specific issue areas. We have considered the issues and causal dynamics surrounding population and we have begun to explore the leverage we may have with respect to population growth. We have also reiterated and elaborated the process by which a model user can develop scenarios alternative to the Base Case (Chap. 5 laid the foundation). Scenarios with new assumptions about model parameters or exogenous variables allow the user to explore the effects of policy choices (usually general policy orientations rather than specific policies). Scenario analysis can help us explore how policy in one area, such as demographics, affects things such as economic growth or energy use as well as population. Remember that these scenarios do not in any sense predict the future. Rather, we use IFs and scenario analysis to help us think about shaping the future in the face of uncertainty.

References

Eberstadt, N. (2000). *Prosperous paupers and other population problems.* Transaction Press.

Eberstadt, N. (2005, January). Global democracy: Four unanticipated trends. *Current*, *469*, 10–15. https://www.readabstracts.com/Education/Global-democracy-Four-unanticipated-trends-Are-there-too-many-workers-The-global-workforce.html

Ehrlich, P. R., & Ehrlich, A. H. (1968). *The population bomb.* Ballantine Books.

Lutz, W., Butz, W. P., & Kc, S. (Eds.). (2014). *World population and human capital in the twenty-first century.* Oxford University Press.

Lutz, W., Stilianakis, N., Stonawski, M., Goujon, A., & Kc, S. (2018). *Demographic and human capital scenarios for the 21st century: 2018 assessment for 201 countries*. Publications Office of the European Union. https://data.europa.eu/doi/10.2760/835878

Notestein, F. (1967). The population crisis: Reasons for hope. *Foreign Affairs, 46*(1), 167–180. https://www.foreignaffairs.com/articles/1967-10-01/population-crisis-reasons-hope

O'Neill, B. C., Kriegler, E., Ebi, K. L., Kemp-Benedict, E., Riahi, K., Rothman, D. S., van Ruijven, B. J., van Vuuren, D. P., Birkmann, J., Kok, K., Levy, M., & Solecki, W. (2017). The roads ahead: Narratives for shared socioeconomic pathways describing world futures in the 21st century. *Global Environmental Change, 42*, 169–180. https://doi.org/10.1016/j.gloenvcha.2015.01.004

O'Neill, B. C., Kriegler, E., Riahi, K., Ebi, K. L., Hallegatte, S., Carter, T. R., Mathur, R., & van Vuuren, D. P. (2014). A new scenario framework for climate change research: The concept of shared socioeconomic pathways. *Climatic Change, 122*(3), 387–400. https://doi.org/10.1007/s10584-013-0905-2

Sciubba, J. D. (2022). *8 billion and counting: How sex, death, and migration shape our world*. W. W. Norton & Company.

Economics

7

Abstract

International Futures (IFs) includes a dynamic general (all economy) equilibrium model (GEM) representing the supply and demand sides of six economic sectors: agriculture, energy, other raw materials, manufactures, services, and information and communications technology. Partial equilibrium physical-unit models of agriculture and energy link closely to the GEM. Two features of the economic model make it quite unique: (1) extensive elaboration of multifactor productivity in the production function, and (2) a full social accounting matrix structure that facilitates analysis of interacting household, firm, and government finance across 188 countries. It also includes a representation of the impact of climate change on economic production. These features help link the model closely in two-way interactions to almost all other models in IFs. This chapter provides selected Base Case forecasts with some comparison to those from other integrated assessment models, summarizes model structure, and includes some exercises to support economic analysis with IFs.

7.1 Introduction

Countries devote much effort to determining policies that will sustain or increase economic growth rates. At first glance, they appear to have been very successful. Between 1900 and 2025, the global GDP grew from just under $4.5 trillion to nearly $110 trillion (each in constant international 2021 dollars at market exchange

Dr. Evan Hillebrand contributed much to this chapter in the previous edition. Deva Sahadevan, Pardee Institute Senior Research Associate, contributed to this chapter.

B. B. Hughes, *Exploring and Shaping International Futures*, World-Systems Evolution and Global Futures, https://doi.org/10.1007/978-3-032-22084-4_7

rates [MER]), an annual average growth rate of about 2.8%.[1] In per capita terms, GDP grew from about $2,000 to just over $13,000 globally. Despite the Great Depression, World Wars I and II, and the collapse of empires, those 125 years saw by far the most rapid economic growth in human history.

In addition to admiring the pace of economic growth, we should recognize the great importance of having reached contemporary global per capita GDP levels. Dramatic social transformations occur in almost all societies as GDP per capita grows across such a range (see Chaps. 8 and 12). Primary school attendance becomes nearly universal and literacy rises well above 50%. Life expectancy jumps sharply toward that in the richest countries. Change in such "quality of life" variables is much less marked above $10,000 per capita than below it (Hughes, 2001). There is also evidence that when a democracy's GDP per capita is above about that same level, it is unlikely to slide back to autocracy or anarchy (Przeworski et al., 1996). Thus, it is very significant to be able to say that the world on average—and 72 of the countries for which we have comparable data—have reached at least $10,000 per capita.

There are reasons to think that global economic growth rates across the twenty-first century could approximate those of the twentieth (although the International Futures (IFs) Base Case suggests that slowing population growth will likely also slow economic growth rates). First, global economic growth has been on a generally accelerating path for more than 200 years. Maddison (1995) pointed out that the "Golden Age," the period of most rapid economic growth in human history at that time, was 1950–1973. Yet, even growth during the much-maligned 1973–1992 period was extraordinarily rapid by historical standards, and it has risen since then.

Second, the very large differentials that built up globally in GDP per capita have created a pattern in which countries at lower-income levels can significantly increase their growth rates by strengthening human capital, improving governance, and adopting technology from richer countries. In fact, the growth rates of GDP (at MER) in countries outside of the generally high-income countries of the Organisation for Economic Co-operation and Development (OECD) have exceeded those of OECD countries every year since 1995 except the COVID-19 impacted years of 2021–2022, as have GDP per capita growth rates since 2000.

Third, the enormous stock of human capital devoted to advancing science and engineering is expected to produce further dramatic improvements in technology—in biotechnology, in microelectronics, in nanotechnology, and in energy production and utilization—that should enhance global economic growth. A major wild card is the emergence of artificial intelligence (AI).

[1] Maddison (2001: 261) and author's calculations. Maddison's numbers use purchasing power parity calculations to account for between-country differences in price levels that are not reflected in aggregations that use exchange rate calculations. Purchasing power parity figures tend to increase the estimates of GDP in the less developed countries compared to measures that use market exchange rates.

It is, of course, also possible that some combination of failures to improve governance, breakdowns in international cooperation, and environmental damage—or all forces combined—could significantly slow global growth.

A major challenge is that despite the spread and acceleration globally of economic growth, it has been extremely uneven and income inequalities among and within many countries have widened. In 1900, the richest countries of the global economy, such as Australia, New Zealand, the United Kingdom, and the United States, had GDPs per capita in the $9,000–$12,000 range (in 2021 international dollars at MER), whereas the poorest were at levels well below $2,000. Today, those same four high-income countries have reached per capita levels near or exceeding $50,000, while 44 in the IFs country set remain below $2,000.

Looking ahead, the World Bank, the United Nations, the International Monetary Fund (IMF), and the OECD regularly produce forecasts of global economic growth, as do several non-governmental organizations like the World Economic Forum and the McKinsey Global Institute. Generally, the forecasts extend just a year or two. But the IMF's regular *World Economic Outlook* reports add value with forecasts about five years ahead, and the IFs project uses those forecasts in its own years across that same horizon before moving to endogenous calculation.

These organizations provide data and forecasts that are sometimes at market exchange rates (MER), sometimes at purchasing power parity (PPP), and sometimes at both. GDPs at MER reflect its value in local currency converted to U.S. dollars at official exchange rates in a specified year (like 2021). GDPs at PPP account for the differential costs of goods between countries and the United States. Because goods in lower-income countries tend to cost less at the official exchange rate than they would in the United States (tourists often appreciate that lower cost), the local economy has more effective value than the conversion rate suggests.

Thus, the GDP at PPP for such countries, adjusted for that higher purchasing power, is almost always higher than the GDP at MER (they are identical for the United States). In Ethiopia, that difference in 2020 was a factor of 2.6 and in India it was nearly 3.0. As such countries become richer and more broadly tied into international trade, the PPP values will converge downward toward the MER values. Therefore, GDP PPP growth rates will generally be slower than GDP MER growth rates. The next section of this chapter will provide more information about such unique features of the IFs economic model. IFs computes GDPs at both MER and PPP. Many figures in this chapter show it at PPP, in part because other long-range forecasting generally uses PPP.

Specialized OECD-based analyses sometimes extend the time horizon across the century by contributing to the Shared Socioeconomic Pathways (SSP) forecasts (see Chap. 1 for introduction to the SSPs) that extend across the century (Dellink et al., 2017), as have analysts within the International Institute of Applied Systems Analysis (IIASA) system (Cuaresma, 2017). Figure 7.1 shows the 2023 release of OECD projections for the five SSPs and for the IFs Base Case forecast (all at PPP). The value from IFs in 2100 is in the cluster of OECD forecasts that sit between the two extremes of SSP3 and SSP5. By 2100, the IFs value is slightly higher than those in the OECD cluster, but that reflects in part IFs use of 2021 dollars and the

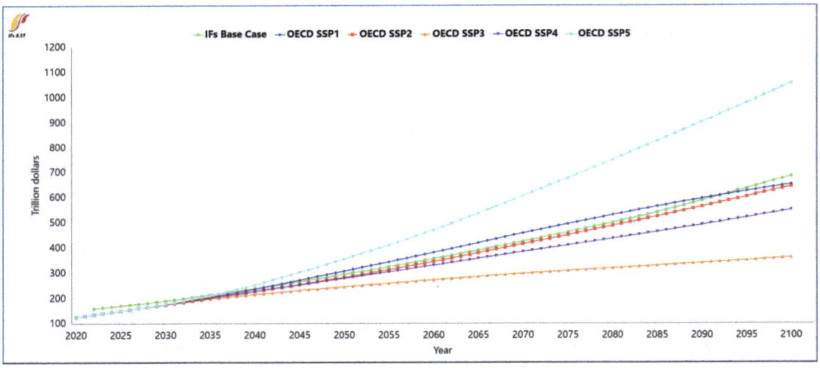

Fig. 7.1 Forecasts of global GDP at PPP by the OECD across the SSPs and by the IFs Base Case. *Note* IFs values are in 2021 dollars and OECD values are in 2017 dollars. *Sources* IFs 8.57 and IIASA database of OECD forecasts (2023 revision) using the model described in Dellink et al. (2017)

OECD use of 2017 dollars. Inflation reduced the value of the dollar between 2017 and 2021 by about 11%.

To put the variation of GDP projections into broader context, Fig. 7.2 shows a second set of 2023 revised GDP projections across the SSP scenarios that sit on the IIASA website, this one undertaken by a team associated with IIASA. The economic model of that team is somewhat less elaborate than those of the OECD and IFs. Note that the entire set ends the century at considerably lower GDP values than do the scenarios in the OECD set. The IFs Base Case values at the beginning and end of the century are higher than all SSP scenarios in the figure. But a conversion of values in IFs from 2021 to 2017 dollars would lower the IFs line by about 11% and place it below SSP5.

To conclude this discussion of GDP forecasts, we must recognize great uncertainty surrounding long-term analysis of it, considerably more so than for population (Burgess et al., 2023). That is an important recognition before we turn later in this chapter to creating and exploring alternative forecasts using IFs. GDP growth affects almost everything we forecast (and, of course, the lives of people around the world).

Moving beyond the aggregate size of the global economy (very important for many reasons, not least of which are economic well-being and the environmental impact of humans) it is important to consider also the changing distribution of GDP across countries and the changing sectoral character of the economy. Figure 7.3 shows the ratio of GDP per capita at PPP in the OECD and non-OECD country sets, comparing the forecast of the IFs Base Case with the SSP2 (middle-of-the-road) scenarios from the OECD and IIASA. As indicated earlier, the great divergence of GDP per capita in the two sets that began with the Industrial Revolution and continued through the twentieth century reversed toward convergence

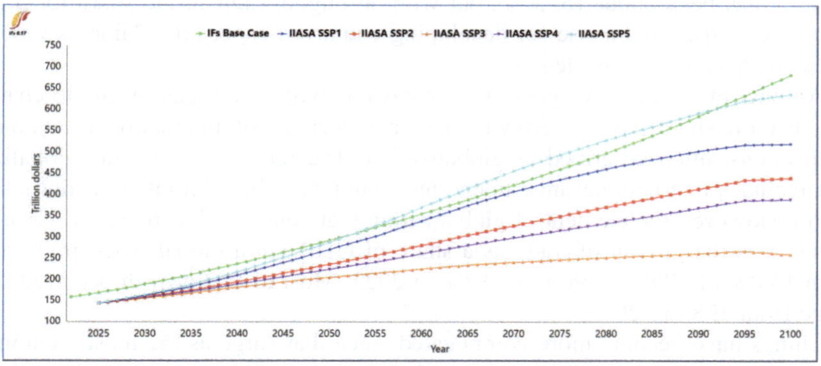

Fig. 7.2 Forecasts of global GDP at PPP by IIASA across SSPs and by the IFs Base Case. *Note* IFs values are in 2021 dollars and IIASA values are in 2017 dollars; inflation reduced the value of the dollar between 2017 and 2021 by about 11%. *Sources* IFs 8.57 and IIASA database of forecasts using model described in Cuaresma (2017)

in roughly the first 20 years of this century. Interrupted briefly by the COVID-19 period, that convergence then resumed. The IFs Base Case forecast suggests a rough stabilization of the ratio until the second half of the century, related in part to the remaining demographic challenges of developing countries but significantly also to slowing of the earlier torrid pace of GDP growth in China. (China is a partner, but not a member of the OECD.)

Turning to the sectoral shifts occurring in the global economy, one of them is in what is being produced and consumed as a portion of the GDP. Whereas the agricultural sector was about 80% of the global economy in 1800 (Pomeranz, 2009), it was 32% in 1960 and is now about 4%. The service sector has replaced most of that share, rising to about 60% globally. Since 1960, manufacturing's share

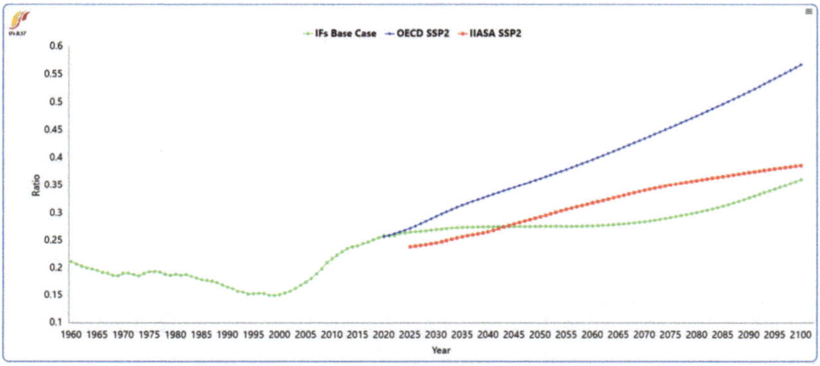

Fig. 7.3 Ratio of GDP per Capita at PPP in non-OECD to OECD countries. *Source* IFs 8.57 Base Case using International Monetary Fund values through 2027

globally has been about 15–20%, but with an ongoing fall in the share for high-income countries and a rise for developing countries, especially China as well as others in Asia to a lesser degree.

Another major issue in economic forecasting, with some relationship to changing sectoral shares, is the growth in recent decades of international economic connections, often described as globalization. The rubric of economic globalization includes international investment and remittances by migrants to countries of origin. However, the element of globalization that generally has received the most attention is the growth of trade as a share of GDP. On a global basis, trade rose from 13.0% of GDP in 1960 to 30.7% in 2022. Most of that growth occurred in a surge from 1985 to 2005.

Things have become more complicated since that surge as the mostly enthusiastic acceptance of economic globalization waned. Looking at exports as a share of GDP, Fig. 7.4 shows that prior to the surge from 1985 to 2005, it was low-income countries for which the share of exports in GDP was the largest. That was related in substantial part to the provision of primary goods (raw materials, energy, and food) from those countries to the rest of the world. The export share of GDP from the currently upper-middle-income countries (including China) was lower than that from low-income countries in the 1960s and 1970s, but that has changed dramatically since the early 1990s. Exports as shares of GDP rose sharply across all country-income groups into the first decade of this century, especially those from the upper-middle-income countries (dominated by China). This had much to do with the growth of manufacturing exports from China and those of services from high-income countries. The pushback from importing countries has been a factor in roughly stabilizing trade as shares of GDP since 2010.

The combination of growth in ratio of trade to GDP and the sectoral expansions of trade in manufactures and services have had domestic impacts for exporting and importing countries, including loss of manufacturing employment in many

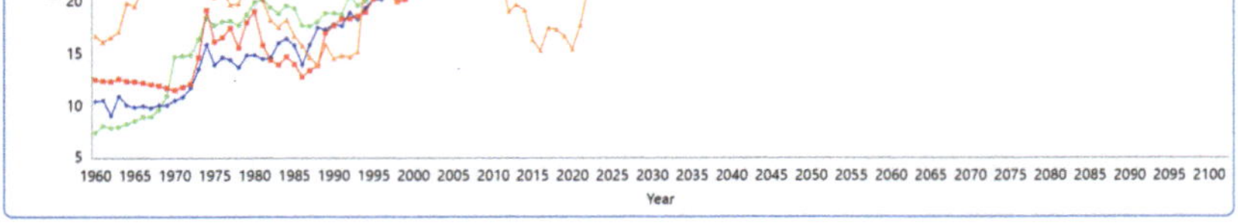

Fig. 7.4 Global exports as a percent of GDP at MER. *Source* IFs 8.57 Base Case using historical data from the World Bank

high-income countries. That is a major factor in the dampening of enthusiasm for globalization.

At this point it is extremely difficult for anyone to have confidence in forecasts of future trade patterns.[2] Figure 7.4 shows that the IFs Base Case forecast for the export share of GDP globally is quite stable. Under the hood of that, we see that the share of exports in GDP of upper-middle-income countries has already fallen substantially since the sharp peak in 2006 of 33%. China reached 34.5% in that year and its ratio declined to 20% in 2024, a reality that might surprise many readers.

The IFs forecast shows the ratio for upper-middle-income countries as roughly stable, and a look at that grouping by country (not shown here) suggests that the GDP and exports of other members will likely grow faster than those of China. Figure 7.4 suggests that export shares of GDP in the lower-middle-income countries and then especially in the shares in low-income countries could grow quite substantially, related in significant part to the spread of industrialization within them as a complement to continued raw materials exports. The forecasted export shares of their GDP seem very high, however, so we should remember that all forecasts are wrong—even if some are useful.

In addition to global economic disparities, sectoral shifts, and trade patterns, there are other important and related global economic changes that have geopolitical implications. Figure 7.5 shows the ratio of GDP in China to that in the United States. It is much more complex than one might expect to understand the economic size of the two countries relative to each other, an important element of their relative positions globally (an issue for considerable attention in Chap. 13). The first insight is suggested by the ratio at MER in the IFs forecast (orange line), the measure used by most analysts as the basis of their relative economic positions globally. That ratio climbs very close to equality by midcentury and then roughly stabilizes. The convergence pattern reflects both the currently faster Chinese GDP growth rate and the slowing of that advantage toward midcentury, in part because of China's already decreasing population and soon to decrease labor force as the one-child policy years and current low fertility rate cut into its growth considerably more than do demographic changes in the United States (see Chap. 6).

The second insight comes when looking at the ratio in terms of GDP at PPP. The OECD, IIASA, and IFs models all suggest that the Chinese economy is already bigger than that of the US at PPP. The IFs MER (green) and especially the OECD PPP (orange) forecasts anticipate considerable growth in the Chinese advantage prior to some reversion in the second half of the century, probably again associated significantly with demographics. In contrast, the IFs PPP (blue) scenario projects near constancy in the Chinese advantage, not growing again because in part because of demographics and in part because of the diminution of the PPP to

[2] The first great period of globalization (1820–1914) was derailed by war and depression. The second great period of globalization is sometimes dated from 1950. See Lindert and Williamson (2001).

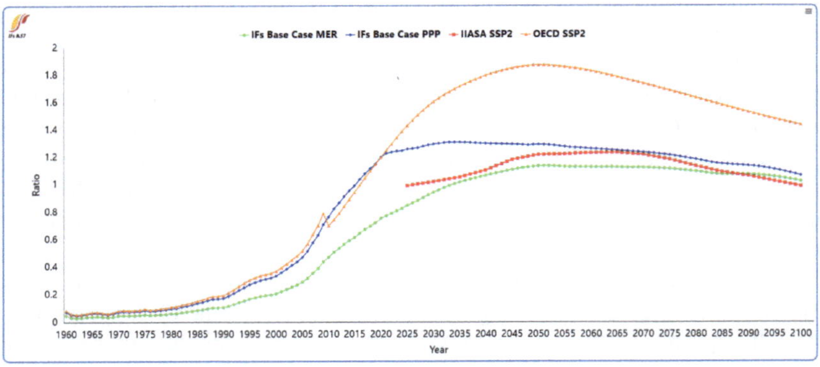

Fig. 7.5 China/U.S. GDP ratio across three models and across PPP and MER representations. *Note* OECD and IIASA forecasts are the 2023 revisions. *Sources* IFs 8.57 and IIASA database of forecasts using models described in Dellink et al. (2017) and Cuaresma (2017)

MER ratio (not shown) within China with continued economic growth. The stabilization of the red line (IIASA SSP2 at PPP) is also in part because the higher PPP advantage in China diminishes with approach of its GDP per capita at MER to that of the U.S, as explained in the discussion of Fig. 7.1. Chapter 13 will turn to related geopolitical forecasting and analysis, complicated by the wide range of such forecasts.

Still another important economic issue is the distributional pattern of incomes *and wealth* generated by the economy within and across countries (Piketty, 2014). A standard measure of such distribution is the Gini coefficient, also used by IFs. Higher Gini values indicate more unequal distribution. Considerable research has found that as countries become richer, the values of Gini tend to decline (Brueckner et al., 2015). The IFs Base Case (version 8.57) forecasts a modest decline in the global population-weighted within-country average values from 0.348 in 2025 to 0.338 in 2100. If we look at a global Gini for all the people in the world rather than a country average, however, there could be a much more meaningful decline if, as in the IFs Base Case, the GDPs per capita of developing countries continue to close the gap with those of high-income countries. Taking that into account, the total global Gini in IFs 8.57 falls from 0.550 to 0.483 in the SSP2 (middle-of-the-road) scenario. This will interact in positive dynamics with decreases in poverty and hunger and rises in education, issues to which Chap. 8 will return.

By no means least in terms of the forward impacts of economic growth are the environmental consequences of that growth. Chapter 11 will turn to those. It is impossible to study the future of the global economy and measures like GDP (as flawed as those aggregate values may be in representing human well-being) without a strong focus on the driving forces of them. These dynamics are the focus of the next section.

7.2 Economic Dynamics and Leverage

Although the IFs population model has some complex dynamics around fertility and mortality, the supply and demand dynamics of the economic model are far more complex. This basic description of the IFs economic model begins by explaining some of the dynamics around the production of GDP, the variable to which the introduction of this chapter gave much attention. That explanation will give special attention to the concept of productivity and the IF's system's use of what is called a Cobb-Douglas production function. In addition, financial flows among agent categories, namely households, firms, and governments, are critical to understanding the economy and the linking of its production to the needs and goals of those agents. It is important to think about those financial flows generally, with some drill-down attention to those within countries and those across them. With some understanding of these elements of the IFs economic model, we will be able to turn to scenario analysis. For information beyond that in this section, see the IFs Wiki, the IFs network diagram, and the Pardee Institute's home page.

7.2.1 Economic Production

Economics, like almost all issue areas in this book, involves an interaction of positive and negative feedback loops. Figure 7.6 portrays important components of those dynamics in IFs, focused on the gross domestic product (GDP), the total production of goods and services that can be delivered to meet the demand of households, firms, governments, and net exports.

Economic production requires three key factors: labor, physical capital (buildings and machines), and productivity of producers. IFs represents production and demand for it (as well as trade) in six economic sectors: agriculture, energy, other raw materials, manufactures, services, and information and communications technology (ICT).

IFs uses a common building block of economic theory and modeling, the Cobb-Douglas production function, to combine capital, labor, and their productivity in the computation of GDP. Some GDP moves through savings (from domestic and foreign sources) to investment in new capital and then to further production. That positive feedback loop has, for most of the world since the Industrial Revolution, very substantially prevailed over the negative feedback loop that decreases physical capital by the depreciation (wearing out) of it.

Additions to capital and labor explain, however, only a portion of economic growth for most countries; improvements in technology and knowledge and organizational efficiencies explain the rest in ways that are still not fully understood theoretically or demonstrated empirically. The economic growth model in the IFs system augments the Cobb-Douglas production function—a theoretical bedrock of "neoclassical growth theory"—with insights from "endogenous growth theory" and extensions to neoclassical growth theory that emphasize the importance of the roles of technology and of human and social capital (Barro, 1998). IFs combines

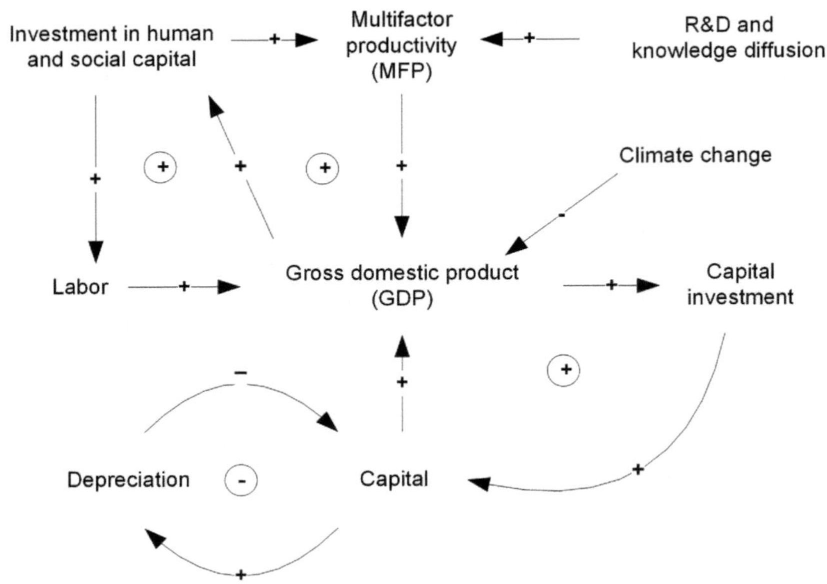

Fig. 7.6 Long-term production dynamics. *Source* Author

assumptions and estimates of the growth of human and social capital with knowledge growth and the diffusion of knowledge into the growth of what economists call multifactor productivity (MFP) or total factor productivity (TFP). Countries that have strong multifactor productivity growth tend to have strong economic growth and improving living standards. In summary, high growth in physical capital and/or labor supply, are not, by themselves, enough to guarantee rapid economic growth.

There are, of course, linkages of production with demand (to which we will return) on which much of day-to-day economic discussion focuses because they give rise to important cycles. Figure 7.7 shows three feedback loops, with the inventory level in the six economic sectors of IFs at the center of them. The first two use the rise and fall of prices as the signal back to producers and consumers that helps equilibrate production and demand. The third links inventory levels and production via capacity utilization. These negative feedback loops promote a rough balance between production and demand, but systemic shocks and lags in the operation of the linkages means that the two will almost never be in exact equilibrium.

On a more technical note, some readers will be interested in the way the general equilibrium economic and partial equilibrium models in IFs (notably agriculture and energy) undertake such pursuit of equilibrium. It involves a variation on what is known in control theory as a proportional-integral-derivative (PID) controller (Hughes, 2019: 144–145) and economic model documentation in the Wiki). That

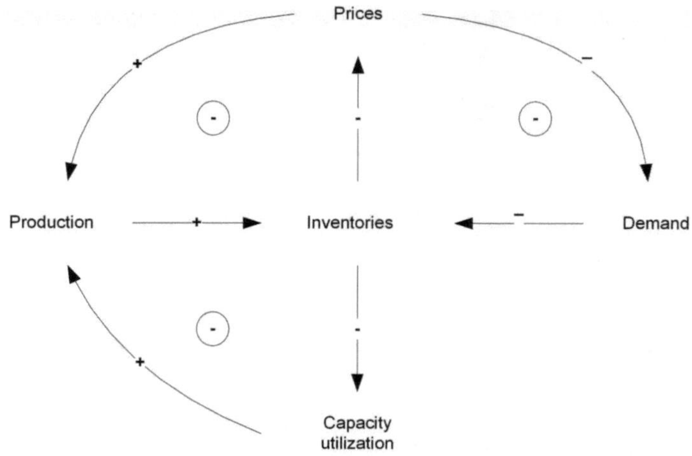

Fig. 7.7 Supply-demand equilibrium dynamics. *Source* Author

is, changes in prices (and capacity utilization) respond to both the size of inventories in any year relative to a target level and the change in those inventories year-to-year.

7.2.2 Multifactor Productivity (MFP)

Staying with the production side of the economy, it is important to understand the dynamics that drive MFP because those dynamics are critical in the real-world economy and they link the supply side of the economic model of IFs to pretty much all other models in IFs. Among the important features of the IFs system is that it includes so many such linkages fully integrated into its calculation of MFP, quite unique among economic models (Hughes & Narayan, 2021).

Roughly speaking, the drivers of MFP in IFs fall into four categories: Human Capital, Social Capital, Physical Capital, and Knowledge. The model computes values for each of those and combines them into the MFP term. Using Nigeria as an example, Fig. 7.8 shows how IFs includes these concepts in a *Development Profile* (you can pull up this for any country or grouping, for any year, by clicking on *Display, Specialized Displays, Development Profile*). The variables shown in each category will normally advance with economic development level as represented in GDP per capita at PPP. When any one of those variables exceeds or falls short of the expected level at any level of GDP per capita, it will make a positive or negative contribution to the rate of change in MFP.

7.2.2.1 Human and Social Capital

Human capital and social capital are not precisely defined terms, nor are they easily measurable. The economists who deal with these concepts in their models think

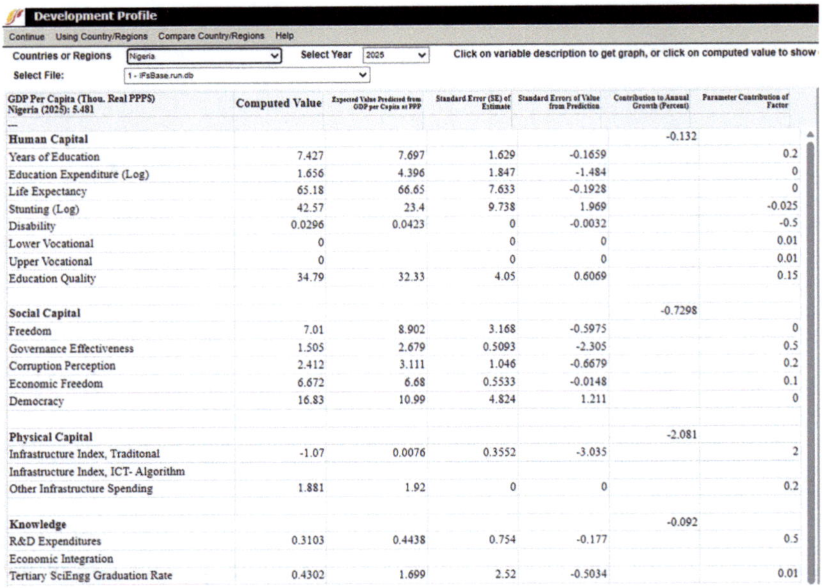

Development Profile

Continue Using Country/Regions Compare Country/Regions Help

Countries or Regions: Nigeria Select Year: 2025 Click on variable description to get graph, or click on computed value to show

Select File: 1 - IFsBase.run.db

GDP Per Capita (Thou. Real PPPS) Nigeria (2025): 5.481	Computed Value	Expected Value Predicted from GDP per Capita at PPP	Standard Error (SE) of Estimate	Standard Errors of Value from Prediction	Contribution to Annual Growth (Percent)	Parameter Contribution of Factor
—						
Human Capital					-0.132	
Years of Education	7.427	7.697	1.629	-0.1659		0.2
Education Expenditure (Log)	1.656	4.396	1.847	-1.484		0
Life Expectancy	65.18	66.65	7.633	-0.1928		0
Stunting (Log)	42.57	23.4	9.738	1.969		-0.025
Disability	0.0296	0.0423	0	-0.0032		-0.5
Lower Vocational	0		0	0		0.01
Upper Vocational	0		0	0		0.01
Education Quality	34.79	32.33	4.05	0.6069		0.15
Social Capital					-0.7298	
Freedom	7.01	8.902	3.168	-0.5975		0
Governance Effectiveness	1.505	2.679	0.5093	-2.305		0.5
Corruption Perception	2.412	3.111	1.046	-0.6679		0.2
Economic Freedom	6.672	6.68	0.5533	-0.0148		0.1
Democracy	16.83	10.99	4.824	1.211		0
Physical Capital					-2.081	
Infrastructure Index, Traditonal	-1.07	0.0076	0.3552	-3.035		2
Infrastructure Index, ICT- Algorithm						
Other Infrastructure Spending	1.881	1.92	0	0		0.2
Knowledge					-0.092	
R&D Expenditures	0.3103	0.4438	0.754	-0.177		0.5
Economic Integration						
Tertiary SciEngg Graduation Rate	0.4302	1.699	2.52	-0.5034		0.01

Fig. 7.8 Development Profile within IFs. *Source* IFs 8.57 screen capture

of human capital in terms of things that make workers more productive, especially schooling, specialized training, and physical health, while social capital refers to the organization or governance of society.

Human capital, in the present version of the IFs model, is represented by values for years of education, education expenditures, life expectancy, stunting, disability, upper and lower vocational education, and education quality. An improvement in any of these is usually good for economic growth. The last column in the *Development Profile* shows the default value for the elasticity of the multifactor productivity term to the various factors that we use to represent human and social capital. Thus, looking in Fig. 7.8 at Nigeria in 2025, the years of adult education are below the "expected value" and will retard annual economic growth (negative human capital contribution to the multifactor productivity term).

Turning to social capital, improved governance for economic development means policies and institutions that increase people's incentives to work, save, and invest. Factors such as consistent rule-of-law, improved property rights, and lower corruption are part of the concept, although translating the general concept into practice is not easy, nor is modeling it. In IFs, we include factors such as indices of corruption and of governance effectiveness, both as indicators of social capital and as leverage points.

7.2.2.2 Physical Capital and Knowledge

The *Development Profile* also lists some useful variables under the headings of "Physical Capital" and "Knowledge." By physical capital we mean infrastructure in distinction from production capital like an automobile factory. Spending on roads and energy supply and communications links has been shown in the literature (Calderon & Serven, 2004) to have a significant impact on economic growth. IFs includes several indices of infrastructure development, calculated in its elaborated infrastructure model (see Chap. 8).

With respect to knowledge the model accounts explicitly for differences in national levels of spending on research and development (R&D) and on the degree of integration into the global economy via trade as a share of GDP. Further, by increasing or decreasing the amount of certain types of investment—including investment in R&D—we increase the efficiency of investment, whatever its aggregate level, and hence spur economic growth. Also, the graduation rate within a society of those pursuing tertiary level education in science or engineering contributes to productivity.

As the IFs project moves forward, it will be important to add variables representing the development and use of artificial intelligence. An earlier Pardee Institute study of the prospects for the advance of AI was undertaken (Scott et al., 2022), but that was shortly before a huge surge in the rate of AI's advance materialized globally.

7.2.3 The Impact of Climate Change

Another factor affecting economic production is climate change. The environment model provides a measure of global temperature increase. In the economic model that increase can adjust the calculation of value added and therefore GDP coming out of the Cobb-Douglas production function with its extended MFP representation. The formulation for economic impact in IFs builds on work done in the integrated assessment model (IAM) community, especially that of Nordhaus (2019) with his Dynamic Integrated Climate-Economy (DICE) model and analysis by Dietz and Stern of that representation (2015: 579–580). The DICE model and therefore IFs include a second order polynomial formulation with its linear and squared terms that link rise in global temperature to annual impact on GDP. Dietz and Stern have argued, however, that temperature rise does not just have an annual impact on GDP (a flow variable) but also has cumulative impact on underlying stock variables such as capital and productivity. IFs includes parameterization that allows exploration of either or both impacts.

7.2.4 Mapping Financial Dynamics: The Social Accounting Matrix

Three primary features of the IFs economic model make it quite unusual in long-term forecasting models, and they facilitate a wide range of studies with IFs. The first is the elaboration of the MFP term discussed above, rather than, as in many models, a focus on only 1–2 driving variables, usually the contribution of education. That elaboration benefits from the close integration of the multiple models in the IFs system, which facilitates broad, policy-focused exploration of variables that can boost productivity.

The second feature is the inclusion of a full social accounting matrix (SAM) that structures monetary flows across the economy, representing such flows not just among economic sectors but also the revenues and expenditures of households, firms, and governments. Further, country-specific SAMs also show intercountry flows, including foreign aid, foreign direct investment, and remittances, plus transfers to and from the World Bank and International Monetary Fund. Using Mexico as an example, Table 7.1a shows the highly collapsed matrix that is called up by *Display/Specialized Displays/Social Accounting Matrix*, providing an overview of the financial flows. The *Switch/Expand SAM* option calls up a much larger table from which Table 7.1b shows only a very small portion. The first six rows of 7.1b show in billion dollars the value of output from each of the six economic sectors in IFs to all the sectors and to unskilled and skilled households (that is, their purchases). The first six columns show the purchases of each sector from all sectors and purchases also of unskilled and skilled labor.

Table 7.1 shows only illustrative elements of the SAM. Its importance in analysis with IFs is that it helps ensure that revenues and expenditures are balanced (no free lunches for any agent category). Savings and borrowing of actors are not represented in the table, but the model helps ensure balances by accounting also for the stocks of those (assets and liabilities), available for display via *Switch/Show Stocks*.

Extensive representations also of stocks and of stock-and-flow dynamics is very important and is a third feature that makes IFs relatively unique. For instance, an issue that has taken on real significance in the 2020s is the growth of the stock of government debt. Globally, government debt was about 40% of GDP in 1990. By 2022 it had grown to 102%. Given inertia in revenue and expenditure patterns, the IFs Base Case shows that growth continuing for at least a time. Concerns have risen quite sharply around the world about the sustainability of the current level of debt, much less growth in it, because interest on that debt burden as a percentage of GDP and government revenue has been growing significantly. Pressures for continued growth include aging populations, with living and medical expenses that governments often support. Yet historically, such high levels of debt have repeatedly led to financial crises. The IFs Base Case represents these stock and flow variables, forces some turnaround in the growth trend of government debt in coming years, and explores the broader implications of such patterns. Forecasting on this issue is very important.

Table 7.1 Social accounting matrix and elaboration of intersectoral flows plus household purchases (columns) and earnings (rows)

Table 7.1a Collapsed matrix

Social Accounting Matrix

Continue Switch Help About

Countries or Regions: Mexico   Select Year: 2025

Select File: 1 - IFsBase.run.db

Supplied by	Used by Sectors	Used by Household	Used by Firms	Used by Capital	Used by Government	Used by ROW	Used by Total	Environment
Sectors	1054	951.2	-21.27	367.2	194.7	550.9	3097	0
Household	736.1	0	278.5	0	225.9	46.93	1288	0
Firms	709	0	0	0	0	64.57	773.5	0
Capital	0	106.7	332.9	0	-69.63	-24.25	345.7	0
Government	0	183.1	144.1	0	0	3.175	330.4	0
ROW	597.6	0	20.19	0	2.713	0	620.5	0
Total	3097	1241	754.5	367.2	353.8	641.3	6455	0
Environment	0	0	0	0	0	0	0	0

Table 7.1b Small portion of expanded matrix with intersectoral flows

SAM	IFsBase.run.db	Sectors						Households	
Mexico	2025	Agriculture	Energy	Materials	Manufactures	Services	ICTech	Unskilled	Skilled
Sectors	Agriculture	25.13426	0.12111	0.23248	6.21921	8.11966	0.11768	28.3339	24.8627
	Energy	0.21932	3.56369	2.23283	5.97954	1.51455	0.20104	50.1205	43.9802
	Materials	0.56772	1.29623	56.8208	70.26778	3.57719	3.0964	19.1026	16.6525
	Manufactures	6.40085	13.15136	15.64911	199.03551	76.63342	17.99971	97.6849	84.896
	Services	12.71032	7.45308	26.56425	97.55402	150.62679	17.41496	216.5758	190.6001
	ICTech	0.34623	0.41573	1.91337	15.95312	18.04849	48.7624	26.1707	23.3387
Households	Unskilled	21.704	20.6119	29.1861	116.7655	183.8005	19.0168		
	Skilled	2.3657	9.275	6.5061	16.9062	216.6912	3.7995		
	Total								

Source IFs 8.58 screen capture

7.2.5 Domestic Financial Flows and Stocks

The flows and stocks associated with the three agent categories—firms, households, and government—have special interest to us. For instance, skilled and unskilled households in IFs obtain income from both labor contributions to production and net transfers from the government. IFs represents portions of that as saved (small in low-income countries) and portions directed toward consumption. IFs further differentiates that consumption spending across the economic sectors.

The GDP of all countries finds its way via the spending, saving, and taxation decisions of households, firms, and governments to four principal end uses: household consumption on a wide range of goods and services; governmental expenditures for defense, education, health, infrastructure, and other purposes; reinvestment in capital stock; and net exports. The reinvestment in capital stocks is normally sufficient to cause those stocks to grow and therefore to increase GDP in future years. Household saving, especially in higher-income countries, is another stock of importance. And as Fig. 7.7 indicated, rising and falling inventory stocks in each economic sector have importance in facilitating recursive computation and pursuit of equilibrium across time.

Countries attempt policy formulation that will increase savings and therefore reinvestment in capital. Although this sounds relatively simple, it is not. Despite substantial governmental efforts to spur additional investment, the portion of GDP that the high-income OECD countries in aggregate reinvested in capital between 1970 and 2020 has been fairly constant (at about 21%). Debates rage over the degree to which interventions by government via policies on taxation, spending, or interest rates can increase savings and investment.

Other debates concern direct governmental expenditures, mostly over how much and with what targets. Those include spending on the public capital infrastructure of roads, highways, and information systems and on human capital via education and health care expenditures. They also include transfer payments to help specific subpopulations, including those disadvantaged in terms of income or health and also frequently aging populations regardless of financial condition.

There are clear trade-offs across expenditure categories and also with respect to revenue raising via taxation or borrowing. The social accounting matrix structure of IFs tracks such expenditures and revenues to ensure that there are no free lunches and has parameters that facilitate scenario intervention concerning various trade-off decisions.

7.2.6 International Trade and Financial Accounts

International financial flows include those associated with trade balance (exports minus imports), foreign investment earnings, financial remittances by migrants, and foreign aid. The current account of a country is the net inflow value of those flows. A negative current account, often driven by a surplus of imports relative to exports, requires offset by a positive capital account balance, which includes the net inflow of investment and foreign borrowing. IFs represents these variables.

7.2.6.1 Trade and Globalization

Trade in goods and services is a large and generally dominant part of current account flows, and it is a part of the balancing of supply and demand within each economic sector (see Fig. 7.9). One pair of negative feedback loops again works around inventories and domestic prices of goods and services. An interacting pair functions through trade balances (exports minus imports) and the exchange rates (or prices) of currencies. This second pair of negative feedbacks is complex because it interacts additionally with the current account. For instance, the figure indicates that foreign aid and remittance receipts improve the current account, raise the exchange rate, and therefore dampen exports. We will elaborate some of the other figure elements in the next subsection.

Trade's ability to spur economic growth and incomes has long been appreciated by governments, but its ability to adversely affect the prospects of some

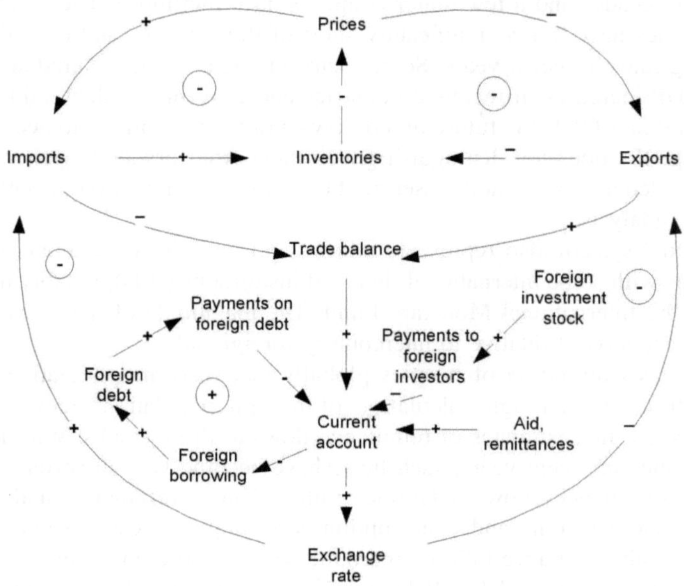

Fig. 7.9 International economic dynamics. *Source* Author

groups within a society has also always led to opposition. In recent decades, globalization—trade, financial, and labor market linkages—has intensified, presumably spurring overall growth because it helps countries better utilize their differing comparative advantages in production factor endowments (labor, capital, and productivity). But it has also sparked a growing political backlash. The backlash is heavily related to disruption of employment patterns as countries specialize and it is also related to security concerns as countries can become dependent on critical imports.

The competing causal analyses around the impact of trade on economies give rise to questions concerning choice of trade policy in the face of uncertainty. How actively should countries pursue free markets? To what degree might protectionism help or weaken local economies? How do states committed to free trade deal with a state that takes advantage of open markets elsewhere but protects its own?

7.2.6.2 Other International Flows and Related Stocks

Beyond trade, the international flows most often discussed are foreign aid, remittances, foreign direct investment, and foreign borrowing. IFs also represents these.

Foreign aid analysis pays much attention to the percentages of GDP that outflows and inflows constitute and, especially for aid recipients, to the portion of government finances tied to those receipts. Targets for aid donations by higher-income countries as a portion of GDP at 0.7% or even 1.0% have been identified across past decades and a few donor countries have met those. The reality is that most countries have fallen significantly short of those values and even decreased their giving rates in recent years. Some recipient countries have “graduated” from aid. As GDP increases in recipient countries and government debt values rise as portions of donor GDP, the future of aid flows will likely continue to become more challenging. IFs does not distinguish government from privately (generally non-profit) provided aid, but a model user might assume a combination of both sources in scenario analysis.

The model system also represents inflows and outflows of country finance in transactions with some international financial institutions (IFIs), notably the World Bank and the International Monetary Fund. The magnitude of these IFI flows is, however, very modest relative to intercountry foreign aid.

IFs forecasts migration of peoples globally (see also the population Chap. 6 on this). It makes a rough calculation of foreign population stocks based on those flows and ties estimates of remittance flows to those stocks. With increased migration rates in recent years, such flows have become very important for some countries. Remittances flow to families rather than governments, and IFs augments household income and consumption accordingly. Because some countries experience both remittance inflows and outflows, IFs represents them separately.

Most users of the model will be focused on what the IFs project refers to as “monadic” flows of trade, aid, migration, etc. That is, the focus will be on country total inflows and outflows, balanced globally, but without specification of the countries to which or from which the flows move (that is, without “dyadic” or “bilateral” elaboration of the monadic values). IFs has developed extensive dyadic representations, however, which switches can call up in the personal computer-installed version of IFs but not the web-based version (see Chap. 13).

As indicated above, current account flows must be offset by capital account flows. IFs represents flows of foreign direct investment (FDI) and portfolio investment, and it calculates net external debt or asset values from those flows. Further, it computes the return on such investment and represents that as another flow in the current account calculation. Figure 7.9 provides a partial view of the dynamics in IFs around financial flows and some stocks beyond those involved in trade.

The issue of external/foreign debt is important and is separate from, but often interacting with, the issue of government debt discussed earlier (government debt can be financed by either or both domestic and foreign entities). Because of significant borrowing across multiple decades by low-income countries, IFs data initializes the level of that debt in its base year at around 50% of GDP; data are somewhat weak in this area. Foreign debt levels are significant also for middle-income countries. Just as government debt levels are currently high and likely to need addressing in coming years, that is true for external debt of many countries. In fact, both sets of debt issues have regularly generated debt crises involving single (e.g., Argentina repeatedly) or multiple countries.

The ability of funds from abroad to improve economic well-being is uncertain in all countries, including high-income ones. It is useful to understand that the inflow and outflow of funding from abroad is closely related to the two other issues of our leverage discussion: savings/investment and trade. A country like China, with a large surplus of exports over imports, achieves that surplus by saving more than it invests domestically. Economists represent that relationship with an identity in the economic accounts of all countries:

$$\text{Savings} - \text{Domestic Investment} = \text{Exports} - \text{Imports} = \text{Foreign Investment}$$

It is not always clear in this relationship where the driving forces lie. Does China export so much relative to imports because it has a surplus of savings relative to domestic investment demand? Or do its large net savings reflect the export-oriented character of its corporations and governmental policies?

Economic policymakers often believe with respect to both domestic and international policy that they have considerable leverage via trade and investment policy over economic well-being and perhaps also over the distributions of income and wealth. At the same time, however, there are very substantial debates over the appropriate application and utility of leverage. The next section will help you use IFs to further explore such issues. You should understand, however, that the answers to all debates that the model in some way addresses are to some extent built into the model. Thus, the model should be used with the purpose of thinking about the issues, arguments, and competing understandings, not with the intention of obtaining "the answers."

7.3 Interventions and Scenario Analysis

Chapters 4 and 5 explained the mechanics of using International Futures and elaborated the process of scenario analysis. That knowledge is essential, and it is recommended that you have mastered the intervention exercises there and/or in Chap. 6 on Population before doing those here.

The first step in scenario analysis is to pick and explore some indicator variables of importance to you. Most users will explore across the many displays identified In *Display/Flexible Displays*. For that you will not need to know the names of variables in IFs, but if you have highlighted one of those displays, a click on the menu's *Explain/Explain List* option will tell you the names of those in that display. Table 4.2 also provided a list of variable names across the models of IFs. This section will give you names of many of those likely to be of special interest with respect to the IFs economic model, along with parameters that facilitate interventions. With respect to interventions, the primary path to exploration involves use of *Scenario Analysis/Quick Scenario Analysis with Tree.*

With respect to economic well-being, you will want to look at worldwide and country or region-specific gross domestic product **(WGDP** and **GDP)** and at the growth rates in those values (**GDPR**). To obtain those values most quickly, go to *Display/Flexible Displays* and pick the *Economy/Production* categories. Those variables show values in billion dollars (2021 prices) at market exchange rates. **GDPP** shows values at purchasing power parity. **GDPPC** and **GDPPCP** provide per capita values. Again, the *Flexible Displays* don't require knowing the variable names, but they can help you become familiar with them.

You may want to look at the variable **HDI**, the United Nations Development Programme's human development index, a well-being measure that combines information on life expectancy, educational attainment, and income. Click on *Display/Flexible Displays/Human Development/HDI—UNDP index* and ask for a *Line Graph*. The resulting picture gives you a snapshot of the projected trends in the HDI by *Country/Regions* or *Groups* in the Base Case and/or alternate scenarios.

As you turn to issues of distribution, and therefore to potential equality, you will also want to look at the domestic Gini index (**GINIDOM**) and the global Gini Index (**GINIFULL**) on which lower values indicate greater equality. If you go to *Display/Flexible Displays/Self-Managed Display* you can see the absolute and relative values of North-South gaps (**NSGAPA** and **NSGAPR**), important indicators in your analysis. Or you could go to the *Display/ Flexible Displays/ Distribution* category to find those.

IFs also includes poverty indicators, most easily available in the *Display/ Flexible Display/Human Development/Poverty* category, where you can explore numbers of people or percentage of population at multiple poverty levels. In *Self-Managed Display* you can find those by using the *Variable Search* option to search on "income" or you can simply scroll down though the list of variables and parameters to find them (remember variable names are all capital letters and parameter names are all lower case).

→ One interesting way to observe changes in the global income distribution is to click on *Display/ Specialized Displays for Issues/ World Map, Lorenz Curve, Gini, Histogram*, then select the *Lorenz* Curve option at the bottom of the screen and ask for GDPPC or GDPPCP. A Lorenz curve will appear showing the initial distribution of world income with a Gini coefficient summarizing the inequality measure across countries. By advancing the *Year* in the box for changing it, changes in these relationships over the forecast period can be examined. *Options/Add Curve* with selection of a different year allows comparisons across time.

Once you have identified variables of interest for development and exploration of alternative scenarios, you can turn to identifying parameters for interventions that will shape your scenario. When we reviewed thinking about dynamics and leverage with respect to economic growth and distribution, we focused on three specific issues: economic growth with special attention to multifactor productivity; domestic financial flows with special attention to government finance; and international financial flows. We will do the same here, starting with developing alternative scenarios of economic growth.

7.3.1 Exploring Economic Growth

In discussing the dynamics of economic growth in IFs, we focused on the elements of the production function: capital, labor, and multifactor productivity.

7.3.1.1 Investment and Capital Stock

Influence over the sector-specific capital stock term (**KS**) in the production function comes primarily from control over the level of investment (**I**). The **invm** parameter is a multiplier like most of the parameters ending in "**m**". Changes to **invm** over time will increase or decrease investment (**I**) and its share of **GDP**. There are many other demands on GDP: portions flow to household consumption (**C**), to government consumption (**GOVCON**), and to net trade (**X** minus **M**). A country that devotes a greater share of its output to investment—whether by government decree or carefully managed incentives—can encourage a higher rate of economic growth, but doing so may lower these other terms.

→ When **invm** is 1.0, the model computes investment according to internal rules. When the value is 1.5, investment will be 50% higher than normal; similarly, a value of 0.5 will reduce investment by 50%. The parameter **invm** is country specific, so if you want to see an entire world in which investment increases, you will want to change it for the World. As with other interventions, the model might fight you somewhat as households and government compete to protect their basic calculations of share of GDP.

→ If you wish to increase a specific sector's share of investment, such as shifting more into the ICT sector, use the parameter **idsm**, dimensioned by country and sector. You can see its sectoral impact by looking at the sector's capital stock (**KS**). Remember that the agriculture and energy sectors are "partial equilibrium models" where their investment needs are determined and sent to the full equilibrium model for reconciliation with other investment demands, so the parameter will not work for them. In the agriculture and energy models the equivalent parameters are **aginvm** and **eninvm**.

Research Possibility. How great might a country's leverage over economic performance be if it could increase investment? Because investment increases come at the expense of consumption, can less-economically-developed countries use this strategy effectively, or does it have short-term implications for the poorer countries (loss of consumption power, declines in quality of life) that are unacceptable? For example, can Africa increase investment without worsening immediate food problems and malnutrition of children (**MALNCHIL**)?

7.3.1.2 Labor Force Size

Population size itself (**POP**) influences the size and associated contribution of the labor force employed by sector (**LABEMPS**). That contribution is especially great when the dependency ratio (shares of pre- and post-labor age population) is low. Note that it is possible to increase the share of the population in the labor force with parameters controlling the ages of entry and retirement (**workageentry** and **workageretire**, or the portion of people of working age who participate in the labor force with a gender-specific multiplier (**labparm**). Change a specific sector's demand for labor with **labdemsm**.

7.3.1.3 Technology and Multifactor Productivity

The productivity or technology stock term (**TEFF**) in the production function is almost certainly the greatest unknown with respect to future economic growth. The theoretical logic behind the IFs production function is that the rates of annual sectoral growth in MFP (**MFPRATE**) have four primary determining elements. The first element consists of the basic rate of sector-specific technological growth in a leading country (**mfpleadr**), historically the United States, although increasingly shifting to China (IFs does not yet have code to make that switch, but the model's logic does not require it). Changes in that parameter, which can be temporally specific, gives the model user direct control over the productivity advance rates around the world (**MFPRATE**). The second element is another "brute force" parameter named **mfpadd,** which allows additions or subtractions for specific countries. Other countries will endogenously catch up with the leader based on a convergence logic using an inverted U-shaped curve across GDPPCP levels (**MFPPREMIUM**), the third element.

The fourth element is the sum of four variables related to human capital (**MFPHC**), social capital (**MFPSC**), physical capital (**MFPPC**), and knowledge capital (**MFPKN**). All four variables are computed with driving logic that draws upon other models in the IFs system, as described in the previous section of this chapter. For instance, **MFPHC** depends on progress in education and health variables from those models. Interventions in those other models drive an elaborated use of IFs that facilitates a focus on policy strategies.

Initial work with IFs often focuses just on **mfpleadr** and **mfpadd.** Those who wish simply to drive alternative economic forecasts and explore forward linkages from them can use those parameters. More elaborate work generally moves to exploration of the drivers of economic growth and broader implications of interventions across the models of the IFs system.

→ To make a large change in global growth from the Base Case, increase the value of **mfpleadr** for all years from the base year through 2100. Use the scenario tree option; click on *Scenario Analysis/Scenario Analysis with Tree* to pull up the tree form. Do a *Parameter Search* on **mfpleadr** and *Load* it into the form. That will call up a set of the economic sectors, perhaps pick *Total*. That will produce a sub-display for changing the parameter. Select the *High* option, ramping up the parameter. The *Run Scenario* menu option allows you to choose a run horizon; try 2100. In analysis of results, how much does average annual global economic growth go up (select global GDP [**WGDP**] to display in this scenario and in the Base Case, display a table, and use the *Percent* option to see annual growth rates). How much does world energy demand (look at **ENDEM** for the World grouping) change? How about world temperature (**WTEMP**) in 2100? We will examine the connections between economic growth and energy and the environment in later chapters.

Research Possibility. Do natural resources limit economic growth? Many people have warned that natural resources pose an insurmountable limit to economic growth. If the world has only a fixed supply of nonrenewable natural resources, how can population, production, and living standards continue to grow over time? Most economists, however, assume that if and when a particular natural resource grows scarce its price will rise, causing consumers to lower demand, encouraging producers to increase supply, and motivating technologists to improve processes for recovery and use and to discover or improve substitutes. You can see how this behavior is embedded in IFs by changing the assumed level of global oil reserves (cut **RESOR**—Saudi Arabia—oil by a large amount using **resorm**) and seeing what happens to GDP growth, energy prices, and energy use by fuel type. In later chapters there will be more examples of the interactions between energy, economics, and the environment.

As artificial intelligence moves rapidly toward artificial general intelligence (AGI) and to artificial super intelligence (ASI), it will become critical that these variables be added to IFs and to the calculation of **MFPKN**. As of this writing, the AI variables in IFs with **aitaskm** and **aimachiqm** parameters are too conservative and are not driving MFP. They need to be reformulated before they can be drivers of **MFPKN** and many other elements of IFs, such as the advance in health and reduction of mortality rates.

7.3.1.4 Climate Change Impact

Increase in global temperature (**WTEMP**) associated with atmospheric carbon dioxide (**CO2PPM** and **CO2PER**) can have an impact on value added (**VADD**) and therefore **GDP**. That impact (carried in the variable **CLIMECONIMP**) is controlled by two primary parameters. The first is **climeconimpsw**, which turns impact on or off (1.0 is on). The second is **climeconimpstk**. That determines whether and how much impact warming has not just on GDP, but on total factor productivity (TEFF). This second parameter is not just a switch (0 is off), but also controls the magnitude of impact, and we suggest a value of 1.0 or more.

There are two other parameters that control the function that relates temperature to economic impact. That function has a linear term controlled by **climeconimplin** and a second order (squared) term controlled by **climeconimpsq** that more significantly increases the impact of greater warming. The default values for both come from literature around the DICE model (including Nordhaus, 2019).

→ Experiment with either or both **climeconimpsw** and **climeconimpstk** and with the parameters controlling the magnitude of impact (the default values of which are quite low, reflecting the work of Nordhaus). Anything that affects the economy will have forward linkages across the entire IFs system. For instance, anything that slows economic growth can affect also the rate of poverty reduction (Moyer, 2023; Moyer & Hughes, 2012; Moyer er al., 2022).

The agriculture model has its own formulation that relates warming and precipitation change from climate change to crop yield, recognizing the differences in national impact related to latitude, an impact that passes through to the economic model. No impact of warming is represented for the energy sector.

7.3.1.5 Use of Exogenous SSP Scenarios

Another scenario feature in IFs is the ability to override the economic growth rates that are computed endogenously in IFs (and 22 other variables, including population and education) with values from other projects. Many were computed across the five SSPs by various modeling projects in the IAM community. To call them up from the scenario tree, look under *Add Scenario Component/World Integrated Scenario Sets/SSP scenarios/SSP Update 2024/Absolute.* They can be run individually or in a batch run (Solórzano et al., 2024).

As of this writing there are also pre-run SSP scenarios already available when you open the IFs system. The pre-run versions of the SSPs are results of the Pardee Institute's translation of the story logics behind the SSPs into parameter changes for IFs. The Pardee forecasts have similar logic to the files that other IAMs have produced and can be loaded exogenously into IFs, as described above. But the Pardee Institute pre-run versions have some advantages. Whereas the stored scenario files from other IAMs are "hard" overrides of the variables they specify, the pre-run scenario representations of story logics allow IFs variables to fully interact to tell a more dynamic story of scenario unfolding. Additional parameter changes can also be added to those pre-run stories, affecting their unfolding across all variables (Moyer et al., 2026 with forecasts through 2150).

7.3.2 Domestic Finance

Many scenarios with IFs involve changes to government finance, including to total government revenues (**GOVREV**) using **govrevm** and government expenditures (**GOVEXP**) using **govexpm**. IFs will, however, maintain a general revenue-expenditure balance over time by computing government debt (**GOVDEBT**) and requiring that it maintain a "reasonable" relationship to GDP over time.

Much analysis makes changes on the expenditure side, which is divided into transfer payments made to households (**GOVHHTRRN**) affected by **govhhtrrnm** and direct expenditures identified as government consumption (**GOVCON**) affected by **govconm**. The transfer payments are divided into those for pensions/retirement payments (**GOVHHTRNPEN**) changed with **govhhtrnpenm** and those for welfare/social security payments (**GOVHHTRNWEL**) changed with **govhhtrnwelm**. The **GOVCON** expenditures are directed into destination sectors (**GDS**), specifically military, health, education, R&D, infrastructure in the categories elaborated in the IFs infrastructure model, other infrastructure, and other (which includes administrative expenses). The destination patterns can be shaped by **gdsm**, which does not change overall **GDS** and **GOVCON**.

In short, there is a great deal of control over total government finance and its targeting in IFs. Sectoral targeting is especially useful when exploring the potential implications of differing societal/governmental priorities. Inside the models of some of those target areas (including education, health, and infrastructure) are consequences of such targeting, such as implications for schooling of children.

Research Possibility. Changing a country's spending on education or health, for example, can boost human capital and thus multifactor productivity growth and economic growth. Can it have offsetting impacts that might be undesirable? See what happens to R&D spending and military spending when you increase the proportion of government spending (**GDS**) that goes to education (use **gdsm**). Are there obvious trade-offs in economic growth or otherwise?

It is not just government financial expenditures that affect economic growth, its distribution, and multiple sociopolitical variables. It is also the character of government and societies more generally. Levels of corruption and patterns of government effectiveness are elements of such character. Chapter 12 on the sociopolitical system and Chap. 8 on human development explore these issues. However, you could explore now with multipliers affecting government corruption (**govcorruptm**), government effectiveness (**goveffectm**), and economic freedom (**econfreem**) to manipulate social change in these areas.

7.3.3 International Finance

In looking at international finance, two useful variables are the current account balance (**CURACT**), which shows the aggregate net inflow of funds from the trade balance (**X-M**), from foreign aid receipts minus outflows (**AID**), and from net remittances received (**XWORKREMITIN- XORKREMITOUT**). The population chapter noted the use of **migrateinm** and **migrateoutm** to build scenarios around migration, which affects net remittances, so this section focuses on trade and foreign aid.

When the current account flow variable is negative, it requires offsetting inward capital account flows and those raise external debt stock levels (**XDEBTRA**); positive values reduce that debt. The equilibration dynamic in IFs that works to control changes in the absolute value of the ratio of foreign debt to GDP adjusts exchange rates (**EXRATE**).

7.3.3.1 Trade

Explore the growth of trade in the Base Case. Look at world trade as a percentage of the world economy (**WTRADE**). If you use *Display/Flexible Displays/Economy/Global Economy/Exports Value—Percent of GDP* you may see that in the Base Case of IFs the forecast value flattens relative to the historical pattern of growth in recent decades.

The advocates of free and expanded trade argue that continued growth in trade as a percentage of the economy carries with it many benefits. Opponents point to costs, including loss of jobs and economic independence. Tariffs are taxes on imports and thereby raise the price and reduce the volume of them. **MTARIFF-TAXR** and **MTARRIFREV** represent import tariff rates and revenues. The user can change rates with **mtariffaxrm**. Various nontariff barriers to trade, such as quotas or quality restrictions on imports (like health standards on food imports), have the same general effect.

→ Does IFs show the benefits of greater trade and fewer trade barriers that economic liberals claim? To explore this question, you can manipulate the protection multiplier (**protecm**) country by country. In scenario tree analysis, click on *Governments/SocioPolitical Systems/Economic-International/protecm/United States* (or just do a *Parameter Search* on **protecm**). Select *Total* or a specific sector. Use the sub-Display that appears to significantly ramp up the parameter value. When the value is 1.0, there is no change in tariff and nontariff barriers to trade. A value of 1.5 effectively raises the world price faced by a country on all its imports. As the value approaches 0.5, suggesting import subsidies, the price of imports is cut in half (the model will not allow you to reduce prices by more than that). You might begin by asking about the consequences in a scenario where the United States moves away from free trade. Simulate the model using this higher level of protectionism. What are your expectations? What are the results? What does it do to economic growth and the structure of the economy relative to the Base Case. You might even look at employment patterns. Consider also the possible economic consequences for China, Germany, or Japan.

If the global economic impact of U.S. protection is less than you expected, one reason might be that trade protection, like military spending, is often subject to action-reaction dynamics. IFs does not build in any automatic reaction by other regions to the U.S. protectionist measures. You must do that through your scenario(s).

→ Introduce a proportionate reaction to the increased U.S. protection by all other countries in the system (representing an overall breakdown in the liberal trading order). You can do this by increasing the protection multiplier for the World grouping. Look again at the results for the economies of the world. You might also experiment with a severe global trade war, with tariffs of perhaps 100% (you need to click on the *Fully Customize* button to push prices up that high). What does this level of protectionism do to economies, including global economic growth?

Export Promotion. Many believe that trade enhances aggregate welfare—and the results of the above scenarios suggest that the structure of IFs mostly carries that causal logic. Commercial liberals recognize that the less developed countries face a particularly demanding world, but they argue that the appropriate response is to use the market intelligently, not to withdraw from it. They point to the economic success of South Korea, Taiwan, and other export-promoting countries as evidence. Go back to the Base Case (on the tree click *Scenario Files/Clear Tree*) and develop a new scenario that incorporates an export promotion policy for a country of interest to you or perhaps for the important country group called the BRICs (which originally consisted of Brazil, Russia, India, and China and uses those first initials, but has grown to a membership of 10).

→ There is an export shift (**xshift**) parameter in IFs that allows you to "force" exports of countries onto the global market. Values of 0 leave exports unchanged from the Base Case, but positive values increase exports (unlike many parameters, this is an additive factor rather than a multiplier). The effects of **xshift** are cumulative. That is, a value of 0.05 adds about five percent to exports for that year and all subsequent years until other factors in the model dampen the increase. Thus, it is best to set the value of **xshift** low (such as 0.02) for all years. Apply your scenario to exports (X) for your selected country or the BRICs. And then look at economic and quality-of-life indicators for that country grouping. Does the scenario have any effect on other countries or groupings?

How does trade improve economic performance in IFs? If there were both a high level of sectoral disaggregation and a detailed representation of factor costs in IFs, the benefits of trade from comparative advantage and specialization would be explicit. IFs does not have that level of detail. Instead, as the ratio of trade to GDP increases, the productivity in all sectors increases (this is controlled by an elasticity of multifactor productivity with economic integration, **mfpeconints**). In essence, this linkage serves as a surrogate for all trade benefits, including efficiency gains through comparative advantage and productivity gains through the import of manufactures, especially of capital goods that bring advanced technology.

Terms of Trade. Even skeptics concerning free trade seldom question that trade confers aggregate economic benefits. Instead, they typically argue that some parties benefit more, perhaps even much more, than others. The key to such claims often lies with terms of trade—the price a country receives for its exports relative to the price of its imports.

In the real world, it is difficult to change the terms of trade. A country cannot simply begin to charge more for that which it exports and expect thereby to earn more. Charging more would likely dampen demand and elicit competing supply; the country might ultimately earn less. It is possible, however, for some countries to shift production and exports from goods of less value to those with more value, a strategy that Japan pursued in the 1960s and 70s and that China is pursuing now.

→ You can explore the North-South terms of trade in the Base Case by looking at the computed variable **TERMTR**. Does it show a clear trend upward (more relative value for exports from the South) or downward (less value for their exports)? There is considerable debate among those who have studied the terms of trade as to how much the trend historically has shifted to the disadvantage of the South, and there is therefore little basis for confidence in any forecast.

We can see how important the terms of trade are for the South, however, by simulating the financial impact of an improvement in them. The terms-of-trade multiplier (**termtrm**) is a multiplier on the value of Southern exports (and proportionately reduces the value of Northern exports). If you change it gradually from 1.0 initially to perhaps 1.2 over 20 years, you can see the impact that a 20% improvement in the terms of trade would have for the South (even if it appears inconceivable that real-world policies might accomplish such a shift).

→ Run the model with such a change and look at the new terms of trade (**TERMTR**). You might want to look with special care at a particular Southern region such as Africa. What happens to the value of exports (**X**) and the cost of imports (**M**) or to the trade balance (**TRADEBAL**)? What happens to the GDP per capita (**GDPPC**) and the human development index (**HDI**) relative to the base case? Does the improvement in export earnings allow Africa to import any more food (**AGM**)? If so, does that have any effect on malnutrition of children (**MALNCHIL**) in Africa? Overall, how does a change in the terms of trade seem to compare with foreign aid (discussed below) in its potential for improving the quality of life for Africans? Why?

If you set the model to run for less than thirty years, you are likely to see results of this scenario very much as you anticipated. If you run the model that long, however, you may see some "counterintuitive" or perverse results—that is, regions that you expect to benefit from terms-of-trade intervention on their behalf can do less well than in the Base Case. Explore the results carefully to see why (you might look at external debt levels (**XDEBTRA**) and exchange rates (**EXRATE**) for some insights). Subsequent discussion of aid will also help you understand such results.

Research Suggestions. How much benefit do countries gain if they open themselves up to imports? Can one country gain that benefit unilaterally or must many countries/regions open markets simultaneously? If all countries and regions open markets, do they share equally in benefits? Are the costs of protection comparable to the gains of free trade?

7.3.3.2 Foreign Aid

The IFs model can be used to explore the impact of foreign aid and other financial flows from North to South. Look at the ratio of GDPs per capita in North and South, that is, at the relative North-South gap (**NSGAPR**) and at the absolute difference in GDPs per capita in the two global regions (**NSGAPA**). The persistence of these gaps has led to many proposals for substantial aid programs from Northern to Southern countries.

→ An aid parameter (**aiddon**) specifies the percentage of GDP that donor countries give as aid (that percentage is unlikely to exceed 1.0 [an aid outflow of 1%] in even the most substantial North-South aid program; values like 0.3 are now common). In this scenario, you will presumably want to change the outflow of aid for many aid-giving countries, possibly the entire OECD grouping, perhaps ramping **aiddon** up to a global target of 0.7 from its initial value over a five- or ten-year period.

The aid given by donor countries goes into a pool that the model then splits among recipients. **AIDREC** indicates the portion of GDP that recipients are targeted to receive as aid; actual receipts in IFs are global aid distributed proportionately to those values, which change as the GDP per capita of countries rise. The parameter **aidrecadd** can change shares received by recipients. **AID** is the amount of interstate aid calculated by IFs in billion dollars; it is negative for donors and positive for recipients.

Once again, before you run the model with increased North-South aid transfers, consider how much impact you expect the scenario to have on the human development index or GDP per capita in both donor and recipient regions. How much might the increased aid narrow the absolute and relative North-South gaps? How much impact will the aid have on different recipient regions? Overall, how much leverage does the model suggest we have with this policy "handle" and how does that compare with your expectations?

We saw earlier that change in terms of trade can produce counterintuitive results in the long run. If you look at the impact of increased aid receipts on Africa in the very long run, you may see similar counterintuitive results for similar reasons. Especially if the model results differ from your expectations, you will want to know more about the basis for the model calculation: *How does foreign aid affect the economies of donors and recipients in IFs?*

First, aid affects government revenues and expenditures. For recipients, it is an addition to government revenues; for donors, it competes with other government expenditures. In both cases, it therefore has secondary implications for other categories of government spending and thereby for the economy. Second, aid affects investment, although the amount of aid that augments investment is debated by scholars. Third, aid is either a credit (for recipients) or debit (for donors) to the external current account balance (**CURACT**). As such, it will affect external debt (**XDEBTRA**), the exchange rate (**EXRATE**), and the availability of imports (which in turn affects economic performance). This last linkage can counterintuitively cause improvements in terms of trade or aid inflows to increase the exchange rate, thereby dampening exports and growth.

Research Suggestions. Experiment with different foreign aid scenarios. Overall, do these policies prove more or less important than you expected? How do you explain the results? To help understand the results, you might want to look at the relative magnitude of arms spending (**GDS**), foreign aid (**AID**), investment (**I**), and trade (**X, M**). And you may want to run the model for thirty years or more for many analyses. How substantially and how rapidly could a large-scale aid program reasonably expect to narrow the North-South well-being gaps?

International Borrowing and Debt. There is very often a substantial loan component to what we loosely call foreign aid. Thus, foreign aid contributes with private flows of capital to the foreign debt that accumulates with interest and restricts the ability of less developed countries (LDCs) to import. In fact, a considerable portion of foreign debt, especially for the poorest LDCs (like many in Africa) is public debt owed by countries to other governments or international financial institutions.

→ IFs computes annual international public borrowing of Southern countries as a portion (**aidlp**) of the aid they receive. IFs accumulates the total foreign debt of regions (**XDEBTRPA**) as a function of both annual trade deficits and such public borrowing. Interest (the rate is **lintr**) can further add to debt. IFs computes the annual repayment of debt depending on the portion that must be repaid each year (**repayr**). When foreign debts and interest rates are high enough, the amount that must be repaid annually can easily come to exceed the amount of new foreign aid, creating a net aid outflow from poorer to richer countries. Explore the values of these parameters and variables in the Base Case of IFs. You might focus on sub-Saharan Africa.

→ Compare the condition of Africa, including the level of malnutrition in children (**MALNCHIL**), under one or more scenarios that reduce or eliminate the loan portion of aid (giving it all as grants by setting **aidlp** to 0.0) or that effectively forgive the outstanding loan (by reducing the repayment rate to 0.0). How much leverage do these policy variables have to improve the condition of Africans? In contrast, what if all the "aid" came as loans?

Research Suggestions. Try developing some best-case and worst-case scenarios for the economies of the developing world, especially the World Bank's low-income country group. They might involve either high or low aid (depending on what you found about the long-term implications of aid). They might involve changes in the terms of trade, export promotion or lack of it, and/or changes in governmental spending patterns. You now have many possible policy levers with which to experiment. How much leverage do you believe there realistically is in addressing the economic problems of low-income countries and with what policies does it lie?

7.4 Conclusion

Economic liberals tell us that market-led (rather than state-dominated) economic institutions and free trade in goods, services, and capital—referred to as globalization—will promote growth and improved living standards. And surely the history of the last 50 years suggests that as technology has advanced, as economic freedom has increased, and as international barriers have come down, there has been extremely rapid world economic growth, a decline in global poverty rates, and even—in the last 40 years—a sharp decline in the absolute numbers of people living in poverty (World Bank, 2024). Critics of liberal economic policy remind us that the precise causal factors behind economic success and failure are not well-understood, that the benefits of rapid global growth have not been spread evenly, and that government regulation and intervention in the economy will always be essential for a variety of reasons, including the augmentation of human capital. Different values and understandings direct our attention to different issues and to different policy levers. Once again, however, you must draw your own conclusions about the relative or combined merits of competing understandings and the leverage we have with respect to our future. Explorations with IFs can help.

References

Barro, R. J. (1998). *Determinants of economic growth.* MIT Press.

Brueckner, M., Dabla Norris, E., & Gradstein, M. (2015). National income and its distribution. *Journal of Economic Growth, 20*(2), 149–175.https://doi.org/10.1007/s10887-015-9113-4

Burgess, M. G., Langendorf, R. E., Moyer, J. D., Dancer, A., Hughes, B. B., & Tilman, D. (2023). Multidecadal dynamics project slow 21st-century economic growth and income convergence. *Communications Earth & Environment, 4*(1), 220. https://doi.org/10.1038/s43247-023-00874-7

Calderon, C., & Serven, L. (2004). *The effects of infrastructure development on growth and income distribution* (World Bank Policy Research Paper 3400).

Cuaresma, J. C. (2017). Income projections for climate change research: A framework based on human capital dynamics. *Global Environmental Change, 42*, 226–236. https://doi.org/10.1016/j.gloenvcha.2015.02.012

Dellink, R., Chateau, J., Lanzi, E., & Magné, B. (2017). Long-term economic growth projections in the Shared Socioeconomic Pathways. *Global Environmental Change, 42*, 200–214. https://doi.org/10.1016/j.gloenvcha.2015.06.004

Dietz, S., & Stern, N. (2015). Endogenous growth, convexity of damage and climate risk: How Nordhaus' framework supports deep cuts in carbon emissions. *The Economic Journal, 125*(583), 574–620. https://doi.org/10.1111/ecoj.12188

Hughes, B. B. (2001). Global social transformation: The sweet spot, the steady slog, and the systemic shift. *Economic Development and Cultural Change, 2*(January), 423–458.

Hughes, B. B. (2019). *International futures: Building and using global models.* Academic Press.

Hughes, B. B., & Narayan, K. (2021). Enhancing integrated analysis of national and global goal pursuit by endogenizing economic productivity. *PLoS One, 16*(2), Article e0246797. https://doi.org/10.1371/journal.pone.0246797

Lindert, P., & Williamson, J. (2001). *Does globalization make the world more unequal?* (NBER Working Paper 8228). National Bureau of Economic Research. https://www.nber.org/papers/w8228

Maddison, A. (1995). *Monitoring the world economy 1820–1992.* OECD.

Maddison, A. (2001). *The world economy: A millennial perspective.* OECD.

Moyer, J. D. (2023). Blessed are the peacemakers: The future burden of intrastate conflict on poverty. *World Development, 165*, Article 106188. https://doi.org/10.1016/j.worlddev.2023.106188

Moyer, J. D., & Hughes, B. B. (2012). ICTs: Do they contribute to increased carbon emissions? *Technological Forecasting and Social Change, 79*(5), 919–931. https://doi.org/10.1016/j.techfore.2011.12.005

Moyer, J. D., Hughes, B. B., Irfan, M., Solórzano, J., Petry, C., Xiong, Y., Hanna, T., Meisel, C., Sahadevan, D., & Sullivan, E. (2026). Broadly quantifying the roads ahead: SSPs for 188 countries to 2150. Forthcoming in *Nature Communications NCOMMS-25-55570B*. https://doi.org/10.21203/rs.3.rs-7104157/v1

Moyer, J. D., Verhagen, W., Mapes, B., Bohl, D. K., Xiong, Y., Yang, V., McNeil, K., Solórzano, J., Irfan, M., Carter, C., & Hughes, B. B. (2022). How many people is the COVID-19 pandemic pushing into poverty? A long-term forecast to 2050 with alternative scenarios. *PLoS One, 17*(7), Article e0270846.

Nordhaus, W. (2019). Climate change: The ultimate challenge for economics. *American Economic Review, 109*(6), 1991–2014. https://doi.org/10.1257/aer.109.6.1991

Piketty, T. (2014). *Capital in the twenty-first century*. Harvard University Press. https://www.hup.harvard.edu/books/9780674430006

Pomeranz, K. (2009). *The great divergence: China, Europe, and the making of the modern world economy*. Princeton University Press. https://doi.org/10.1515/9781400823499

Przeworski, A., Alvarez, M., Cheibub, J. A., & Limongi, F. (1996). What makes democracies endure? *Journal of Democracy, 7*(1), 39–55. https://muse.jhu.edu/article/16730

Scott, A. C., Solórzano, J. R., Moyer, J. D., & Hughes, B. B. (2022). The future of artificial intelligence. *International Journal of Artificial Intelligence and Machine Learning, 2*(1), 1. https://doi.org/10.51483/IJAIML.2.1.2022.1-37

Solórzano, J., Hughes, B. B., Irfan, M. T., Xiong, Y., Kwasi, S., Ibrahim, A., Makhmatova, S., Hanna, T., & Moyer, J. D. (2024). Protocol for making forecasts exogenous in the International Futures model. *STAR Protocols, 5*(4), Article 103414. https://doi.org/10.1016/j.xpro.2024.103414

World Bank. (2024). Poverty, prosperity, and planet report 2024: Pathways out of the polycrisis. Washington, DC: World Bank. https://doi.org/10.1596/978-1-4648-2123-3

Human Well-being

8

Abstract

Eight of the 17 Sustainable Development Goals (SDGs) have direct linkages to human well-being. Those goals help organize presentation in this chapter of selected International Futures (IFs) Base Case long-range modeling and forecasts for poverty, nutrition, education, gender equity, water and sanitation, informal economy size, and income distribution. IFs produces forecasts for 188 countries, but this chapter shows century-long graphics for them grouped primarily across the four country-income categories of the World Bank. Exploration of possible futures for human well-being variables is dependent on their linkages within IFs to essentially all other models in the system. For instance, the demographic and economic models of IFs provide much of the foundation needed for the exploration of human well-being, but so do the models of agriculture, energy, the environment, and sociopolitical systems. This chapter also explores linkages of human well-being to the education, health, and infrastructure models of IFs, the structures of which are summarized in this chapter rather than elaborated in chapters of their own. Exercises at the end of the chapter can help readers develop and explore a wide range of alternative scenarios.

Mohammod Irfan (Pardee Institute Senior Scientist), José Solórzano (Pardee Institute Lead System Developer), Taylor Hanna (Pardee Institute Associate Director of Development Analysis), and Keith Gehring (Korbel School Teaching Associate Professor) are co-authors on this chapter.
Yutang Xiong (Pardee Institute Lead Data Operations Manager) provided underlying data support for it and recent versions of the larger IFs system.

B. B. Hughes, *Exploring and Shaping International Futures*, World-Systems Evolution and Global Futures, https://doi.org/10.1007/978-3-032-22084-4_8

8.1 Introduction

Individuals and their societies constantly strive for improvements in human well-being. Chapter 2 documented how successful the efforts have been over past millennia, centuries, and decades. While earlier success was often localized and reversible, more recent gains are largely global, with very considerable momentum in the rates of progress.

The 17 Sustainable Development Goals (SDGs), developed in the 2012–2017 period and refined through 2020 (see Chap. 1), provide elaborated statements of goals that very roughly fall into three categories. The first is improving well-being; the second is pursuing it in an environmentally sustainable world; and the third is enhancing the pursuit with capable and responsible domestic and international institutions. As of this writing, we cannot know whether the target dates for the goals (many of which will not be met) will simply be extended beyond their 2030 horizon (Fuso Nerini et al., 2024) or whether they will be reworked or replaced. It is almost inconceivable that such goals will simply be abandoned.

The interconnected model set within the International Futures (IFs) system provides the capability of exploring scenarios for human futures with variables and dynamics that cut across all 17 SDGs and many of their specific targets and indicators (Hanna et al., 2024; Hughes, 2025; Hughes et al., 2021; Moyer & Bohl, 2019; Moyer & Hedden, 2020). Attention to those goals, targets, and indicators can help the readers of this volume in their use of IFs and support their presumed interest in advancing human well-being in a sustainable world.

The futures of global demography and economics, the focus of the preceding two chapters, have very direct importance for human well-being and, therefore, for the first six SDGs as well as a few others within the full set. This chapter will survey those goals and selected targets associated with them that scenarios with IFs, especially those with varying demographic and economic characters, significantly affect. Specifically, we look here at goals focused on poverty and hunger reduction and at advances in the areas of health, education, gender equality, and access to safe water, safe sanitation, and electricity.

Subsequent chapters in Part IV (Sustainable Development) focus on the futures of agriculture (with a critical role in providing nutrition and greatly affecting the environment, including forestation patterns); of energy (access to which is another important human need with large-scale environmental impacts); and of the environment generally (deterioration of which negatively affects human well-being).

Part V (Sociopolitical Impetus) then provides an opportunity to explore futures with IFs that are very closely related to the final two SDGs, which are drivers (or impetuses) of change across the SDGs tied closely to human development and sustainable development as well as important in themselves. Because there is no organizational structure that can clearly link each IFs model to specific SDGs, please accept some arbitrariness in that of this volume.

Chapters 1 and 3 indicated the role that the SDGs have served in many analyses with IFs and in explanations of its structure, and Figs. 1.1 and 3.9 showed the entire set of SDGs. As refined through March, 2020, the United Nations (UN) has established multiple target variables (typically 8–12) for each goal and one or more indicators identified for each target variable, a total of 251 distinct indicators. No world model, including IFs, represents all of those, but IFs represents more than 100 indicators that replicate official ones precisely or closely.

In this chapter, we first identify the SDGs that relate most closely to human well-being, then turn to information about some of the models and modules in IFs that are especially useful in their analysis of life conditions. Finally, we identify some of the target variables and indicators in IFs that are available for your scenario analysis as well as parameters on which you can build those scenarios. Ongoing development and use of IFs continues to enhance such analysis, and the regularly updated IFs Wiki can elaborate and update information provided here.

8.2 The SDGs and Human Well-being Goals

SDG 1: "End poverty in all its forms everywhere." A 2012 conference in Rio de Janeiro called for the Goals. After working group elaboration of them, the UN agreed on their statement in 2015. The first target for the first goal was defined in 2017 as eradication by 2030 of "extreme" poverty.

With analytic leadership from the World Bank, extreme poverty came to be understood in indicator terms as living on less than $1.90 per day at 2011 prices (in this volume we represent such price years as 2011$), widely used in global economic analysis at that time. That value was in turn understood to be linked to an earlier concern for basic security, including calorie access, by those living on less than $1.00 (1996$)—or $1.25 in the 2005$ used by the Millennium Development Goals (MDGs) that preceded the SDGs (see Chap. 1). In 2022, the World Bank moved to $2.15 in 2017$ and in 2025 it went to $3.00 in 2021$. Changes in the IFs system attempt to keep up with these changes, and version 8.53 moved to 2021$ and measurement of poverty at related levels. IFs has the capacity in 2021$ to calculate poverty numbers and rates at any level a user specifies (*Specialized Displays/Poverty Level Display*).

Over time, the extreme poverty measure, although measured globally, has come to be the target that low-income countries seek to move beyond. For its lower-middle-income and upper-middle-income country categories, the World Bank set poverty levels at $4.20 and $8.30, respectively, in 2021$. To ensure conceptual clarity across countries, those poverty levels are understood to be assessed at purchasing power parity (PPP), accounting for consumer price levels of each country.

This elaboration of the evolution of specification and measurement of poverty is an important foundation for understanding and developing forecasts of poverty in IFs. For instance, it is important for analysis of all SDG goals to understand (1) the changing values of goals/targets across time; (2) the distinction between economic values measured at market exchange rates (MER) from those using purchasing power parity (PPP); and (3) the distinction between household income and consumption.

In IFs, significantly because of changes in global and regional goals over time and across subpopulations, there can be multiple variables for analysis of poverty, including legacy goal values from different base years of the international dollar in earlier IFs versions. Version 8.57, available at the time of this writing, uses 2021 international dollars (2021$), as will versions in future years until the World Bank again changes its international dollar measures. Thus, poverty values associated with targets for each country-income grouping calculated in these versions will reflect 2021$ calculations.

Figure 2.6 shows the tremendous progress made globally since 1820 in reducing the percentage of population living on less than $1.90 per day and the increase from negligible levels to more than 50% of humans globally living on $5 or more (all in 2011$). Figure 8.1 shows the more recent progress and century-long IFs Base Case forecasts across income-level groupings in the numerical reduction of those in poverty at the $3.00 2021 PPP$ level. Most of the reduction occurred first in upper-middle-income countries and more recently in lower-middle-income countries, while poverty in low-income countries has continued to grow numerically.

It is commonly recognized that no society will likely eliminate poverty or hunger for absolutely every citizen or completely reach other SDG targets, such as having every citizen complete secondary education. A variety of physical and mental illnesses and other factors, including personal lifestyle and social policy choices, will prevent true universality in attaining most SDGs. Therefore, analysis with IFs generally explores targets three percent less than universality, e.g., reducing poverty and hunger to under 3% and increasing secondary education

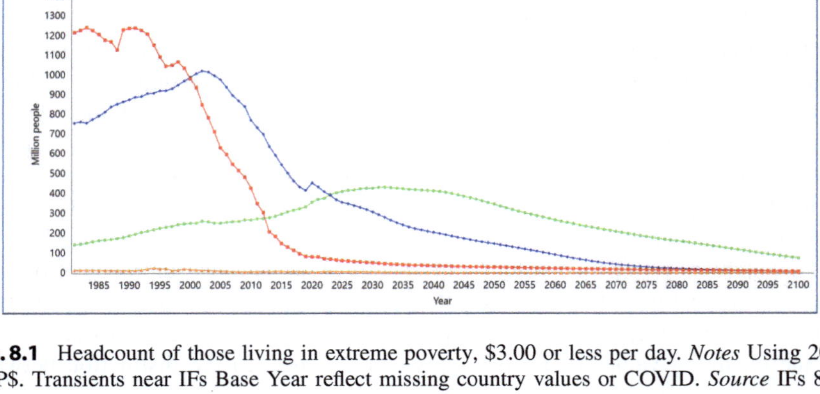

Fig. 8.1 Headcount of those living in extreme poverty, $3.00 or less per day. *Notes* Using 2021 PPP$. Transients near IFs Base Year reflect missing country values or COVID. *Source* IFs 8.57 Base Case using historical data from the World Bank Poverty and Inequality Platform

completion to 97%. In Fig. 8.1, low-income countries reach that 3% target in the 2090s. No income groups reach exactly zero.

A subsequent section of this chapter explains the computation of values for poverty and other SDG-related variables in IFs and therefore can assist in analysis of how progress toward the SDGs might be accelerated.

SDG 2: "End hunger, achieve food security and improved nutrition, and promote sustainable agriculture."

The first target and indicator related to SDG 2 focuses directly on ending undernourishment for all people. Achieving that requires attention to total calorie availability and its distribution around the world. The second target turns to malnutrition more generally and a key indicator is child stunting. Attention to improved nutrition can also address issues of either missing nutrients (not represented in IFs) or obesity (represented in IFs). Here we focus on average calorie levels and child stunting.

Figure 2.16 shows the significant progress made historically in average calorie availability, including in sub-Saharan Africa beginning in the 1990s. Figure 8.2 shows the continuation of such progress globally in the IFs Base Case. Country averages of or below 1,800 generally indicate very significant undernutrition. Figure 8.2 shows that in the 1960s most of the world was very near that level (see also Moyer et al., 2020).

Children under 5 years of age are a key demographic for those wanting to reduce undernutrition because prolonged child hunger results in stunting. Figure 8.3 shows the stunting rate in recent years, again across World Bank income categories. Note that stunting has not reached zero even in high-income countries but at least has stabilized at relatively low levels. For low-income and even the lower-middle-income categories it remains far from target levels.

SDG 3: "Ensure healthy lives and promote well-being for all at all ages." The first three targets of this goal focus on (1) maternal mortality; (2) health of newborn

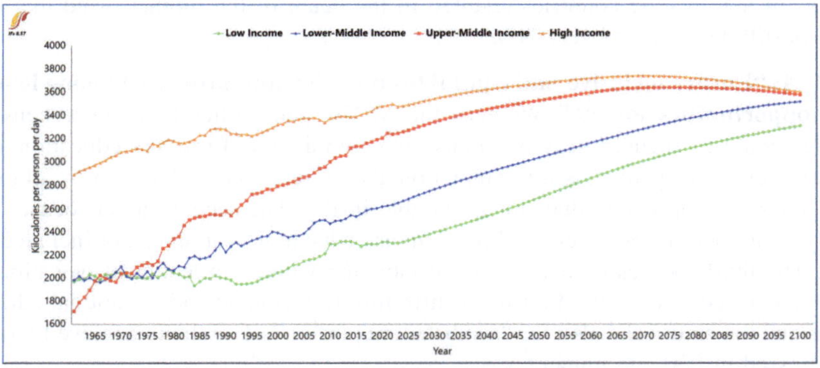

Fig. 8.2 Average kilocalories per capita. *Note* Transients mostly reflect irregular country reporting. *Source* IFs 8.57 Base Case using historical data from FAO Food Balance Sheets

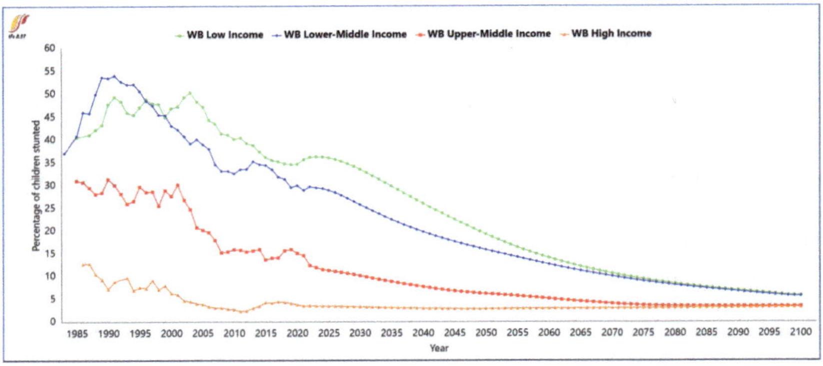

Fig. 8.3 Percentage of stunting among children age 0–5. *Note* Uses 5-year moving average to smooth irregular historical data. *Source* IFs 8.57 Base Case using historical data from the World Bank

children and those under 5 years of age; and (3) ending epidemics of many communicable diseases, including major killers like AIDS, tuberculosis, and malaria. The maternal mortality target explicitly calls for reduction below 70 per 100,000 live births; the under-5 mortality target calls for reduction to or below 25 per 1,000 live births; and the neonatal target is 12 or less per 10,000 live births.

Progress on these variables has been quite remarkable, as illustrated in Fig. 8.4 for maternal mortality. Yet, even though the high-income and upper-middle-income countries have, on average (population-weighted), reached the target already, the trajectory of historical progress for low-income and lower-middle-income countries, reflected in the IFs Base Case scenario, suggests that not only will the target not be met by 2030, but that it is unlikely to be reached even by 2050. The better news is that even in low-income countries the death rate per thousand live births will likely reach or be very close to the target by 2060. It is appalling to see that as recently as 1990, approximately one of every 100 live births in low-income countries resulted in the death of the mother—and that the rate is still near one in every 300.

SDG 4: "Ensure inclusive and equitable education and promote lifelong learning opportunities for all." Whereas the MDGs and earlier less-comprehensive global goal sets focused heavily on ensuring completion of primary education, the first target of this goal calls for completion also of secondary education by all girls and boys, an important educational variable in IFs. The second target focuses on early childhood development and pre-primary education, variables not included in IFs. The third focuses on access for all men and women to technical, vocational, and tertiary education; the fourth and fifth turn to technical and vocational education; and the sixth calls also for "a substantial portion" of adults to achieve literacy (addressed in IFs) and numeracy.

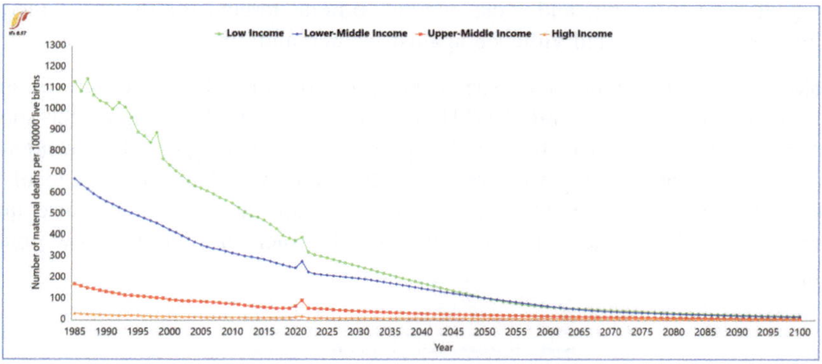

Fig. 8.4 Maternal mortality ratio (deaths per 100,000 live births). *Source* IFs 8.57 Base Case using historical data from the World Bank

Here, again, however, true universality is a goal unlikely to be reached by almost any society. Even high-income countries today average only about 90% secondary completion by of-age children (somewhat more complete at higher ages). Figure 8.5 shows that, on the Base Case trajectory, even the average gross (irrespective of age) enrollment in upper secondary education of youth in low-income countries falls far short of universality by 2050. In that country set, female enrollment may also continue to lag male enrollment until far into the century. Not shown here (but in IFs), the education gender gap has closed in much of the world. In fact, in high-income countries women have pulled well ahead of men in tertiary enrollment, raising different gender parity issues.

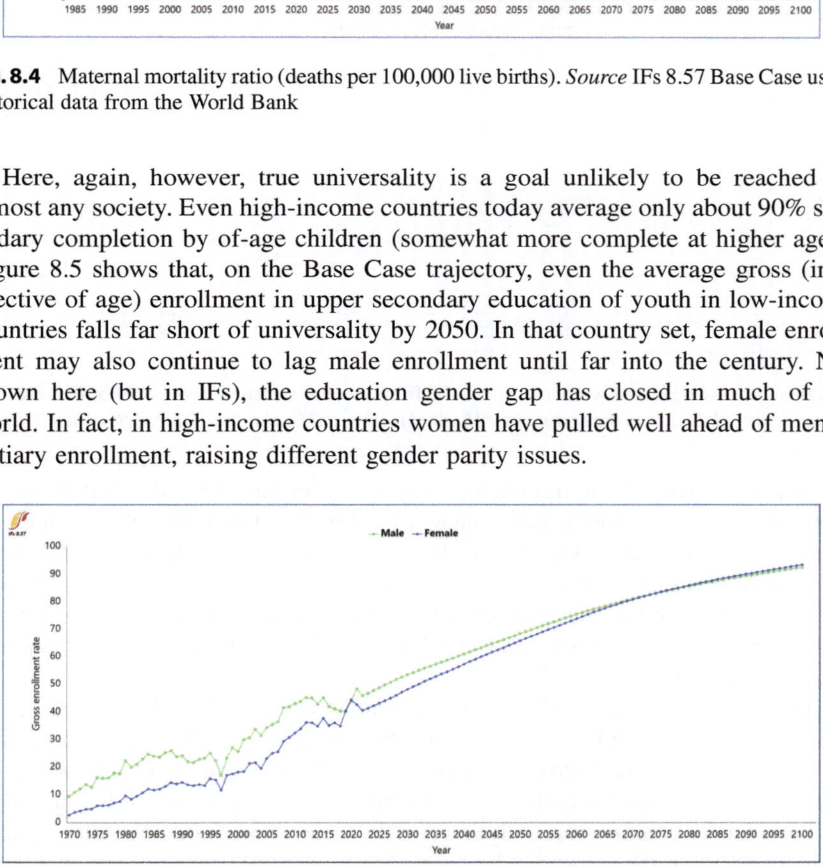

Fig. 8.5 Gross secondary enrollment percentage of of-age males and females in low-income countries. *Source* IFs 8.57 Base Case using historical data from UNESCO Institute for Statistics

SDG 5: "Achieve gender equality and empower all women and girls." The discussion above concerning education for girls and women was already relevant to this goal. The two first, and most general, official targets for this goal involve ending discrimination and violence against women and girls.

Looking first at the important issue of violence, IFs includes a measure of societal violence against women (SVDTHSWOMEN) and children. Figure 8.6 shows violence against women to be a global phenomenon, but especially prevalent in low-income countries. Given the relatively flat pattern globally at the end of the last century, the decreases that began near 2000 appear hopeful, but also quite uncertain. We do not have a similar measure for men that would allow gender comparison.

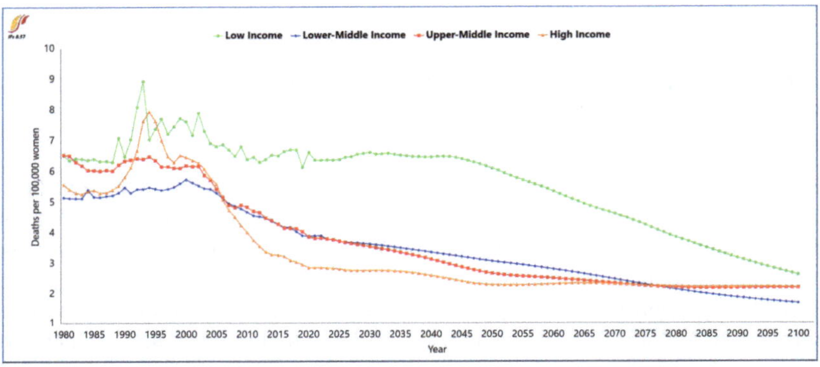

Fig. 8.6 Annual deaths of women per 100,000 from societal violence. *Source IFs 8.57 Base Case using historical data from the Institute for Health Metrics and Evaluation, Global Burden of Disease*

Turning to gender discrimination, many measures have been developed and IFs represents several of them. For instance, the IFs population model computes the Gender Development Index (GDI), the ratio of the Human Development Index for women and men. The HDI is a geometric mean of three sub-indices measuring health with life expectancy, education with a combination of schooling and adult educational attainment, and standard of living with gross national income per capita at PPP.

IFs also computes a Gender Empowerment Measure (GEM) that represents the ratio of women to men in senior/management positions, in parliaments, and earned income. In IFs, it is one driver of multifactor productivity in the economic model. Contemporary values for high-income countries are just a bit above 0.7 and those for low-income countries are only about 0.3. A newer measure of gender equality in the field and in IFs is the Gender Inequality Index (GII), which looks at equality losses from reproductive health (including the maternal mortality rate), education and parliamentary representation, and labor participation.

Nearly all drivers in IFs of the GDI, GEM, and GII are widely improving the status of women, thereby generating improvement in Base Case forecasts. Except for the GDI, however, the measures have a limited historical data foundation. Therefore Fig. 8.7 displays only the GDI.

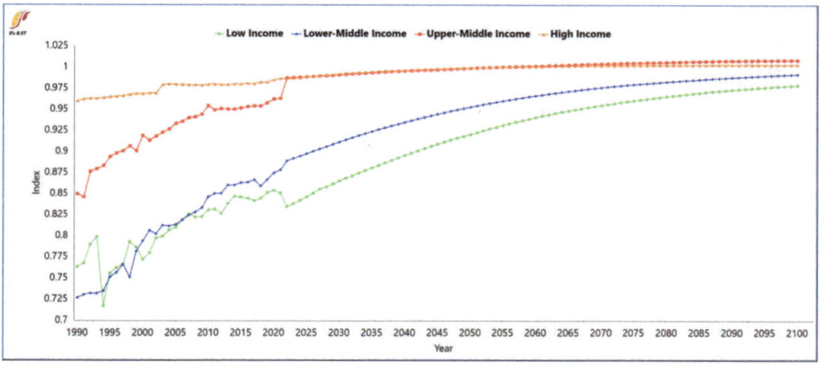

Fig. 8.7 Gender Development Index. *Note* Transients, including in the 2022 base year, reflect both missing data and some transients in it. *Source* IFs 8.57 Base Case using historical data from the United Nations

UN Women, a group within the UN focused on gender equality, has developed a Women's Empowerment Index and a Global Gender Parity Index, which together provide a considerably richer picture of gender equality. In collaboration with UN Women, the IFs project (Hanna et al., 2023) has also added measures of unpaid domestic work (UNPAIDLABOR) and women in managerial positions (WOMENINLEADERSHIP).

Other SDG 5 targets focus on specific gender issues such as forced marriage, genital mutilation, pay for domestic work, and equal opportunities for participation in leadership. Data on indicators for these targets are seldom extensively available across countries and time.

SDG 6 "Ensure availability and sustainable management of water and sanitation for all." The first target of SDG 6 focuses on safe affordable water and the second on adequate and equitable sanitation and hygiene, with a specific reference to ending open defecation. For some time, analysts have used "ladders" to identify different quality levels of access to drinking water and sanitation. Ladders with five steps have now replaced earlier three-step sets.

For reasons of basic health, societies tend initially to put more weight on access to safe water than to safe sanitation. The upward steps on the drinking water ladder are surface water, unimproved, limited, basic, and safely managed. For sanitation the steps are open defecation, unimproved, limited, basic, and safely managed. The first two targets for SDG 6 involve reaching safely managed levels on both variables. Figure 8.8 shows the percentage of populations in the World Bank income categories with safely managed water access historically and in the Base Case scenario of IFs.

Important note to model users: The transients within Fig. 8.8 in the IFs base year of 2022 reflect the absence of data for many countries, illustrating an important issue when looking at history-plus-forecast graphics. When values for the base year are missing for a country, IFs generally uses cross sectionally-estimated

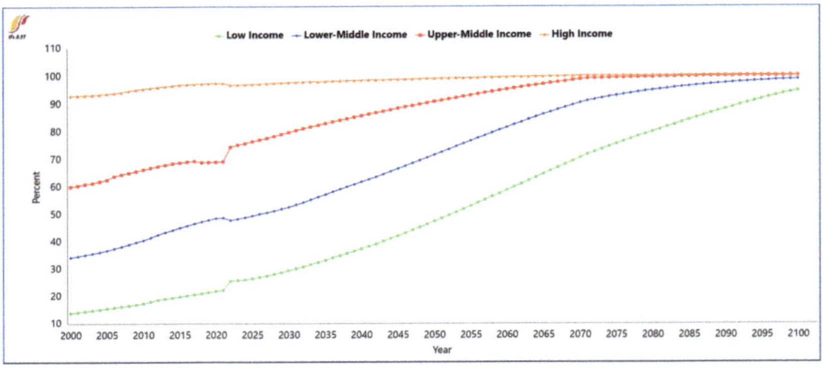

Fig. 8.8 Percentage of total population with safely managed water. *Note* Transients in the IFs base year reflect much missing historical data. *Source* IFs 8.54 Base Case using historical data from World Health Organization

functions (often against GDP per capita at PPP) to calculate values. For access to safe water, values for many countries are missing, including some with large populations. That greatly affects the population-weighted access rates of the base year.

Although upper-middle-income countries will likely be approaching the levels of high-income countries by 2050, only the high-income set is shown as passing the 97% level of near universality by then.

The first six SDGs, discussed to this point, are all clearly related to assessments and enhancement of human well-being. While most of the remaining SDGs are important in analyzing environmental quality and/or domestic and global agency for pursuit of all the goals, two additional goals, SDGs 8 and 10, are directly connected to human well-being.

SDG 8: "Promote sustained, inclusive and sustainable economic growth, full and productive employment, and decent work for all." Although the goal has relevance for people in all countries, the importance of economic growth for advance toward other SDGs is especially great in low-income countries.

The first target of SDG 8 calls for at least 7% annual growth in GDP for the least developed countries. It does not specify GDP at MER or PPP. Figure 8.9 shows IFs Base Case forecasts for growth of GDP at MER for the World Bank income groupings. Despite highly irregular and generally slow low-income-country growth historically, the IFs Base Case does show a rise into the range near 7% by the 2040s.

Even were the GDP growth rate in low-income countries to exceed that in high-income countries by the large amounts shown in Fig. 8.9, the absolute gaps in GDP per capita would continue to grow, rising from $48,000 in 2025 to $132,000 in 2100 (not shown). Figure 8.10 provides more insight, however, by indexing the GDP per capita values for the four groupings from the year 2020. Although the value for the low-income grouping grows more slowly than that of the high-income

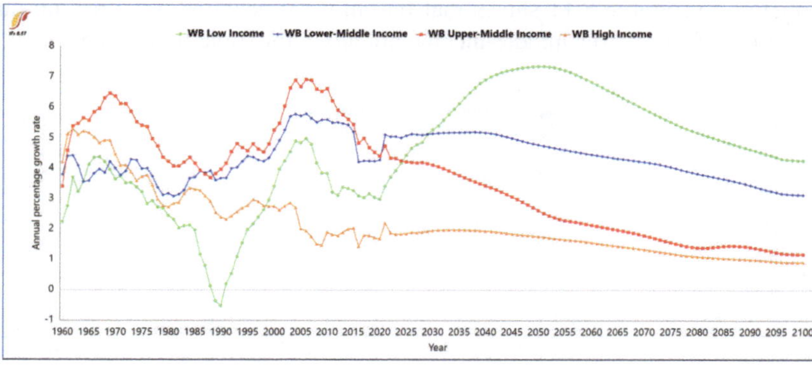

Fig. 8.9 Annual growth in GDP at market exchange rates. *Note* Using 10-year moving average to somewhat smooth historical fluctuations. *Source* IFs 8.57 Base Case using historical data from the World Development Indicators

grouping through about 2035 (because in large part of rapid population growth), the middle-income groupings grow more rapidly. And by 2040 and thereafter, growth of the low-income group begins to clearly exceed that of the high-income and middle-income groups. Overall, the path through 2100 in the IFs Base Case is one of convergence in the ratios of GDP per capita even in the face of divergence in absolute values.

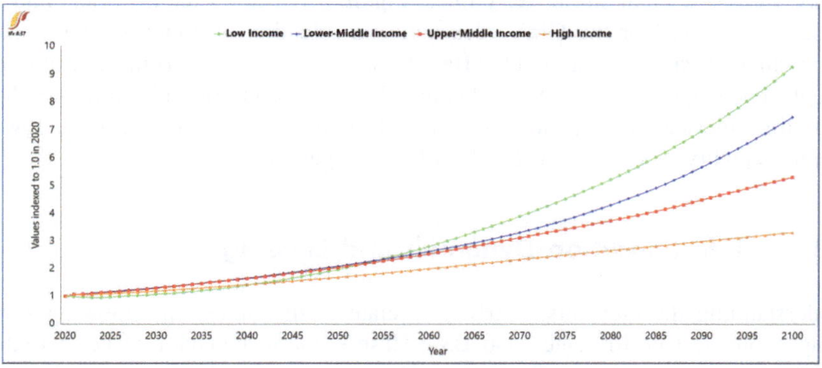

Fig. 8.10 GDP per capita at purchasing power parity indexed from 2020.
Source IFs 8.57 Base Case

The third target of SDG 8 calls for reduction in the size of the informal economy as a portion of the total economy. Although it provides income for individuals who might not otherwise have it, wages and job security are almost always low in the informal sector. Figure 8.11 shows that the informal economy may remain quite sizeable in low and lower-middle-income countries for some time to come.

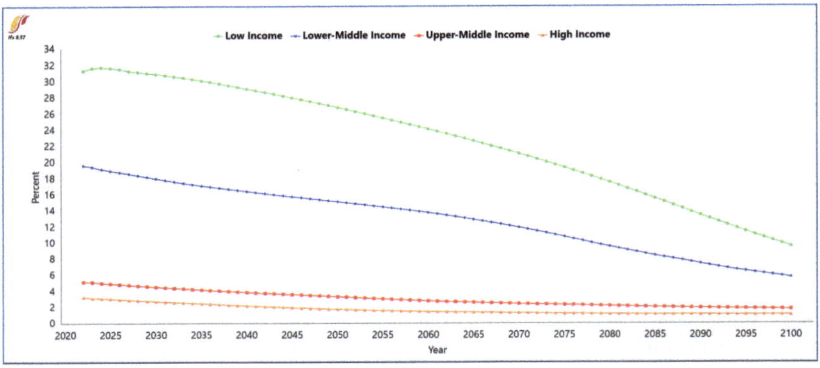

Fig. 8.11 Informal economy as a percent of the total economy.
Source IFs 8.57 Base Case using data from multiple sources, including Elgin and Oztunali (2012), Schneider and Enste (2013), and United Nations Economic Commission for Europe (2008)

SDG 10 "Reduce inequality within and among countries."
A key measure of inequality is the Gini index. Higher Gini values represent less equal income distributions and have implications for a variety of human well-being variables, including poverty rates. Figure 8.12 shows the historical and IFs Base Case values for the Gini index across a selection of countries with a very large range of values. Although the Base Case scenario maintains initial conditions largely unchanged over time, the historical data of countries like three of those in Latin America suggest the possibility of significant change, including toward greater equality. Scenarios in IFs also allow exogenous change.

8.3 Life Condition Dynamics and Leverage

Understanding the elements in IFs that relate to the SDGs and knowing something about past performance and Base Case scenario behavior of key variables and indicators are only the first steps. The next step is learning more about the dynamics within IFs that underlie movement toward or away from the goals. For information beyond that in this section, see the IFs Wiki, the IFs network diagram, and the Pardee Institute's home page.

Because of the highly integrated nature of global system elements in both the world and IFs, those dynamics often cross multiple interlinked IFs models. For instance, it is impossible to understand the future of poverty (SDG 1) without exploring the implications for that future of demographics, economic growth, and patterns of income distribution and inequality. The subsections of this section

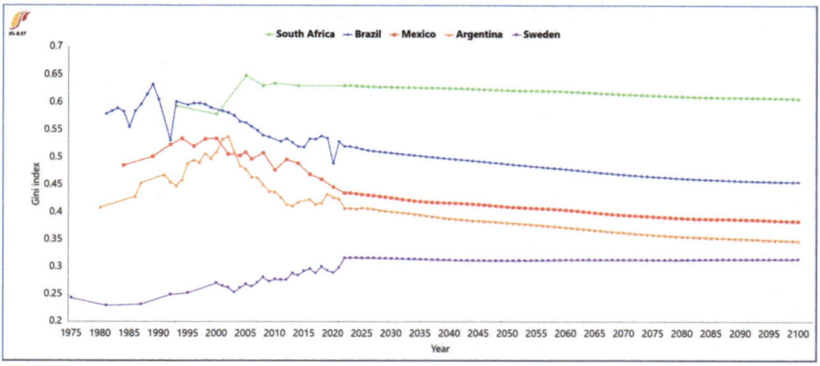

Fig. 8.12 Country-specific Gini values. *Source* IFs 8.57 Base Case using historical values from the World Development Indicators

emphasize those integrated connections. Key resources that support the discussion here exist online in the Network Diagram for IFs and its Wiki, especially the pages explaining various models.

8.3.1 Poverty and Its Foundations in Economic Growth and Distribution

The three immediate foundations of the IFs algorithmic calculation for poverty numbers and percentage rates are (1) the average household consumption per capita; (2) an assumed shape of national income and consumption distribution; and (3) a specified poverty line (e.g., $3.00 per person per day). Figure 8.13 indicates how those elements combine to compute the percentage of population in poverty. The assumed probability distribution function (PDF) has a log-normal shape, which means that the log of the distribution variable (here, income or consumption) is normally distributed around the distribution mean. It has a single peak but is not symmetrical, with the side to the left of the peak moving quite sharply to 0.0 but the side to the right of the peak having a long tail (right-skewed distribution). The median value under the curve, identified in the figure as "Per capita income" but applied in the model to per capita consumption rather than income, is therefore somewhat to the right of the peak. Along the x-axis, any specific per capita value can be identified as the "Poverty line" and the percentage share of the area under the distribution to the left of the line is the percentage of the population calculated to be in poverty.

The log-normal assumption roughly fits the reality of income distribution in most countries, but there are some with quite different distributions, including, for instance, a bimodal one with peaks for the very poor and very rich subpopulations. Data for income distributions across countries are quite skimpy so IFs applies the log-normal assumption for all countries, using the Gini index to shape it.

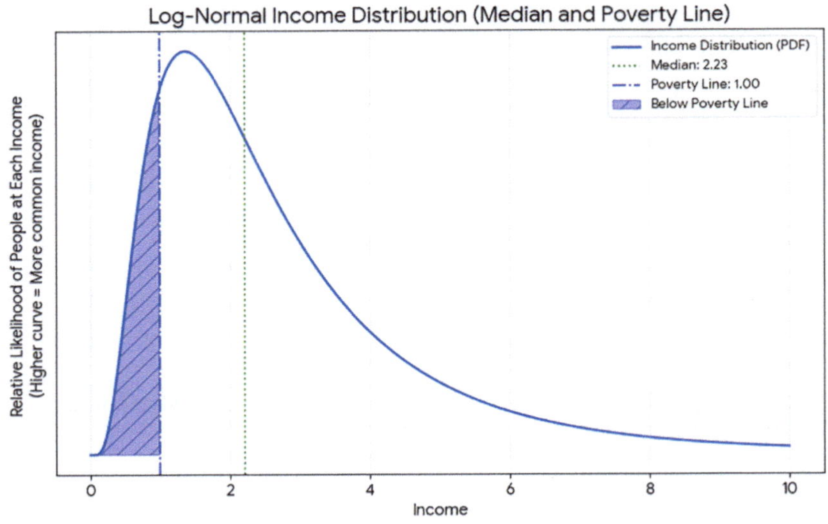

Fig. 8.13 Log-normal income distribution and poverty portion. *Source* Mohammod Irfan

The lower the peak and more spread the distribution, the more equal the income and/or consumption patterns are. As we saw in Fig. 8.12 and the discussion around it, we can use the Gini coefficient to summarize information related to distributional patterns. Figure 8.14, called a Lorenz Curve, is the foundation of Gini values. From left to right, that curve shows the cumulative portions of national income (or consumption or whatever is being distributed) that accrues to portions of the national population from the fraction with the least on the left to the total population on the far right, which of course receives 100% of whatever is being studied. A diagonal upward-sloping line (green in Fig. 8.14) would show that each portion of the population obtained the same share of the total and is known as the line of equality. The further the actual distributional pattern (blue line) falls below that equality line, the greater the area A is relative to the total area under the equality line (A + B) and the larger the Gini and inequality are (with a maximum of 1). As we saw in Fig. 8.12, a country like South Africa might have values that exceed 0.6.

Although common discourse often links poverty to household income level or even to GDP per capita, the World Bank and the SDGs define it in terms of "living on less than" various specific daily dollar values. It is thus more properly understood in terms of consumption level, as it is in IFs. In IFs, household income reflects earnings from labor adjusted by flows to and from governments and firms. At a global level it is quite close to 60% of GDP. Given the split by households of income into savings and consumption, the consumption share is less than that, somewhat more than half of GDP globally. The share has declined somewhat over time with the growth of income and the extension of life expectancy, both supporting higher levels of savings, especially in high-income countries and some of

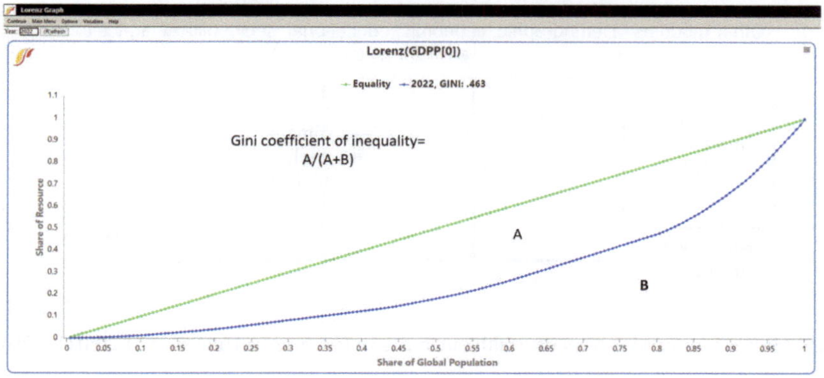

Fig. 8.14 The Lorenz curve basis of the Gini coefficient. *Source* IFs 8.57 Base Case

the upper-middle-income countries, most notably China. The next major section of this chapter will discuss the control a model user has over forecasts of household consumption, its distribution, and therefore of poverty levels.

8.3.2 Nutrition and the Absence of Hunger

The next chapter will elaborate the agricultural model, which provides much of the foundation for nutrition, although trade of agricultural goods is such that consumption can be very different from what is produced in a country. In general terms, in IFs the demand for (consumption of) calories and protein is fundamentally tied to income, with some responsiveness also to food prices. As incomes increase, the consumption levels in IFs rise, with saturation levels for calories and protein consumption. Crops, meat, and fish all contribute to both calories and proteins. Calculations of undernutrition for both children and the broader population are tied in IFs to calorie consumption, with the latter also responsive differentially to levels of income in skilled and unskilled households.

8.3.3 Health[1]

Because IFs contains elaborate models for both health and education, full chapters could be devoted to each. For the purposes of this volume, we instead provide discussion of them in this and the next subsections of this chapter. See the IFs Wiki for extended documentation.

[1] Randall Kuhn, José Solórzano, and Cecilia Peterson led the development of the IFs health model.

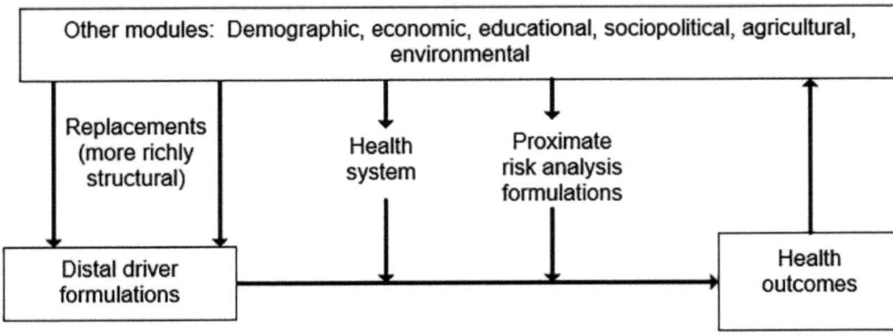

Fig. 8.15 Categories of drivers of health outcomes in IFs. *Source* Hughes et al., (2011a, 2011b: 38)

The health model builds heavily on extensive work in the Global Burden of Disease (GBD) project (Mathers & Loncar, 2006). It represents 15 causes of mortality and morbidity in the three GBD general categories, namely communicable diseases (diarrheal diseases, malaria, respiratory infections, HIV/AIDS, and other communicable); non-communicable diseases (malignant neoplasms, cardiovascular diseases, chronic respiratory diseases, digestive diseases, diabetes, mental health, other non-communicable); and injuries and accidents (traffic accidents, other unintentional injuries, intentional injuries).

There is tremendous variation in the IFs representation of the drivers of mortality and morbidity across the 15 causes; the driver variables come from most other models of IFs, including those for population, economy, agriculture, education, infrastructure, environment, and sociopolitical systems. Figure 8.15 suggests the top-level complexity of driver systems. Within IFs it is important to understand the distinction between distal and proximate drivers of change in a dependent variable like health outcomes. Such dependent variables are immediate functions of the proximate drivers. Proximate variables are driven by other variables, which are therefore more distant/distal relative to the dependent variable.

There is a very central distal driver formulation with roots in GBD analysis in which GDP per capita, educational attainment of those 25 years of age or older, and technological change (via parameterization of the constant term of the function) play roles for many of the 15 causes of morbidity and mortality. However, this formulation is subject to significant augmentation or total replacement by formulations and algorithms for some of the causes. Also, because of its pervasive importance, smoking was added in IFs to the basic distal formulation for some of the causes. The coefficients linking the distal causes to mortality and morbidity were estimated for multiple age categories and are specified in IFs across its 5-year age groupings.

Other elements that represent causality are labeled super-distal (such as policy-driven or human behavior change-driven influences on health) or proximate risk factors (including obesity and air quality). Hence the ultimate formulations across the 15 causes vary dramatically. Figure 8.16 emphasizes how drivers vary

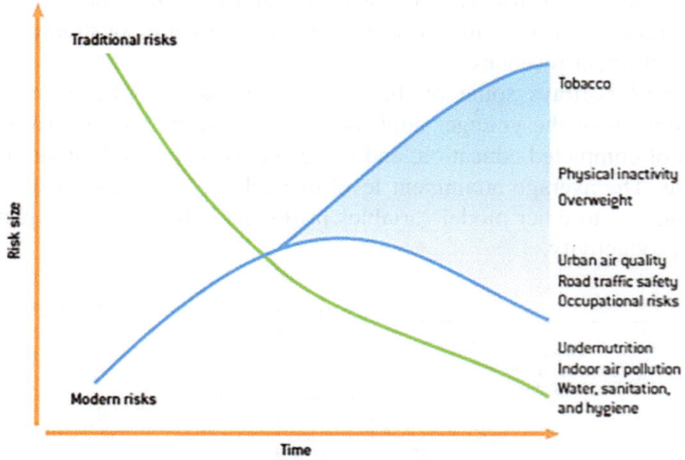

Fig. 8.16 The transition in health risks with development. *Source* Hughes et al., (2011a, 2011b:73); see also the World Health Organization (2009: 3)31. Reprinted with permission

not only across mortality causes but evolve with a combination of economic, socioeconomic, and technological developments from some risk factors labeled "traditional" and some called "modern." We know, of course, that the figure is a gross simplification, with the impact of tobacco now decreasing in what might be called a post-modern-risk development phase, and that even obesity might be increasingly brought under control. Similarly, environmental risk factors can change character from household focused (e.g., cookstoves), to broader society (urban air pollution), to global (global warming). In short, the causal systems, combining statistically based formulations and structural representations, vary greatly across, and even within, the 15 mortality causes. In addition, interventions across IFs models more broadly can also affect health forecasts.

8.3.4 Education[2]

The education model of IFs (Irfan, 2017) represents grade-by-grade progression across elementary and lower- and upper-secondary levels, as well as total years of tertiary education. The entrance, progression, grade repetition, and completion patterns are influenced strongly by changing demographics and the evolving educational capacity of societies, which, in turn, is significantly shaped by economic growth, changes in school enrollment patterns, and government expenditures. Educational expenditures are affected by the demands from enrollment patterns and, because of the labor intensity of education, the average level of GDP per capita.

[2] Mohammod Irfan developed the IFs education model.

The expenditures must also compete with demand for other spending, including that on the military, health, and infrastructure, and they are constrained over time by total government revenues.

Figure 8.17 portrays some of these key relationships. The years of educational attainment of the younger adult population cohorts are directly shaped by their years of completed education, and those cohorts carry their attainment levels through life. The average attainment level of adults is a variable with significant forward linkages to other model variables in IFs, including mortality patterns and economic productivity.

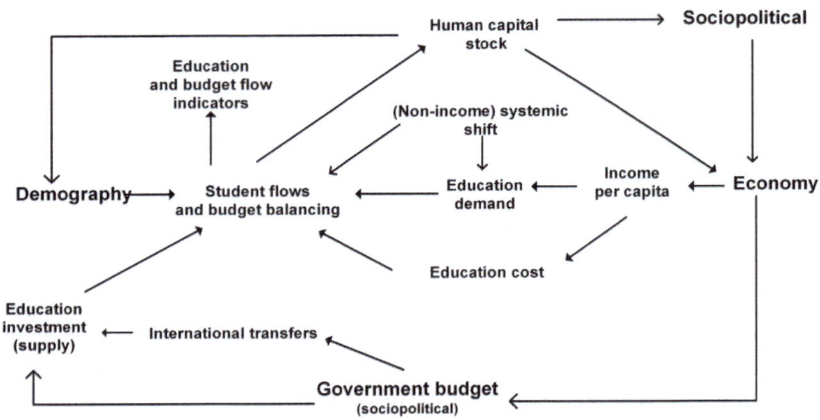

Fig. 8.17 High-level representation of dynamics in the IFs education model. *Source* Dickson et al. (2010: 60) and broader model documentation in the Wiki

The model dynamics also reflect base year and historical data on patterns of enrollment by sex and the changes in those patterns across time, which are affected by cross-sectionally-estimated functions for sex-specific intake and outflow rates with GDP per capita at PPP. Enrollments in lower-income countries for girls often increase more rapidly than those for boys, with national patterns generally evolving toward the higher enrollment and completion patterns now common in high-income countries. Figure 8.18 provides elaboration of the student flow and attainment stock structure of the model.

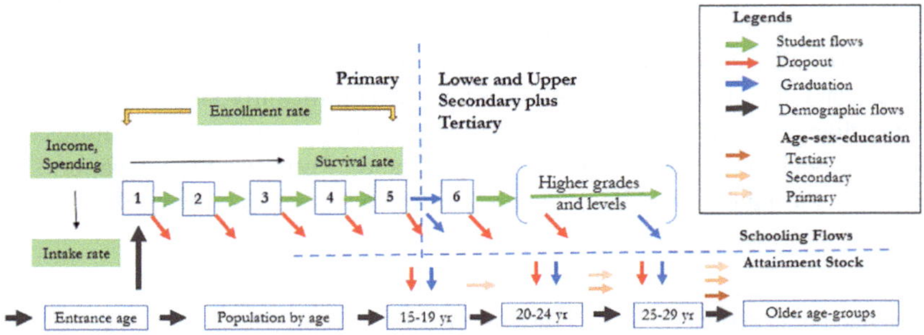

Fig. 8.18 Enrollment and attainment in the IFs education model.
Source Mohammod Irfan

8.3.5 Gender Equality

The earlier discussion about how girls are now catching and then passing boys in educational enrollment and completion illustrates how the study of potential advance toward gender equality is spread across many models within IFs rather than being embedded in a separate model. The health model similarly addresses important issues for women, including maternal mortality and broader life expectancy, as well as computing measures such as the rates of violent deaths for children and women. The economic model includes attention to labor participation rates of women and the gender empowerment measure. The population model computes the sex-specific Gender Development Index as well as important variables like the age of marriage and violent deaths of women (and children). In summary, gender equality is open in IFs for exploration in many ways (Hanna, et al., 2023).

8.3.6 Access to Water and Sanitation, Electricity and Other Forms of Infrastructure[3]

IFs includes a broad infrastructure model (Rothman et al., 2014), not just attention to water and sanitation (watsan or WASH). In addition to representing the changing availability of different levels of access up to safe and improved water and sanitation, it represents access to electricity (partly in contrast to the direct use of solid fuels for cooking and other purposes); access to safe roads (Moyer et al., 2017), including paved ones; access to various forms of information and communication technology, including mobile telephones and broadband services; and in association with the IFs water and agriculture models, access to irrigation.

Infrastructure is particularly difficult to model because, unlike poverty, health, or education, there is no single or small set of sources to turn to for basic data and analysis. There tend to be different intergovernmental and policy analysis institutions for each infrastructure type.

Figure 8.19 shows basic causal structures in IFs for watsan. Demand tends to be linked at least in part to GDP per capita (normally at PPP). Supply requires finance, and a large share of infrastructure globally is governmentally financed. That puts financing demand for each type of infrastructure in competition with each other type and infrastructure in general in competition with other demands

[3] Dale Rothman and Mohammod Irfan led the development of the IFs infrastructure model.

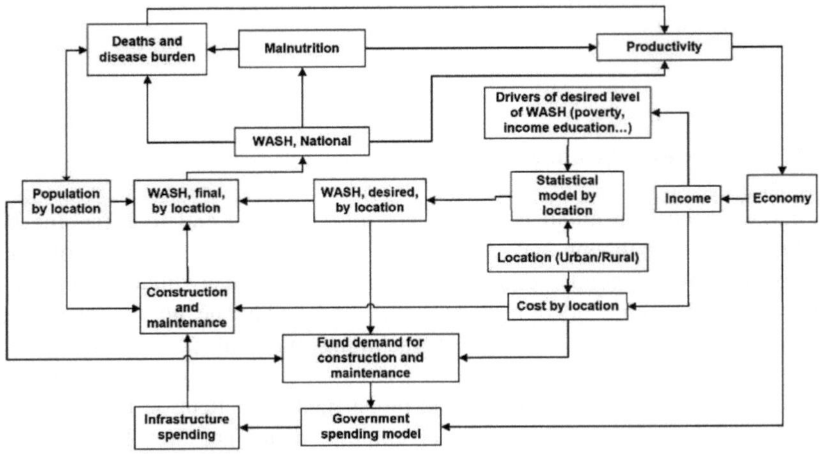

Fig. 8.19 High-level representation of water and sanitation (WASH) in IFs. *Note* Parameterization differentiates the steps of the WASH ladders. *Source* Mohammod Irfan

for governmental direct consumption spending, including that for health and education, as well as in competition for government transfer payments, such as those that reduce income inequality and poverty.

Only some of the infrastructure model relates directly to pursuit of the SDGs. Much of it and other models within IFs treat their subject matter more broadly.

8.4 Interventions and Scenario Analysis

In its survey of undertaking scenario analysis with IFs, Chap. 5 made the distinction between building scenario runs by specifying the values of just one or two parameters versus building and using scenario files involving multiple parameter changes in a larger story. In their discussion of scenario interventions to explore alternative demographic and economy futures, Chaps. 6 and 7 generally focused on small sets of parameter changes to explore impacts of specific interventions in those models of IFs.

The discussion of human well-being in this chapter has, however, required broadening our focus to multiple and interacting models within the IFs system. In fact, many readers will be interested in scenarios that cut across nearly all IFs models in the simultaneous pursuit of the SDGs. Thus, this section will briefly outline an existing scenario, an alternative global future named *SDG Push*, that does just that with a focus on the human development SDGs before we turn to specific variables and parameters that narrow the focus to specific issues/goals, moving us closer to the realm of much policy analysis. At the time of this writing, the SDG Push is one of the “pre-run” scenarios immediately available for analysis.

8.4.1 Use of the SDG Push Scenario

Recognizing the importance of the SDGs and whatever set of global goals might succeed them, and in a joint project with a team at the United Nations Development Programme early in the COVID-19 period, a team at the Pardee Institute developed what came to be labeled the SDG Push scenario, to be juxtaposed to the IFs Base Case (Hanna et al., 2024). The list below shows core elements of that scenario, focused heavily on the human development SDGs. This scenario could well be of interest throughout your use of IFs and this volume.

There are ten major intervention points in the SDG Push scenario:

1. Agriculture and calorie availability progress (increase yield, reduce losses, increase calorie availability/consumption distribution)
2. Educational emphasis (increase enrollment/completion at all levels, enhance STEM, spend more)
3. Health emphasis (improve cookstoves, reduce ambient air pollution, spend more)
4. Water and sanitation emphasis (advance to safer levels on watsan ladders)
5. R&D and ICT emphasis (spend more, increase ICT access)
6. Infrastructure emphasis (expand electricity access, use water efficiently, spending more generally)
7. Character of governance improvement (advance democracy and effectiveness, reduce corruption)
8. Transfer payment emphasis (increase social welfare and pension spending)
9. Wage equality for women (focus directly on movement to equality)
10. Climate change slowing/stopping (advance renewable technology, improve energy usage efficiency, use carbon tax, reforestation)

If the *SDG Push* is no longer available pre-run, you should be able to find it under *Scenario Development and Change/Scenario Files/Open* and then run it yourself. See the *Building More Extensive Scenario Analyses* section of Chapter 5 for explanation on how to load and run scenario files. If it is not there, you may need to approach Pardee Institute personnel. If you load that scenario file into the scenario tree of the Scenario Analysis form, you will be able to see the parameters that the scenario changes and the magnitudes and temporal patterns of those changes. Those can supplement the discussion below.

The subsections below identify some of the variables and intervention points that link many of the elements in the SDG Push list above to the issues in the SDGs and of general interest in exploration of alternative futures.

8.4.2 Poverty and Its Foundations in Economic Growth and Distribution

The *Flexible Displays* form provides access to many tables and graphics related to poverty, which SDG 1 calls upon us to eliminate and which we know interacts closely with economic growth and distribution, on which SDGs 8 and 10 more specifically focus. Options in the *Human Development* category list of that form, including **POVEXTREME**, **POVMIDLOW**, and **POVMIDHIGH** (key poverty variables in 2021$, representing $3.00, $4.20, and $8.30 levels) are among the variables of interest in showing numbers of people and percentages of population living in poverty.

Note: When you highlight a *Flexible Displays* list, if you choose the *Explain/ Explain List* option on the form you will see the variable name(s) that the list will display. Those names will not always show up in tables and especially graphics, but can be useful when you use *Self-Managed Display*.

You will not see any scenario intervention points identified in the list of 10 above that refer specifically to poverty. As earlier discussion of poverty pointed out, the key drivers of poverty are the level of household consumption per capita (**C** over **POP**), with a flexible display in the *Economy/Consumption* category, and the domestic income distribution display (**GINIDOM**) in the *Economy/Distribution* category.

There are many variables and dynamics in IFs that affect the level of household consumption per capita and income distribution. At the most basic level, anything that affects GDP and GDP per capita will have implications for them. Thus, nearly all the intervention points discussed in the Population and Economics chapters will affect per capita consumption.

Most discussions of poverty focus heavily on GDP growth. Chapter 7 mapped out many of the drivers of change in GDP and the intervention points around which scenarios for that change can be built. Among those points it briefly discussed the contribution to GDP change over time made by multifactor productivity (MFP). To elaborate that, human and social capital (**MFPHC** and **MFPSC**) and infrastructure or physical capital (**MFPPC**) influence change in multifactor productivity terms (**MFPRATE** and **MFPTOT**). Along with capital and labor, MFP shapes value-added by sector (**VADD**) and its national sum in **GDP**. Chapter 7 discussed direct exogenous manipulation of MFP with the **mfpadd** and **mfpleadr** parameters in the *Exploring Economic Growth* subsection. Subsequent subsections here will return to some of the variables and parameters in the health, education, and infrastructure models that underlie the endogenous dynamics of multifactor productivity variables.

Household consumption **C** is not a fixed proportion of GDP, even if that proportion globally tends to be about 50–60% and in most countries that proportion changes fairly slowly. It depends quite heavily on the level of disposable household income (**HHINCDIS**) and then on the part of that which is directed toward consumption rather than savings. Household disposable income depends heavily on earnings (that portion of value-added directed to households as opposed to firms), but also significantly on taxes paid to and transfer payments from governments as well as, for many countries, on worker remittances from abroad (**XWORKREMIT**). Among the parameters that can influence those are multipliers on government revenues (**govrevm**) and expenditures (**govexpm**). Household income can be increased very directly by transfer payments for pensions (**GOVHHTRNPEN**) and welfare (**GOVHHTRNWEL**), influenced by **govhhtrnpenm** and **govhhtrnwelm.** The model does not allow indefinite increase in government debt and therefore restricts the degree to which the expenditures of government outstrip revenues.

Not only does the household disposable income per capita variable (**HHINCDISPC**) affect consumption and poverty levels, it also determines change in the domestic income distribution represented in the Gini index variable (**GINIDOM**). Household transfers from the government can be directed with **govhhtrnpenm** and **govhhtrnwelm** differentially to unskilled or skilled households, thereby changing the Lorenz curve in IFs and the calculation of **GINIDOM**.

While in scenarios it is possible to reshape patterns of household consumption shares across household type, and therefore change GINIDOM via control over flows within the social accounting matrix (SAM), it is also possible to explore more directly the impact of such changes. The parameter **ginidomm** allows multiplier values above 1 to increase **GINIDOM** and inequality and values below 1 to decrease them.

→ Although you may wish also to explore the implications of interventions involving productivity or government transfers, at least explore the impact of changing GINIDOM on the future of poverty rates, probably focusing especially on low-income countries.

8.4.3 Nutrition and the Absence of Hunger

The second SDG calls for the elimination of hunger. That requires that societies have sufficiently high levels of daily calories per capita (**CLPC**) and a distribution of those that allows at least basic levels for everyone. In IFs, that distribution across household types is related to disposable income (**HHINCDIS**), which can be affected by government transfer payments with **govhhtrnwelm**. Chapter 9 will turn to the IFs modeling of agriculture and the computation of CLPC from crops, meat, and fish separately and in total. There is a calculation also of protein per capita (**PROTEINPC**), but unlike CLPC it does not enter the calculation of child and general population malnutrition (**MALNCHIL** and **MALNPOPP**).

In IFs, the calculation of CLPC and related total food-types (crops, meat, and fish) depends on national demand in metric tons (**FDEM**) and in grams per capita (**GRAMSPC**). Food demand is tied to GDP per capita at PPP (**GDPPCP**) via a cross-sectionally estimated function. Across time, changes in GDPPCP and FDEM affect the basic calculation of CLCP, which is also modified by food prices and which users can modify directly with **clpcm.**

After initial calculation, adjustments to CLPC, PROTEINPC, and GRAMSPC can occur when there are crop, meat, and/or fish shortages, which reduce or eliminate food stocks (**FSTOCK**). Shortages can be related to domestic crop yield patterns (**YL** with the multiplier **ylm** influencing it), production of crops, meat, and fish (**AGP**), or agricultural trade patterns (**AGX** and **AGM**). Agricultural losses, including waste by consumers (**AGLOSSCONS**), also affect food availability. Shortage impacts begin with a reduction in grams per capita, which is then translated into new values for FDEM and recalculations of CLPC and PROTEINPC.

To address food supply, calorie intake levels, and undernutrition, the model user can directly affect production patterns via yield and cropland multipliers (**ylm** and **ldcropm**) and can reduce food loss in production (**aglossprodm**) between farm and table (**aglosstransm**) and/or consumption (**aglossconsm**).

Increasingly, obesity rates (**HLOBESITY**) are rising, even as undernutrition pressingly remains to be addressed. The obesity rate multiplier (**hlobesitym**) directly changes its prevalence, which like undernutrition, affects health, our next topic.

→ Among other variables, including undernutrition that scenarios focus on, reduction of stunting of children (**HLSTUNTCHL**) from undernutrition receives much attention. Explore scenarios to reduce that for a country or grouping thereof. Consider interventions that focus on increasing yield, increasing GDP, or deceasing population growth. How does IFs assess their relative and combined leverage? Given that all models are somewhat crude representations of reality, how would you evaluate the results you find?

8.4.4 Health[4]

SDG 3 calls for ensuring healthy lives. Earlier discussion of health, including that around Fig. 8.16, emphasized the considerable variation in proximate causal dynamics underlying the 15 different categories of mortality (**DEATHCAT**) and morbidity in IFs. That variation can complicate analysis with the IFs health model (Hughes et al., 2011a, 2011b; Kuhn et al., 2016), even as it usefully connects that analysis to most of the other models in the IFs system and, therefore, to appropriate driving variables. Some users will want to directly affect mortality, however, addressing questions such as "what if it *were* possible to largely eliminate malaria or cancer or heart disease?" It is possible to use multiplicative parameters that change mortality overall (**mortm**) or by cause of death (**hlmortm**). Further, **hlmortcdchldm** changes the rates of death from all communicable diseases for children age 5 and younger, while **hlmortcdadltm** does the same for adults age 15–49.

In addition to mortality multipliers, the Social Accounting Matrix (SAM) structure's government finance module provides another useful handle for affecting mortality patterns without looking at specific death-cause drivers (see again the discussion of the SAM in Chap. 7). In addition to transfer payments that can directly affect poverty patterns, governments provide finance for a wide range of issue areas. In IFs, those are identified as military, health, education, R&D, infrastructure types elaborated in IFs, other infrastructure, and a residual category. The variable specifying the size of those direct expenditures by category is **GDS**, and the model user can change that for any destination category using a multiplier (**gdsm**), subject to normalization of all spending across categories to total government consumption (**GOVCON**). That multiplier is useful not just in scenario analysis around health, but also in the education and infrastructure models. Total government consumption can itself be affected by a multiplier (**govconm**). But government consumption or transfer expenditures without increased revenues will increase debt and ultimately lead to expenditure restrictions or to additional taxes and revenues. It is possible to increase those revenues more directly with a multiplier (**govrevm**), but doing so will direct more household and firm income to the government, reducing other uses of that income by those agents and even potentially increasing poverty.

Following the lead of the Global Burden of Disease project, the distal driver formulation in the health model theoretically also can facilitate scenario analysis attention to death causality without the requirement of analysis related to proximate drivers. Mortality from almost all immediate causes is influenced by GDP per capita at purchasing power parity (**GDPPCP**) and years of adult education (**EDYRSAG25**). Because smoking rates (**HLSMOKING**) affect so many communicable and noncommunicable health conditions, the IFs project added them to the general distal driver formulation for selected mortality types and developed a

[4] Randall Kuhn, José Solórzano, and Celicia Peterson led the development of the IFs health model.

module to forecast their possible future values and/or to alter them for scenario analysis (**hlsmokingm**). In addition, technological change, represented parametrically as a beta on a time term (**hllongvtybeta**) is very powerful and is represented as a driver of health outcomes in the GBD approach and in IFs.

It is difficult, however, to undertake scenario analysis by changing coefficients in the distal formulation because those are statistically estimated in multivariate analysis by type of mortality, sex, and age in seven age groupings (<5, 5–14, 15–29, 30–44, 45–59, 60–69, and 70+) which are spread in IFs across its 5-year cohorts. IFs does, however, include parameters to change the beta term and, therefore, assumptions about technological advance. See the IFs Wiki for information on those parameters.

Going beyond direct parametric intervention with respect to all or specific types of mortality, beyond affecting all of health through private and government spending, and beyond very complicated efforts to change the distal function, much scenario analysis around health in IFs turns to the proximate drivers of the various causes of ill-health and death.

For example, diabetes and cardiovascular disease are driven significantly by body mass index (**HLBMI**), calculated in a function with CLPC but changeable with a multiplier (**hlbmim**). HLBMI is used to calculate the obese portion of population (**HLOBESITY**), also modifiable with a multiplier (**hlobesitym**). Diarrheal disease is a function of [lack of] access to safe water and sanitation, which can be changed through parameters in the infrastructure model described earlier in this chapter. Respiratory diseases, both for populations under-5 and over-30, are a function of the percentage of population identified in the infrastructure model as using solid fuels (**ENSOLFUEL**), adjustable by a multiplier (**ensolfuelm**). Respiratory disease is also tied to atmospheric convection (**ENVPM2PT5**), still again with a multiplier (**envpm2pt5m**) for scenario analysis. For each of these disease categories, country values are compared with typical values from functions involving the distal drivers GDPPCP and EDYRSAG25 in order to determine the degree to which the proximate driver may raise or lower mortality rates.

The health model computes a number of other variables that are of interest in exploring possible health futures. Given life expectancy of each cohort in the longest lived society(ies), it is possible to compute for each country the millions of years of life lost for males or females due to each health condition (**HLYLL**). Using the approach of the GBD for computing morbidity from mortality patterns, it is also possible to calculate years of living with disabilities (**HLYLD**). When mortality is reduced in a society, the ratio of disability years to years lost can rise, for instance living with heart disease or cancer. The sum of the two variables is a measure of the cost of mortality and disability/morbidity relative to conditions in the most healthy societies and is called disability-adjusted-life-years-lost, or DALYs (**HLDALY**). Although even the most aggressive health-focused scenario cannot reduce DALYs per capita to zero, humanity will presumably continue to pursue movement in that direction.

→ Pick a country or set of them and a type of mortality. Develop one or more scenarios that significantly reduce mortality relative to the Base Case scenario, using distal drivers, governmental financial support for health, and/or proximate interventions. Does each provide leverage and, if so, how much?

8.4.5 Education[5]

SDG 4 directs our attention to ensuring inclusive and equitable education and lifelong learning opportunities for all. In contrast to the focus on primary completion (**EDPRICR**) in the earlier Millennium Development Goals, most attention in recent years has been on efforts to push completion of secondary education toward universality, measured in IFs with a variable representing graduation from the upper secondary level (**EDSECUPPRGRATE**). We can imagine that in future years more attention will also be turned to completion of tertiary education (**EDTERGRATE**), but most countries are still quite far from universal secondary completion. Possibly, in societies where upper secondary and tertiary graduates are not finding good employment, more attention may be on the percentage of upper secondary aged children enrolled in vocational education (**EDSECUPPRVOC**).

The representation of education flows within IFs involves key student flow variables at all levels: intake rates, survival rates to the final grade, graduation or completion rates, and transition rates to higher levels of education. These give the user of the model very considerable control over relative emphasis on different levels and gender balances throughout the educational process. Illustrating this at the primary level, there are parameters influencing primary intake rates (**edpriintnm**), gender parity for primary intake (**edprigndreqintdn**), grade specific dropout rates (**edpridrpgrdm**), grade specific enrollment rates (**edprienrgrdm**), and primary survival rates (**edprisurm**). For more complete lists of variables and parameters across all levels of education, look under Guide to Scenario Analysis in the IFs Wiki and pick the education model. For more elaboration of the model itself, use the Understand IFs option from the Wiki's main page.

The education model tracks students by grade through the upper secondary level. Especially in societies without high portions of "of-age" children who enroll and progress through levels, there may be considerable numbers of children enrolled who are above "normal" age for a given grade level. The model therefore distinguishes between net (of-age) and gross (all-age) enrollment rates (e.g., **ENPRIENRN** and **ENPRIENRG** at the primary level).

Even years of education without completion are important. The stock-flow character of the education model gives much attention to the years of education attained by adults above the ages of 15 or 25 (**EDYRSAG15** and **EDYRSAG25**). Those years of attainment have implications for variables across many social systems

[5] Mohammod Irfan developed the IFs education model.

represented in IFs (e.g., for fertility rates, economic productivity, and health conditions). The model tracks the numbers of years of attainment of adults across their lifespans. The student flow parameters described earlier in this section affect school completion and dropout rates, which in turn determine the years of educational attainment of young adults—attainment that they carry with them as they age.

At the policy level, there is interest in education quality as well as completion of educational levels and years of attainment. **EDQUALAG15** is a general measure of that based on data that is tied to adult test scores. IFs distinguishes between quality in math, reading, and science. We recognize, however, that representation of quality across countries and time is so challenging that confidence in such forecasts will be low.

Both progression through educational levels and providing quality education require finance. IFs represents total government spending for education (**GDS**) and the shares allocated across different levels (**GDSED**). The earlier discussion of health noted that changed weight can be given to any target of government spending with a multiplier (**gdsm**). GDSED is similarly affected by **gdsedm**. But that health discussion also emphasized that the SAM structure of IFs makes sure that social accounting is maintained; thus, additional government spending on education is not without the costs of (1) reduced spending elsewhere, and/or (2) increased debt for a while and then automatic government revenue rise within IFs. The user can also make a scenario intervention to increase revenue (**govrevm**) relative to budget-balancing adjustments on the expenditure side.

The complexity of the education model with its many variables and parameters means that learning to undertake scenario analysis with it can require considerable time. There are, however, existing scenario files that can simplify your exploration of education futures and their implications for other sociopolitical systems, including demographics, economics, and health. For instance, an installation of IFs includes scenarios used in each volume in the Pardee Institute's earlier Potential Patterns of Human Progress series. To find, load, and run the educational volume's Normative Scenario, look for the *Add Scenario Component* option under *Scenario Analysis/Quick Scenario Analysis with Tree.* Scroll down to *World Integrated Scenario Sets* and *select Pardee PHPP Vol Series Main Cases.*

→ Progression across multiple levels of the education system through at least the upper secondary level and acquiring some tertiary education requires 11–16 or more years. Thus, for a decade or more, flow-side interventions will have limited impacts on attainment stocks with their forward linkages. Understanding this, experiment with alternative scenarios. In addition to usage of prepackaged scenarios, among the easiest might be specification of additional education spending.

8.4.6 Gender Equality

Although there is no gender model per se within IFs, variables related to gender exist throughout IFs. IFs computes the Gender Development Index (**GDI**), the Gender Empowerment Measure (**GEM**), and the Gender Inequalities Index (**GII**). The population model not only differentiates people by sex and age, it computes variables related to violence against women and children. These include deaths of women and children from societal violence (**SVDTHSWOMEN** and **SVDTHSCHIL**). The parameters **svdthsadd** and **svmulm** affect these. The health model differentiates morbidity patterns by sex and age, and the education model directs much attention to the lower enrollment and completion rates by girls historically (and even today in many developing countries) as well as the change and even reversal of those patterns that is now occurring, especially in higher-income countries (see earlier discussions of health and education model variables).

8.4.7 Infrastructure Including Watsan[6]

IFs includes an infrastructure model focused heavily on extent of provision and percentage access patterns for selected infrastructure types. Its coverage relates directly to SDG 6 on water and sanitation, already discussed in this chapter. But coverage also relates to SDG 7 on energy, SDG 9 on infrastructure more generally, and SDG 12 on sustainability. The types of infrastructure represented in IFs and the associated parameters to directly influence their development are (see Abu et al., 2024 on water and sanitation):

- Electricity access percentage distinguishing urban and rural in **INFRAELECACC** and **infraelecaccm;** generation of electricity per capita in **INFRAELECGENCAP** and **infraelecgencapm**; **INFRAELECTRANLOSS** and **infraelectranlossm** represent the rate of loss in transmission and distribution
- Safe water access percentage in the older 3-step ladder of Unimproved, Other Improved, and Piped (**WATSAFE, watsafem**)
- Safe water access percentage for households with urban-rural distinction on newer 5-point ladder of Surface, Unimproved, Limited, Basic, and Safely Managed (**WATERHH**, **waterhhm**)
- Safe sanitation access percentage in the older 3-step ladder of Other Unimproved, Shared, and Improved (**SANITATION**, **sanitationm**)
- Safe sanitation access percentage with urban–rural distinction on the newer 5-step ladder of Open Defecation, Unimproved, Limited, Basic, and Safely-Managed (**SANITHH, sanithhm**)
- Waste water collection access percentage (**WATWASTE, watwastem**)
- Waste water treatment access percentage (**WATWASTETREAT, watwastetreatm**)

[6] Dale Rothman and Mohammod Irfan led the development of the IFs infrastructure model.

- Road density in kilometers per 1,000 hectares (**INFRAROAD**, **infraroadm**), as well as percentage of roads paved (**INFRAROADPAVEDPCNT**, **infraroad-pavedpcntm**) and a rural access index that indicates the percentage of rural population within two kilometers of an all-season road (**INFRAROADRAI**).
- Information and communications technology (ICT) per hundred people, namely fixed telephone lines (**INFRATELE**, **infratelem**); fixed broadband subscriptions (**ICTBROAD**, **ictbroadm**); mobile subscriptions (**ICTMOBIL**, **ictmo-bilm**); and mobile broadband subscriptions (**ICTBROADMOBIL**, **ictbroad-mobilm**).
- Area equipped for irrigation in 1,000 hectares (**LANDIRAREAEQUIP**, **landi-rareaequipm**)

In addition to the multipliers for parametric control of future infrastructure, there are parameters available for almost all of the variables listed above that specify a target value and the number of years before the target is reached. For instance, parametric control for access to safe water on the 5-point scale is given by **waterhhsafetrgtval** and **waterhhsafetrgtyr**.

Infrastructure development and maintenance cost money (**INFRAINVESTNEW** and **INFRAINVESTMAINT**, both dimensioned by country and infrastructure type). Depending on computations of change in each infrastructure type within a region over time or patterns affected by multipliers or target values, there will be financing demand for each infrastructure type (**INFRABUDDEMNEW** and **INFRABUDDEMMNT**).

Large portions of infrastructure development and maintenance are publicly financed as computed in **INFRAINVESTNEWPUB** and **INFRAINVEST-MAINTPUB**. Together those must sum to **GDS, infrastructure**. The requirement that the sum of new public investment and maintenance spending must equal **GDS, infrastructure** means that the **gdsm** parameter, dimensioned across all of direct government spending, can be used to change the availability of such governmental funds and the patterns of change in physical infrastructure. As stated earlier, all such government spending (including military, health, education, and R&D) sums to **GOVCON**, which is computed within the SAM structure of IFs and thus constrained by government revenue and expenditure patterns. The model reconciles demand for, and supply of, **GDS, infrastructure** in that structure, which can push it higher or lower. The infrastructure model also calculates a value from GDPPCP, albeit crudely, for spending in the "other" infrastructure category (**GDS[InfraOther]**).

In summary, alternative scenarios for the future of infrastructure by type generally use the multipliers or target values and years, but the total extent of infrastructure's advance (or deterioration) can be shaped by interventions affecting government finance.

→ Pick a country or country grouping. Identify a type of infrastructure of interest, perhaps one that you believe should become stronger than shown in the Base Case scenario. Structure and explore interventions to accomplish that.

Explore also some of the impacts that the resultant infrastructure patterns may have for other variables, including economic advance (through productivity) and health (through proximate variables affecting mortality). Do not forget that changes in government spending patterns, if generated, will also affect other variables. Overall, evaluate the benefits and possible costs of your interventions.

8.5 Conclusion

This chapter completes Part III of the volume with its heavy focus on human development via attention in Chaps. 6–8 to population, the economy, and human well-being. The chapters of Part IV direct more of our attention to environmental sustainability. We stress still again, however, the relationships that link all the issues we are exploring. Human development drives environmental change, and environmental change greatly affects patterns of human development.

References

Abu, T. Z., Achore, M., Irfan, M., Musah, I., & Yussif Azzika, T. (2024). The past, present, and future of Ghana's WASH sector. An explorative analysis. *Water Security, 23*, Article 100185. https://doi.org/10.1016/j.wasec.2024.100185

Dickson, J. R., Hughes, B. B., & Irfan, M. T. (2010). *Advancing global education.* Patterns of Potential Human Progress Series, vol. 2. Paradigm Publishers and Oxford University Press, Boulder, CO, and New Delhi, India. Merger into Routledge. https://doi.org/10.4324/9781315636153

Elgin, C., & Oztunali, O. (2012). *Shadow economies around the world: Model based estimates.* No 2012/05, Working *Papers*, Bogazici University, Department of Economics. https://ideas.repec.org//p/bou/wpaper/2012-05.html

Fuso Nerini, F., Mazzucato, M., Rockström, J., van Asselt, H., Hall, J. W., Matos, S., Persson, Å., Sovacool, B., Vinuesa, R., & Sachs, J. (2024). Extending the Sustainable Development Goals to 2050—A road map. *Nature, 630*(8017), 555–558. https://doi.org/10.1038/d41586-024-01754-6

Hanna, T., Hughes, B. B., Irfan, M. T., Bohl, D. K., Solórzano, J., Abidoye, B., Patterson, L., & Moyer, J. D. (2024). Sustainable development goal attainment in the wake of COVID-19: Simulating an ambitious policy push. *Sustainability, 16*(8), 3309. https://doi.org/10.3390/su16083309

Hanna, T., Moyer, J., Azcana, G., Bhatt, A., Duerta Valero, S., & Meagher, A. (2023). *Forecasting time spent in unpaid care and domestic work.* UN Women Technical Brief. https://data.unwomen.org/publications/forecasting-time-spent-unpaid-care-and-domestic-work

Hughes, B. B. (2025). Analysis of integrated global SDG pursuit: Challenges and progress. *Sustainability, 17*(15), 6672. https://doi.org/10.3390/su17156672

Hughes, B. B., Irfan, M. T., Solórzano, J., Yang, V., & Moyer, J. D. (2021). Estimating current values of sustainable development goal indicators using an integrated assessment modeling platform: "Nowcasting" with International Futures. *Statistical Journal of the IAOS, 37*(1), 293–307. https://doi.org/10.3233/SJI-200715

Hughes, B. B., Kuhn, R., Peterson, C. M., Rothman, D. S., & Solorzano, J. R. (2011). *Improving Global Health.* Patterns of Potential Human Progress Series, vol. 3. Paradigm Publishers and Oxford University Press, Boulder, CO, and New Delhi, India. Merger into Routledge. https://doi.org/10.4324/9781315634012

Hughes, B. B., Kuhn, R., Peterson, C. M., Rothman, D. S., Solórzano, J. R., Mathers, C. D., & Dickson, J. R. (2011b). Projections of global health outcomes from 2005 to 2060 using the International Futures integrated forecasting model. *Bulletin of the World Health Organization, 89*(7), 478–486. https://doi.org/10.2471/BLT.10.083766

Irfan, M. T. (2017). *IFs education model documentation.* Frederick S. Pardee Center for International Futures, Josef Korbel School of Global and Public Affairs, University of Denver. https://korbel.du.edu/pardee-resources/ifs-education-model-documentation/

Kuhn, R., Rothman, D. S., Turner, S., Solórzano, J., & Hughes, B. (2016). Beyond attributable burden: Estimating the avoidable burden of disease associated with household air pollution. *PLoS One, 11*(3), Article e0149669. https://doi.org/10.1371/journal.pone.0149669

Mathers, C. D., & Loncar, D. (2006). Projections of global mortality and burden of disease from 2002 to 2030. *PLOS Medicine, 3*(11), Article e442. https://doi.org/10.1371/journal.pmed.0030442

Moyer, J. D., & Bohl, D. K. (2019). Alternative pathways to human development: Assessing trade-offs and synergies in achieving the Sustainable Development Goals. *Futures, 105*, 199–210. https://doi.org/10.1016/j.futures.2018.10.007

Moyer, J. D., Bohl, D. K., Petry, C., Scott, A., Solórzano, J. R., & Kuhn, R. (2020). The persistent global burden of severe acute malnutrition: Cross-country estimates, models and forecasts. *Global Transitions, 2*, 167–179. https://doi.org/10.1016/j.glt.2020.07.004

Moyer, J. D., Eshbaugh, M., & Rettig, J. (2017). Cost analysis of global road traffic death prevention: Forecasts to 2050. *Development Policy Review, 35*(6), 745–757. https://doi.org/10.1111/dpr.12263

Moyer, J. D., & Hedden, S. (2020). Are we on the right path to achieve the Sustainable Development Goals? *World Development, 127*, Article 104749. https://doi.org/10.1016/j.worlddev.2019.104749

Rothman, D. S., Irfan, M. T., Margolese-Malin, E., Hughes, B. B., Moyer, J. D. (2014). *Building global infrastructure.* Patterns of Potential Human Progress Series, vol. 4. Paradigm Publishers and Oxford University Press, Boulder, CO, and New Delhi, India. Merger into Routledge. https://doi.org/10.4324/9781315635743

Schneider, F., & Enste, D. H. (2013). *The shadow economy: An international survey* (2nd ed.). Cambridge University Press. https://doi.org/10.1017/CBO9781139542289

United Nations Economic Commission for Europe (Ed). (2008). *Non-observed economy in national accounts: Survey of country practices.* United Nations. https://digitallibrary.un.org/record/630778

World Health Organization. (2009). *Global health risks: Mortality and burden of disease attributable to selected major risks.* World Health Organization.

Part IV

Sustainable Development

Food and Agriculture

9

Abstract

The agriculture model in International Futures (IFs) is a temporally dynamic (with annual time steps), partial equilibrium model that represents crop, meat, and fish production, consumption, and trade for 188 countries. It does so in millions of metric tons. Crop demand differentiates food, animal feed, and industrial use categories. The agricultural model converts physical production and trade variables to monetary values in billion U.S. dollars at 2021 prices that override the computations for agriculture in the general all-sector equilibrium economic model (see Chap. 7) unless the IFs user specifies otherwise. Calculations in the economic model determine the portion of household income devoted to food consumption, and that determines food demand and therefore calorie and protein consumption levels. A module in the agriculture model represents changing land use in crop, grazing land, forest, urban, and other categories. A separate water model dynamically represents municipal, industrial, and agriculture use (in the context of water supplies), with agriculture usage determining irrigation extent. This chapter presents century-long forecasts of food demand and undernourished children, outlines the model structure for food and agriculture, and provides exercises to assist the model user with scenario analysis.

Professor Richard Chadwick of the University of Hawaii made important contributions to conceptualizing the revision of this chapter in the 4th edition and to the review of behavior in the IFs agriculture and energy models.

B. B. Hughes, *Exploring and Shaping International Futures*, World-Systems Evolution and Global Futures, https://doi.org/10.1007/978-3-032-22084-4_9

9.1 Introduction

Famines in China in 1928–1930 and again in 1959–1961, in the Soviet Union in 1932–1933 and again in 1946–1947, in Iran in 1942–1943, in Bengal in 1943, in Java in 1944–1945, in Biafra (Nigeria) 1967–1970, and in the Democratic Republic of the Congo 1998–2000 each claimed millions of lives. While most of these, and very numerous other famines, have resulted from domestic conflict and political turmoil, they have also been shaped by the condition of underlying agricultural systems.

Chapter 12 discusses a representation in International Futures (IFs) of the probability of intrastate conflict, but we do not believe we can meaningfully forecast specific conflict events and their related outcomes, such as famines. We can, however, explore changing global patterns with respect to production and consumption of food and of agriculture more generally. As we saw in Chap. 8 on Human Development, doing so helps us understand nutritional patterns around the world. Figures 8.2 and 8.3 showed the average consumption of calories and the prevalence of child stunting across the World Bank's income groupings. And a focus on agriculture will help us when we turn to the future of the environment in Chap. 11 because issues of carbon emissions, deforestation, and water use are all closely tied to the agricultural system.

As we explore agricultural futures in this chapter, we will never lose touch with the implications of national and global agricultural systems for the kind of undernutrition that affects hundreds of millions of people. Focusing much of our discussion is the second of the Sustainable Development Goals (SDGs), the elimination of hunger by 2030. Can this goal possibly be met? The answer that nearly all forecasters (and the International Futures Base Case) are providing, even setting the target to be 3% or less rather than zero hunger, is that it is all but impossible globally by that year. According to the IFs Base Case, we may lower the global rate from about 9% in 2020 to about 7% in 2030, finally approaching that 3% target in 2050. Yet, even the 2050 level in low-income countries could still be about 12%, largely due to high levels some places in South Asia and sub-Saharan Africa.

Globally, about 90 million children were undernourished in 2020 (Fig. 9.1), as were a total of about 500 million people (down from over 900 million in the late 1980s). Inadequacies in calorie intake are responsible for millions of premature deaths every year even as the impact of obesity is rising.

Despite very considerable progress, there is an ongoing debate about our ability to feed ourselves in the future. To put the problem in perspective, an increase in the global population from 8.2 billion in 2025 to somewhat more or less than 10 billion by the end of this century means that we will need to increase global food supplies by about 25% just to maintain current average dietary levels. To also bring global average calorie availability from about 3,000 in 2025 to 3,500 (approximately the level in the high-income countries today), we would need food supply increases of more than 40%.

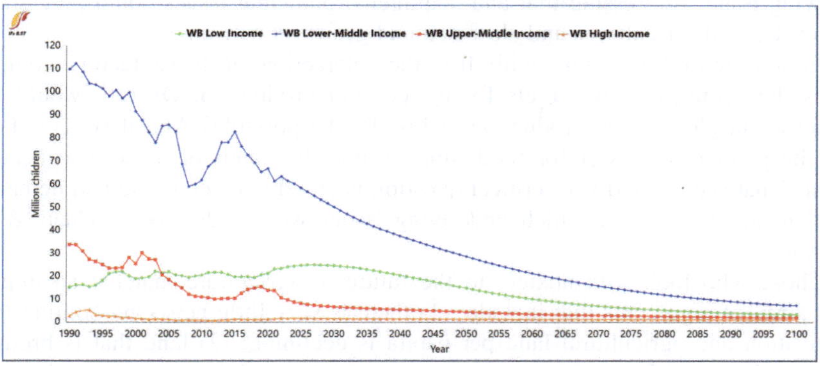

Fig. 9.1 Forecast of undernourished children. *Note* Using a 5-year moving average to reduce historical data irregularities. *Source* IFs 8.57 Base Case with historical data from World Development Indicators

Further, if increasingly well-to-do consumers have their preference, such an increase in calories could come in large part through the addition of more grain-fed meat, considerably further increasing the need for crop production to feed those animals, and from fish, placing additional demand on oceans and aquaculture, both of which face likely constraints. Still another issue is that the percentage waste from raw production through transport and storage, and at the table, is greater for meat and fish than for crops.

Given all these issues, Fig. 9.2 shows potential growth in global demand (interacting with production) over time after 2022. It suggests more than doubling of meat and fish demand and supply as well as about 70% growth in crop demand through 2100.

Cutting the other direction, however, is a rise in obesity levels. In many countries, even with current globally-average calorie consumption levels, that is a

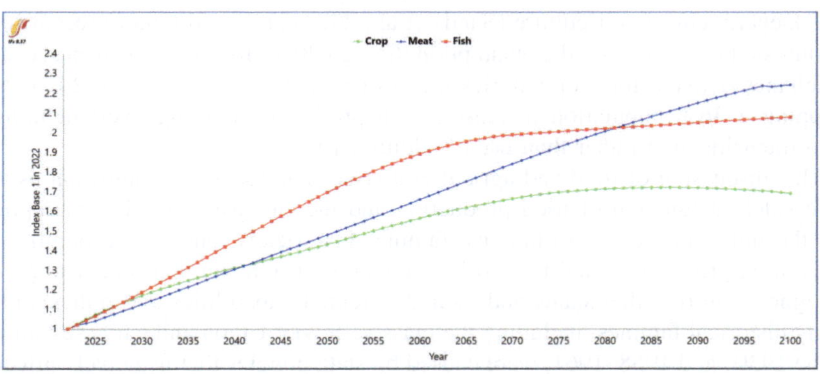

Fig. 9.2 Indices of growth in global crop, meat, and fish production. *Source* IFs 8.57 Base Case

growing issue. Addressing that and health/environmental issues could potentially lower demand for calories and slow demand growth for meat.

In long-term forecasting with IFs, the interaction of these factors complicates the setting of any targets for agricultural production. Overall, would the increases in global food production of Fig. 9.2 be possible? And if so, could we get the purchasing power for food supplies into the hands of those who need it most? That second and very critical question is largely an economic and demand-side inequality one, for which underlying issues were addressed in Chap. 8 on economics.

Those who look with anxiety to the future of agriculture marshal an important array of arguments that include: (1) there is very little new land coming into cultivation and agricultural land per capita is declining; (2) land that is brought into production often comes at the expense of tropical forest and can be land of marginal quality for agriculture; (3) aquifers around the world are being drawn down, streams are heavily used, and sites for placement of new irrigation reservoirs are dwindling; (4) fertilizer and other chemical usage have reached levels that often bring few marginal returns and that contribute greatly to pollution; and (5) ocean fishing is so intense that fish stocks are falling. Chapter 11 returns our attention to environmental issues linked to agriculture (and energy), but this chapter must also be attentive to them.

Those who look more optimistically to the future rely heavily on one major argument: agricultural technology continues to advance, with the pace of change in genetic crop engineering still quite strong. Moreover, a substantial portion of the world has not yet adopted even current state-of-the-art agricultural technology. This argument is buttressed by the reality that the UN Food and Agriculture Organization's (FAO) index of total global food production increased by a factor of 3.97 between 1961 and 2023 (1.63 just since 2000), often in the face of previous pessimistic forecasts.

The UN FAO, mentioned above, and the International Food Policy Research Institute (IFPRI) have been two of the premier data and forecasting organizations with respect to food. In earlier years, they made longer-term forecasts (IFPRI, 2001; UN FAO, 2000), but in recent years they have made fewer forecasts. The U.S. Department of Agriculture (Sands et al., 2023) produced a more recent forecasting study of global food consumption through 2050. Its income-driven scenario anticipates demand for crop calories increasing by 47% from 2011 to 2050 (outstripping a 36% population increase), with production coming overwhelmingly from intensification rather than use of additional land.

This discussion of food and agricultural issues has focused on nutrition as if it were solely a function of food production and income-driven food consumption. But the initial references to massive famines historically, and our recognition of differential progress around the world, points out once again that forecasting cannot ignore the role that states and their differential capabilities and policies play. Many historical famines, including the massive Soviet Union and Chinese famines of the 1930s and 1958–1961, were caused by state policies that negatively affected farmers or by domestic and/or international conflicts. While IFs can forecast general dynamics and potential for such issues, it cannot anticipate specific domestic or international conflict events and their very adverse implications.

We will need in the rest of this chapter to better understand the dynamics that IFs does represent and how scenario interventions can help us explore a broad range of alternative food and agriculture futures.

9.2 Food and Agriculture Dynamics and Leverage

Agriculture (see more in the Wiki and the Getting Started page) is integrally bound to the broader population and economic systems, as is energy (the topic of the next chapter). Thus, their portrayals in IFs have much in common with, and interact closely with, the causal systems we saw in Chaps. 6 and 7. Food demand (see Fig. 9.3) is a function of population size, the income available to that population through GDP, and the share of income spent on food (affected by GDP per capita).

The supply side of the agricultural system consists largely of processes and dynamics that seek to satisfy demand. Like most systems driven by goal-seeking, the primary loops are negative or equilibrating ones. Inventories and prices, which respond to inventory levels, are at the center of almost all the controlling loops. In the key long-term loop, prices help determine whether there are apparent profits to be made in food production and thus whether capital investments will be made. Capital will flow both into land development and into equipment and materials that increase yield per hectare of land. Production intensity will respond even more rapidly to inventory changes. The production system in agriculture is like that in the larger economy (see Fig. 7.6), except that land availability plays a key role.

Additional loops seek to maintain shorter-term inventories at desired levels relative to demand through changes in imports and exports. In addition, price changes affect demand itself, but food is so essential to human well-being that the responsiveness (elasticity) of food demand to prices is not great.

There is a great deal more complexity in the representation of food and agriculture in IFs than Fig. 9.3 can show (see also the Wiki). For instance, as incomes grow, food demand increasingly shifts from crop to meat, necessitating the use of grazing land and/or of feed crops to support livestock herds. Also, the level of agricultural technology is much more important to the overall system than its simple portrayal in Fig. 9.3, including technology's possible reduction of losses in the food system. Also, a key question surrounds the upper limit on yield improvements (for example, photosynthesis limits), parametrically specified in IFs. And even the dire impacts of domestic and international conflict can be at least partially represented via a focus on the linkage between their probability and GDP (see the Economics chapter). Further, the full agriculture model includes the calculation of calories and proteins; representation of land types and changes in their acreage (Chap. 11 returns to this); treatment of livestock herd size and meat production; wild fish catch and aquaculture; and more.

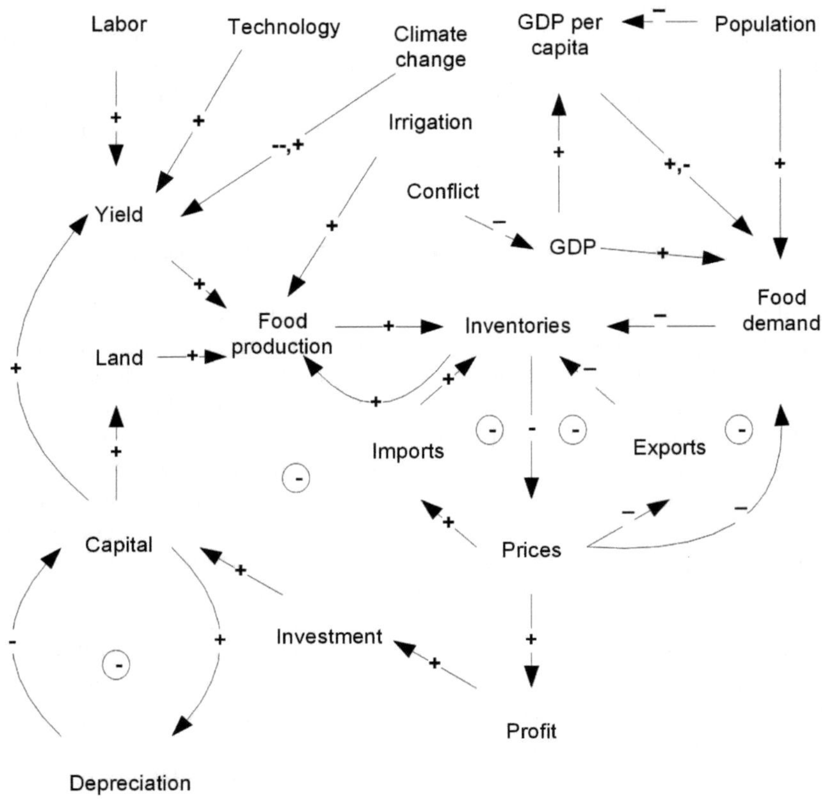

Fig. 9.3 Core food and agriculture dynamics in IFs. *Note* Conflict linkage to food production is not represented in IFs, but climate change is. *Source* Author

Nonetheless, the simplified image in Fig. 9.3 of the global agricultural system indicates the importance of our attention to leverage on the supply side. In the most basic terms, food production depends on the amount of land under cultivation and the productivity of that land. For most of human history, increase in the amount of land was the more important driver, a fact that obviously predisposed leaders and peoples to invasion and conquest of their neighbors as well as to more extensive cultivation of their own domains.

In recent years, and as we look forward, the productivity or yield of that land has become hugely important and a key variable in thinking about the future. Those yields in turn depend on a combination of increased investment in agricultural equipment; on heavier or smarter use of fertilizers, pesticides, and herbicides; and, especially, on improvements in the genetic stock of the grains that we grow. Beginning most notably with the Green Revolution success that Norman Borlaug had with Mexican wheat in the 1950s, new strains of grain slightly more than doubled the world's cereal production per hectare of land between 1961 and the turn of the century. Expansion of irrigated land is another factor that has interacted with other advances and increased production.

A major concern going forward is the impact of climate change. It increasingly affects both temperature and precipitation patterns. Both of those have implications for crop yield. Although dependent on the global location and local conditions and patterns of climate change, almost all the world is being affected adversely and that impact is growing steadily.

Debates surround both the continued possibility of productivity improvements and the best ways to pursue them. The private market obviously provides a critical foundation for productivity as farmers and the companies that sell supplies to them strive for increased yields and profits. The global community has also played an important role by building a global network of agricultural research institutions modeled on the initial successes of Borlaug and the Ford Foundation in Mexico. The Consultative Group on International Agricultural Research brings together states, international governmental organizations, and non-governmental organizations in a partnership that, in 2025, supported 15 research centers across all continents. Those centers in combination with vast private efforts have not only contributed to improvements in grain production; they also nearly quadrupled global meat production between 1961 and 2021, nearly doubling it in per capita terms.

Governments and voluntary organizations also sometimes turn to food aid for countries in special need. For instance, substantial transborder efforts helped reduce deaths in the African famines of 1972–1974, 1984–1985, and 1992–1993. Although long-term failure of production to match population growth and periodic droughts combine to cause famine on the African continent, political failures almost invariably contribute to such famine. For instance, a new government in Ethiopia after 1974 and a civil war in that country disrupted food production, as did a nearly complete breakdown of civil order in Somalia during 1992, and as have long civil wars in the Sudan during the 2020s. Again, IFs does not have such drivers of food shortages in its structure and Base Case forecasts, but the next section of this chapter will identify parameters that can be used to create scenarios including production shocks.

We turn next to a consideration of fish. Although the world's fish catch is less than 5% by weight of the world's grain production, it is nearly half by weight of global meat production. Fishing is therefore a critical global source of protein. Ocean fish catch has been relatively stable since the mid-1980s, and the FAO reports that about 70% of the world's fisheries are either fully exploited or overexploited. Thus, human leverage is increasingly applied to managing fisheries rather than expanding ocean fish catch. The Third United Nations Convention on Law of the Sea established 200-mile exclusive economic zones for coastal states, thereby providing both the incentives and the means for those countries to manage many coastal fisheries. Yet, state subsidies to their fleets, exacerbating competition among ships from individual countries and global competition within the open ocean, result in continued overfishing.

One bright spot in world fish production is a rapid expansion of aquaculture or fish farming, especially in China. The total production of aquaculture reached 50.6 million metric tons in 2020, continuing a rapid pattern of increase that began in the early 1980s with production of only about 3 million tons. Chapter 11 returns to the issue of ocean fish catch constraint and aquaculture.

This discussion suggests significant possibilities for human leverage with respect to food supply. Farmers operating in free markets are central actors. States often intervene in those markets, sometimes distorting prices and, therefore, incentives. There has been a persistent tendency in the post-World War II period for governments in economically developed countries to support prices above market levels to protect the incomes of farmers. Governments in economically less developed countries have often held prices below market levels to protect the diets of consumers. The former policies have created surpluses and the latter have contributed to shortages. At the same time, international research efforts have contributed to long-term supply increases and food aid has helped guarantee short-term food adequacy. A key question for the next century is: will our private and public leverage be sufficient to feed growing and more food-demanding populations?

9.3 Interventions and Scenario Analysis

As always, you would do well to begin looking for indicators of interest with respect to food and agriculture in the Flexible Displays. Perhaps check under the Agriculture and Health (for nutrition) categories and look at the Specialized Display on Development Goals, especially Goal 2 (Hunger). Within the IFs Wiki you will find sections on agriculture and water relevant to this discussion.

The development of scenarios may also take you to *Self-Managed Displays*, where you can explore model results and parameter interventions by name. Variables of potential interest to you in this issue area include agricultural production (**AGP**); crop or grazing land area (**LD, Crop** or **Grazing**); yield per hectare of land (**YL**); agricultural demand (**AGDEM**); food demand (**FDEM**); and agricultural imports and exports (**AGM** and **AGX**). Many variables in this model have separate values for crops, meat, and fish. When undertaking scenario analysis around food and agriculture, it can be useful to differentiate interventions primarily related to production from those focused on consumption.

As you move to interventions, proceed to *Scenario Analysis* for intervention options and already packaged interventions and scenarios. Or search within it on words like food, agriculture, and yield for intervention points. You may find important levers under the *Technological Change* and *Environmental Uncertainties* trunks of that (banyan-like) scenario tree, as well as under the agent categories (notably *Households* and *Firms*).

9.3.1 Production

The most direct intervention on the supply side is the use of **ylm** to increase yield. Changes are basically assumptions about technological advance. They put no requirements on inputs of land, labor, capital, or intersectoral flows, so are to a large extent "free." In fact, more yield without other production inputs can somewhat reduce the requirements for such inputs. Without **ylm**, increased production of more crops requires more cropland (**LD, Crop**) and/or agricultural capital (**KAG**).

→ The yield multiplier (**ylm**), which in the scenario tree is under *Technological Change* (but probably is most easily found by just searching on the name), can introduce assumptions of accelerated technological advance in agriculture. Its normal value is 1, but an increase to 1.1 over 10 years following the model's base year (or even better, starting in the year of your model use) will gradually increase agricultural production by 10% relative to the Base Case with the same production inputs. If you believe that rapid technological breakthroughs in crop genetics through biotechnology are probable, this could be an appropriate scenario. Similarly, a decrease to 0.9 would reduce agricultural production by 10% relative to the Base Case and might reflect an assumption of significant environmental deterioration. Pick a country or grouping and experiment with **ylm**. You may wish, for example, to look at sub-Saharan Africa and to see the impact of alternative yield assumptions on malnutrition. You might also want to look at the global implications of technological advance.

What impact does a technologically optimistic scenario have on world agricultural production (**AGP, World**) and price (**WAP**)? How about the world total of malnourished children across all regions (**MALNCHIL**) or the global human development index (**HDI**)? In this and many other scenarios, the model will to some degree "fight" the change—increases in agricultural productivity through technology will increase inventories of food and put some downward pressure on food prices and that can, in turn, direct some resources away from agricultural production.

Another approach on the supply side is to increase investment in agriculture (**IDS, Agriculture**) with the parameter **idsm** (**Agriculture**). An increase in that will shift investment otherwise going to other economic sectors, so if you wish a general increase in investment across sectors, use **invm**.

One portion of increased agricultural investment will enhance **KAG** and therefore yield per hectare. The agriculture model uses much of any investment increase to add to **LD, Crop** within the constraints of an ultimately arable potential (**landarablepot**). The land submodel in IFs divides land use into crop, grazing, forest, other, and urban/developed. Expansion of cropland is mostly at the expense of **LD, Forest.** Alternatively, a multiplier on cropland (**ldcropm**) can increase cropland without an increase in investment. And a land withdrawal factor (**ldwf**) determines how much land is withdrawn from other usage with population growth, including urbanization's impact.

It is also possible to change crop production via increase in irrigated land area. The IFs water model distinguishes agricultural, industrial, and municipal use of water (see Chap. 11 on the environment for elaboration). Within agricultural use, the focus is on the area irrigated (**LANDIRAREAACTUAL**). To become irrigated, land must be equipped with canals and/or piping (**LANDIRAREAEQUIP**). Multipliers can increase the equipped area (**landareaequipm**) or the irrigated area (**landirareaactualm**). In hard coding of the model, irrigated land produces 3.25 times the yield per hectare of nonirrigated land. Changes in irrigated area obviously also can affect water usage and pressure on supplies.

You may wish to explore the implications of changes in **ylmax**, a parameter under *Technological Change* that sets a maximum on the yields attainable in each country or region. Few agronomists believe that, even with the best technology, yields per hectare can increase indefinitely. In addition, you may wish to investigate the implications of various population scenarios on food availability.

Another issue on the production side of agriculture is the impact of climate change. The environment model calculates the variable **ENVYLCHG.** It represents the combined percentage reduction (or in a very few cases increase) in crop yields because of climate change's impacts on country-specific temperature (**ENVTPCHG**) and precipitation (**ENVPRCHG**), in degrees Celsius and percent, respectively. Interventions that affect these variables involve changing carbon dioxide emissions related to land use and therefore of net deforestation (discussed above) or of fossil fuel use (see the energy model in Chap. 10).

9.3.2 Demand and Consumption

On the food consumption side, it is useful to focus on indicators that summarize the condition of human nutrition. As in our discussion of human development (Chap. 8), consider calories per capita per day (**CLPC**); the number of malnourished children (**MALNCHIL**); the portion of children malnourished (**MALNCHP**); the portion of children who suffer severe acute malnourishment (**MALNCHPSAM**); the percentage of the total population malnourished (**MALNPOPP**); and the overall Human Development Index (**HDI**). Those who have used early editions of IFs may remember that the model can make estimates of starvation deaths (**SDEATH**), but we have moved emphasis to the more common focus on numbers of undernourished children and total population, where there is a database for us to draw on. We have no data on starvation because almost no official statistics indicate that someone dies from it, but rather from respiratory infections or other conditions related to undernutrition.

Nutrition-related variables are affected most directly by food demand (**FDEM**). It might be tempting to believe that changes in production of food will pass through directly to food demand. Unfortunately, however, food demand is overwhelmingly determined by household income (**HHINC**) and the portion of it directed to consumption in total and to food's share of that (**CS**). Changes in agricultural production can affect that in part through price changes, but IFs represents the strong tendency instead for them to change the level of agricultural exports (**AGX**) and imports (**AGM**).

It is possible, however, to combine interventions on the supply side with the usage of direct multipliers on calorie demand. Usage of that parameter for a particular food category such as fish (**clpcm, Fish**) will shift calorie demand to that food category from others. If you wish to change total calorie demand, use **clpcm, Total**.

9.3.3 Other Food and Agriculture Issues

Much potential food that reaches maturity in the fields never reaches the table because it is lost in the production process (e.g., to rodents), in the harvesting process or during inadequate shipping and storage (especially cold storage for meat, fish, and fruit), and at the consumption level. You can see estimates of the portion that regions lose in such ways (**AGLOSSPROD, AGLOSSTRANS, AGLOSSCONS**) and change assumptions about those losses over time with loss multipliers (**aglossprodm, aglosstransm, and aglossconsm**).

Turning to livestock and meat, herd size (**LVHERD**) and meat production (**AGP, Meat**) are ultimately driven by meat demand (**FDEM, Meat**). The care of the herd requires land use (**LD, Grazing**) and/or grain feed demand (**FEDDEM**), subject to a multiplier (**feddem**) and to a productivity parameter representing the conversion of grain into meat (**livhdpro**). IFs does not distinguish among meat-producing animals (e.g., cattle versus poultry) but represents total meat producers in terms of weight.

Turning to fish, demand (**FDEM, Fish**) is tied heavily to income and also to dietary patterns, changeable with **clpcm**. Historically, supply was entirely wild caught in freshwater or marine sources (**AGFISHCATCH**), but it now includes comparably sized aquaculture (**AQUACUL**). The base year values for wild-caught and aquaculture fish come from the IFs preprocessor and are changeable with **fishcatchm** and **aquaculm.** Both catch and aquaculture values are subject to loss rates changeable with **aglossprodm**.

Global raw wild fish catch has been roughly stable for some time and therefore is not computed by a function (in distinction from aquaculture). However, losses of raw catch have been somewhat reduced in recent years, so Base Case values of **AGFISHCATCH** available for food grow, albeit very slowly. Yet, the assumption of constant raw catch value in the Base Case of IFs is problematic because oceans are overfished. Unfortunately, IFs does not yet contain any accounting for ocean fish stocks that could be used to endogenously change **AGFISHCATCH**, should those real-world stocks continue to run down globally. Therefore, in scenario analysis you might decrease ocean catch over time with **fischcatchm**, even as **AQUACUL** continues growth.

9.4 Conclusion

In this chapter, we have considered the causal dynamics surrounding food and agriculture and we have begun to investigate potential leverage to affect them. Remember that our ultimate goal is facilitation of choice in the face of uncertainty. Given your analysis of food and agriculture, where do you think our policy efforts should be greatest?

References

International Food Policy Research Institute. (2001). *2000–2001 IFPRI Annual Report*. https://hdl.handle.net/10568/157978

Sands, R. D., Meade, B., Seale, J. L., Robinson, S., & Seeger, R. (2023). *Scenarios of global food consumption: Implications for agriculture*. Economic Research Report Number 323. https://doi.org/10.22004/ag.econ.338943

United Nations Food and Agriculture Organization (UNFAO). (2000). *World agriculture: Towards 2015/30*. FAO. http://www.fao.org/docrep/004/y3557e/y3557e00.htm.

Energy 10

Abstract

The energy model in International Futures (IFs) is a temporally dynamic (with annual time steps) partial equilibrium model that represents production of oil, gas, coal, hydroelectric, nuclear, solar, wind, geothermal, and other forms of renewable energy for 188 countries. Physical volumes of supply and demand are computed in billion barrels of oil equivalent; the graphical user interface allows conversion to other physical units. The model converts production and trade variables to monetary values in billion U.S. dollars at 2021 prices that override energy sector computations of the general equilibrium economic model unless the IFs user specifies otherwise. The economic model determines inter-sectoral flows of energy. This chapter presents century-long forecasts of the changing relationship of energy demand to GDP and the changing pattern of energy production by type. Of special interest is the growth pattern of the renewables share of the total. The model provides the basis for computation of carbon emissions and atmospheric concentration that the environment model (Chap. 11) tracks. The chapter concludes with exercises to assist the model user with scenario analysis relating to energy and its interaction with other models in the IFs system.

10.1 Introduction

Energy is essential to human economies and our well-being. Across the twentieth century and the first quarter of the current one, the global energy system experienced two notable transformations. First, global energy consumption per capita increased enormously—by a factor of about eight; it now approximates 12 barrels

Dr. Evan Hillebrand wrote this chapter for the fourth edition.

B. B. Hughes, *Exploring and Shaping International Futures*, World-Systems Evolution and Global Futures, https://doi.org/10.1007/978-3-032-22084-4_10

(1.7 tons) of oil equivalent per person each year. Second, the dominant sources of energy changed from biomatter and coal at the beginning of the twentieth century to oil and gas by the 1970s, which provided about 70% of global energy in 1975 (coal was still most of the rest).

There were two "energy shocks" in the 1970s, a decade in which real U.S. oil import prices jumped from $24.64 in 1970 to $132.77 in 1980 (both in 2025 prices).[1] That led many to the mistaken conclusion that fossil fuel shortages were imminent and that an immediate transition to renewable energy was necessary. In fact, as of this writing, known reserves of oil and gas would satisfy current levels of their global demand for more than half a century, and those for coal would cover current demand patterns for more than a century. Such values for known or proven reserves are conservative figures based on contemporary geological and engineering data and existing economic and operating conditions. Estimated ultimately recoverable resources are hugely larger (Bundesanstalt für Geowissenschaften und Rohstoffe).

In late 2025, after much fluctuation following the energy shocks of the 1970s and before the surge in 2026 resulting from the Iran war, the per-barrel price was $62.68 in current dollars. Although about half the 1980 price per barrel, it has been maintained much higher than its 1970 price, and well above production costs, by the marketing control of the Organization of the Petroleum Exporting Countries (OPEC). Founded in 1960, most OPEC members subsequently gained control of their national resources from colonizing countries and foreign companies and therefore acquired significant pricing power in the 1970s.

Fossil-fuel resource constraints and cartel pricing have not proven to be the only or, currently, even the major motivations for a shift toward renewable energy. Instead, environmental pressures—both very poor urban air quality and a sharp climb in atmospheric carbon dioxide with associated global warming—plus cartel control of prices and supply, are motivating it. Rapid falls in the cost of electricity produced by wind, solar influx, and geothermal heat, tied partly to significant learning-by-doing technological advances, have also triggered the early stages of a renewable transition. Chapter 6 (Clarke et al., 2022) of the Intergovernmental Panel on Climate Change (IPCC)'s AR6 report reported that between 2015 and 2022, the cost of solar photovoltaic had declined by 64% (page 630) while onshore and offshore wind costs had fallen by 18% and 40%, respectively (page 634). Those costs have continued to fall sharply.

This new energy transition is in its early stages. In 1960, oil, gas, and coal accounted for 96% (IFs database) of world commercial energy, not counting biomatter usage, and by 2024 that had declined only to about 89% (ExxonMobil, 2025: 6). In the meantime, nuclear energy had grown from almost nothing to about 4%, hydropower was fairly stable at about 3%, and wind and solar had each climbed from near zero to a bit under 2% each (combination of International

[1] The history of oil prices provided here comes from U.S. Energy Information Agency (EIA 2023). Data at https://www.eia.gov/outlooks/steo/realprices/.

Energy Agency (IEA) and ExxonMobil data). If the total is expanded to include bioenergy, ExxonMobil suggests that renewables accounted for about 19% overall in 2024.

Scenarios of global energy futures vary in their expectations about the same variables that have changed quite dramatically in the past: the overall level of energy demand and the general pattern of energy types that will satisfy that demand. In the remainder of this section, we will consider what forecasters are saying about each of these aspects of the energy system. In addition, analysts of energy futures now look with special attention at the environmental consequences of energy use, a subject to which we return in the next chapter.

Beginning with the future demand we might expect for energy, anticipating that depends largely on our expectations for the growth of GDP (see again Chap. 7) and the units of energy demand per unit of GDP. Figure 10.1 shows historical patterns in the IFs Base Case for each of the World Bank's income groups. Note that there has been, and may continue to be, significant decline in the demand per $1,000 of GDP at PPP in high-income countries. That reflects a combination of improved efficiency of usage and a transformation of those economies from significantly industrial to overwhelmingly service sectors. That pattern also characterizes upper-middle-income countries where rapidly industrializing China has heavy weight. Less developed countries may experience some rise in demand per GDP unit before a similar decline. Overall, the IFs Base Case suggests a bit more than a doubling of global energy demand between 2020 and 2100, while GDP approximately quadruples. On a per capita basis, the demand in high-income countries is expected in that scenario to be relatively stable, while that in upper-middle-income countries grows from about half of that to nearly the same. Less developed countries move toward that level but remain substantially below it by century end.

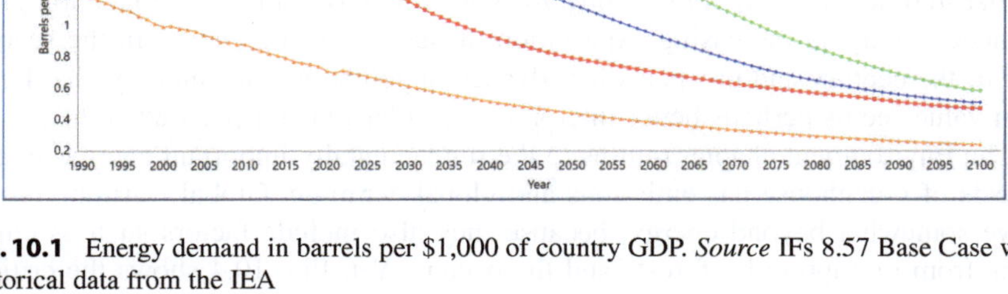

Fig. 10.1 Energy demand in barrels per $1,000 of country GDP. *Source* IFs 8.57 Base Case with historical data from the IEA

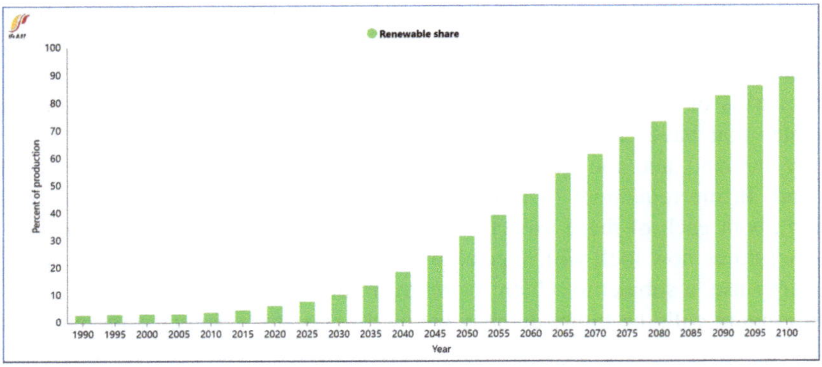

Fig. 10.2 Renewable energy as a percentage of global energy production. *Source* IFs 8.57 Base Case with historical data from the International Energy Agency

Turning to patterns of supply, forecasts by many experts suggest quite sharp rises in wind, solar, and geothermal production by midcentury and beyond. ExxonMobil (2025) projects a rise of those three, plus hydro, nuclear, and bioenergy, to 33% of global energy by 2050, with a reduction in fossil fuels to 67%. The IEA in its Stated Policies Scenario suggests a higher climb of "Clean Energy," also including nuclear and bioenergy, to 42% (fossil fuels at 58%) and possibly/desirably in its Announced Pledges Scenario to 57% and 43%, respectively. This pace of transition still falls far short of a Net Zero Emissions scenario with 90% of energy demand met from the clean category by 2050 (IEA, 2024: 24–25).

Figure 10.2 shows growth of the global renewable energy share to 31% by midcentury in the IFs Base Case (combining the hydro, solar, wind, geothermal, and other renewable categories in IFs). That is very close in total to the forecast of ExxonMobil. Its forecast, however, suggests that bioenergy will reach 10%, and IFs does not explicitly include a bioenergy category. Instead, IFs is considerably more optimistic about solar and wind, forecasting a combined total of 26%.

Particularly as we look beyond 2050 to century end, it is impossible to be confident about the future of non-fossil energy sources in the mix. For example, the IFs Base Case forecast of energy production by energy type (see Fig. 10.3) shows solar energy becoming 18% by 2050 (that seems on the surface reasonable and similar to other sources) but reaching 74% in 2100. Given the rate of technological advance, including increasing expectation of nuclear fusion energy in the second half of the century and prospects for advance in geothermal technology, that long-term value seems perhaps better interpreted as solar plus unspecified "other."

The Environment chapter returns to the fundamentally important energy-related aspects of greenhouse gas emissions and global warming. Global warming issues move somewhat beyond energy, because they also include factors such as emissions from or capture by forests and the oceans. Yet, Fig. 10.4 shows the critical role of greenhouse gas emissions (GHG), foremost among them carbon dioxide,

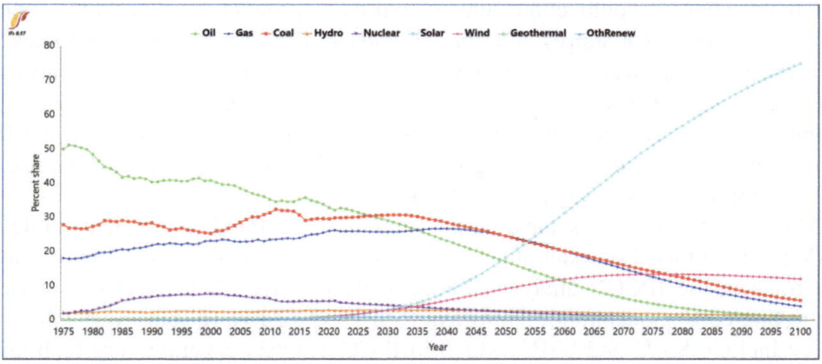

Fig. 10.3 Percentage contributions to global energy production. *Source* IFs 8.57 Base Case with historical data from the IEA

in efforts to hold global warming below 2.0 °C or even 1.5 °C. It shows how far the path of "implemented policies" is from either of those scenarios.

Another significant implication of the shift to green energy is that the expansion of hydroelectric, nuclear, wind, and solar power interacts with the supply and demand for electricity. Although some green energy forms, like geothermal (another big unknown as the century unfolds), can produce directly useful heat,

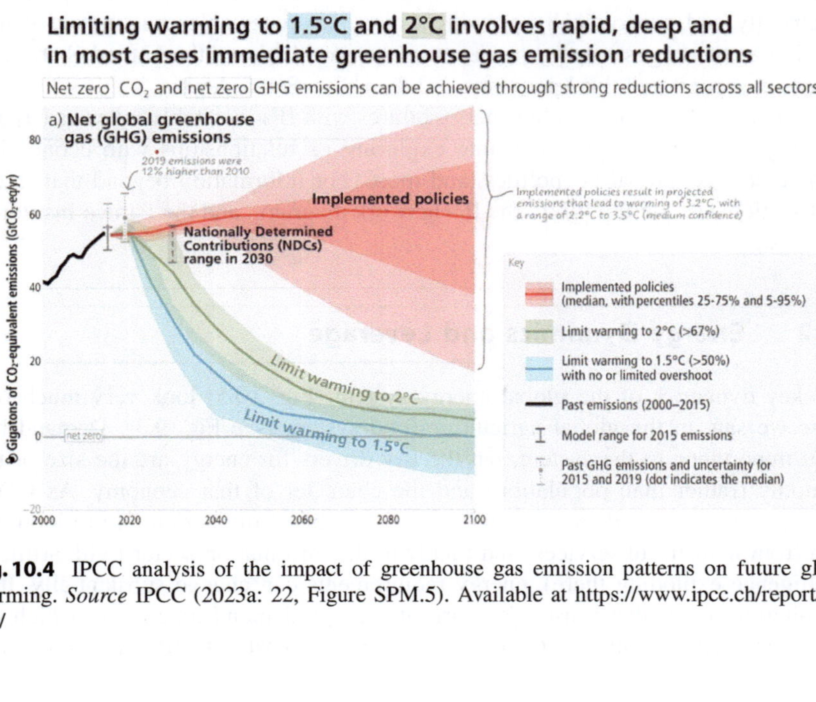

Fig. 10.4 IPCC analysis of the impact of greenhouse gas emission patterns on future global warming. *Source* IPCC (2023a: 22, Figure SPM.5). Available at https://www.ipcc.ch/report/ar6/syr/

and many sources could contribute to hydrogen production, electricity generation is dominant as a supply-side energy carrier. The demand side reinforces the increasing role of electricity via demand from data centers and massive artificial intelligence complexes plus the growing share of electric vehicles. Both ExxonMobil and the IEA anticipate a very rapid rise in electricity production; the IEA notes that demand for it has been growing at 2.7% annually since 2010 compared to 1.4% for all global energy (IEA, 2024: 39).

Many energy forecasts have come from the integrated assessment modeling (IAM) community. For example, IAM models (of which IFs is one) have played key roles in the IPCC's analysis. Those analyses have often explored scenarios in the Shared Socioeconomic Pathways (SSPs) set of five (discussed in Chap. 1). Those include SSP3 and SSP5, which in IPCC analysis generate very high GHG emissions; SSP2 with CO_2 emissions that remain much like current values until midcentury; and SSP1 with declines to near-zero emissions by 2050–2070 (see reports from the IPCC's sixth assessment round in 2023).

Yet, modeling that is done using IAMs to study energy futures in the SSP scenarios typically relies upon exogenous specifications of GDP and population growth, two of the most critical drivers of those futures. In contrast, the energy model in IFs has endogenously close relationships with other elements of the global system, including those in the population, economic, environment, and even geopolitical models. All these factors will continue to play important roles in shaping change in global energy futures and helping us understand it. For instance, they all have shaped and will continue to affect the evolving impact of China on those futures. Even though it still relies very heavily on coal, the huge push by demographically and economically growing China into renewables has moved it into position as the largest energy producer in the world, above the United States and Russia in second and third positions and far above Saudi Arabia.

In short, analysis of possible energy futures with IFs inevitably takes us beyond thinking only about energy and into exploring its relationships with economics, climate change, global geopolitics, and more. For information beyond that in the next section, see the IFs Wiki , the IFs network diagram , and the Pardee Institute's home page.

10.2 Energy Dynamics and Leverage

The key dynamics of the global energy system (Fig. 10.5) look very much like those we saw in the global agriculture-food system (see Fig. 9.3). Demand has great importance to the system, but the key drivers for energy are the size of the economy (rather than population) and the character of that economy. As GDPs per capita increase and as economies shift over time from agriculture to industry, then from industry to services, and finally to the information sector (with artificial intelligence exploding there), energy requirements can change substantially. It is a limitation of IFs that it uses an aggregate energy demand function in which the energy intensity per unit of GDP declines with increasing GDP per capita but is

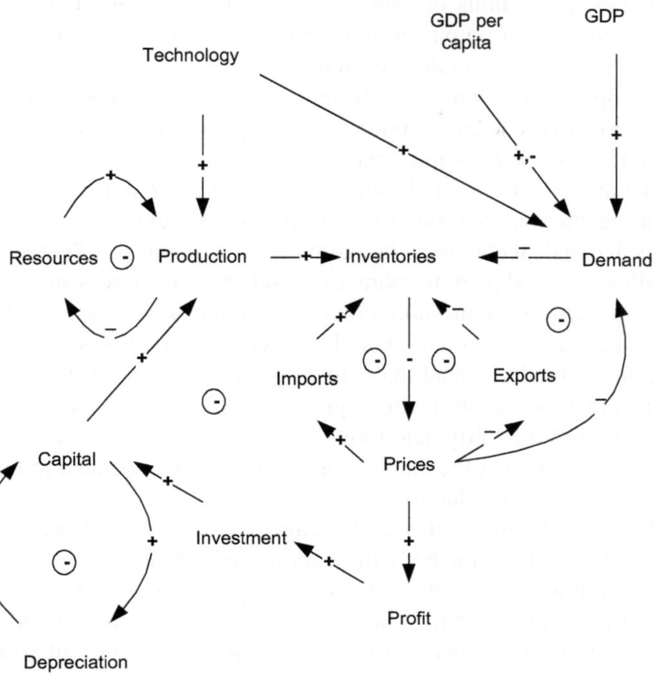

Fig. 10.5 Energy system dynamics. *Source* Author

not responsive to sectoral changes. However, scenario interventions can explore such issues.

Most of the energy system representation in IFs seeks to supply energy demand through a set of negative feedback loops. As with agriculture, prices (or more directly, inventory stocks) play a key role in those loops, including the loops that control investment and those that determine levels of energy imports and exports. Especially given the transition to renewable forms of energy that is underway, it is important to represent multiple energy types in the energy system and to allow the system to shift among energy supply forms as relative costs and sociopolitical decisions unfold over time.

Prices affect energy demand substantially more than they affect food demand (this difference is captured in parameter values, not basic model structure). Perhaps the most important insight to which the energy price increases of the 1970s gave rise is the degree to which energy consumption increases can decouple from economic growth. Prior to 1970, the belief and reality were that energy use rose mostly in lockstep with GDP.

In meeting demand, the model assumes very large possibilities for substitutability across energy types in the long run. In the Base Case scenario, the model also assumes a considerable amount of technological change in the utilization of energy—that is, fewer units of energy are assumed to be needed in the future to achieve the same rate of output. Automobiles, for example, now require much less gasoline to drive a given number of miles.

Energy supply also depends significantly on the profit opportunities available to current and potential energy producers. The supply types represented in IFs with differing and time-variable capital requirements to produce are oil (differentiating conventional and unconventional sources), natural gas, coal, hydroelectric, nuclear, solar, wind, geothermal, and other (unspecified) renewable sources of energy.

As with demand, technology is assumed to continue to advance on the supply side, allowing producers to extract oil and other resources more cheaply and transform and transport them more efficiently. In the Base Case, it is assumed that technological advance will continue to be faster for renewable sources of energy—e.g., wind, photovoltaic, geothermal—than for traditional fossil fuels. The baseline assumption is that the various technological rates of change embedded in the model continue, at least in the fairly near term, at current trend rates. The model user can specify alternate patterns of technological advance and associated reductions in the capital costs of energy production.

To explore energy futures, the key leverage points are thus (1) economic growth; (2) technology advances on both the supply and demand sides; (3) investment decisions, both in terms of energy sources and by geography; and (4) disruptions in supply by various actors. There is also framing of uncertainty around the ultimate resources of oil and gas, although those scenarios are of less interest in a world moving away from fossil fuels. A useful study of energy futures with IFs focused on scenarios around the renewable energy transition was completed during 2025 with IFs 8.32 in partnership with the United Nations Development Programme and Octopus Energy (Sahadevan et al., 2025).

10.3 Interventions and Scenario Analysis

Indicators of importance across demand and supply in the IFs energy model include energy demand **(ENDEM)**; the ratio of energy demand to GDP **(ENRGDP)**; energy production **(ENP),** specified across energy types; and energy prices **(ENPRI)**. IFs also represents the global pattern of energy production **(WENP)** and prices **(WEP)**. You can look at the many available displays in the *Energy* category and subcategories of *Flexible Displays* to develop an overview of the projected energy transition in the Base Case and other scenarios.

Using *Self-Managed Display*, it is possible to explore values of important parameters that control energy scenario analysis. It is also possible to see those values and change them using *Scenario Analysis /Quick Scenario Analysis with Tree*, where you can call up a few existing packaged energy scenarios.

There is considerable potential for leverage in exploring alternative scenarios of energy futures for countries, country groupings, and the world. On the demand side, there are four key parameters. The first is a direct multiplier on energy demand **(endemm)**; it helps examine the effects of assuming that energy demand rises faster or slower relative to GDP growth than the embedded rate. A second parameter is the long-term price elasticity of energy demand **(elasde)**. Note that this elasticity has a negative sign (as prices rise, energy demand falls). Despite many attempts to estimate this parameter based on data about prices and demand, its value still remains conjectural. The value used in the model might be considered by some analysts to be too high for the short-term effect but too low for the long-term effect. Changing the price of fossil energy directly by means of increasing taxes **(carbtax)** can have a quite dramatic impact on energy demand. You can also change the rate of assumed growth in energy use efficiency **(enrgdpr)**.

On the energy supply side, production by type **(ENP)** depends on the capital stock in each energy category (coal, oil, solar, etc.) in each country and on the capital/output ratio **(QE)** for that energy type, with **enpm** and **qem** allowing direct changes in them. The capital/output ratio is heavily influenced for fossil fuels by the level of reserves **(RESER)** and ultimate resources **(RESOR)**. The parameters linking reserves and profits to investment decisions, exploration, and changes in production are based on historical trends, but they are subject to considerable uncertainty.

In addition to the use of **enpm,** you can force a particular growth rate for an energy form via a parameter controlling energy production growth rate **(eprodr)**. The normal value for that parameter is zero, allowing the model to compute energy production by region and energy type endogenously. Nonzero values, such as 0.035, force annual production growth rates, such as 3.5%. See the Help system for access to the Wiki and more explanation of specific parameters.

→ The Base Case shows renewable energy sources growing rapidly in importance (see **WENP** across all energy types or use *Flexible Display* to call up that production). What happens if the assumption is too optimistic? If you cut the value of **enpm** across the century for one or more renewable energy types from 1.0 to 0.2 for the World grouping (you do this in the *Scenario Tree*), what is the impact on production patterns and economic output? You may prefer considerably more optimistic assumptions for renewable energy and so increase the parameter value. Note that even though oil and coal use rises in a scenario of less renewable growth, energy efficiency (lower values of **ENRGDP**) also responds to rising energy prices, ameliorating some of the negative economic and environmental consequences of any slowdown in the shift to renewable fuel sources.

Although these parameters will force the model to produce more or less of a given energy type, the model also includes parameters controlling specific aspects of the energy system that give rise endogenously to different production levels. For instance, the ultimate availability of fossil fuel resources and the known reserves will affect energy production futures. In the Base Case, look at the values over time for known global oil and gas reserves **(WRESER)**. Known reserves increase with discoveries (until the assumed limit of ultimate resources is approached) and decrease with production. Global oil reserves stabilize in the first half of the twenty-first century, and that can constrain production. Even if you assume technological improvements in oil production (via **enpm**), the constraint of total reserves remains. Because oil and gas tend to be less expensive than some other energy forms (such as nuclear, coal, or hydro), constraint on their production can put upward pressure on world energy prices **(WEP)**.

Some observers argue that pessimists have routinely underestimated the amount of oil and gas in the earth (for instance, even by predicting that the United States was running out at the beginning of the twentieth century, before the discovery of the big Texas fields and well before the impacts of fracking technology). Perhaps the assumptions about the world's ultimate oil and gas reserves **(WRESOR)** in the model are still too conservative.

→ You can introduce more optimistic assumptions about world energy resources by changing the value of the resource multiplier (**resorm**). Try doubling it for oil and gas resources of the world. Simultaneously increase the global discovery rate of that oil and gas by doubling the discovery rate multiplier (**rdm**). What impact does that have on world reserves and prices? Does it affect the share of energy production coming from renewable sources? Does it affect world GDP per capita (**WGDPPC**) or quality of life (**WPQLI**)? How about the **GDP** of OPEC?

→ It is also possible that the Base Case assumptions are too optimistic for oil and gas, including given the possibility of high carbon taxes (**carbtax**). Experiment with changes in that parameter.

In earlier years, such scenarios in IFs had larger impacts than they do now because the expectations for growth of renewable energy and less need for fossil fuels have made assumption changes on the fossil side less impactful.

→ Some might also question the assumptions in the Base Case concerning the cost of future production of energy. Technological progress could increase the production of energy possible with the same investment levels. You could explore the implications of decreasing (or increasing) the costs of energy production on a region and energy-type specific level. The multiplier on energy capital costs **(qem)** allows you to do that by energy type. Higher values raise the capital costs of energy production and lower values decrease those costs. This is a good way of investigating the implications of possible changes in the cost of renewable energy technologies. (As with other multipliers, you should leave the initial value at 1.0 and interpolate to a new value over time.)

Still another parameter on the production side of energy is a multiplier on investment in energy **(eninvm)**. If you believe that governmental tax or subsidy policies might lead to an increase or decrease in energy investment, this parameter can help you introduce such changes. The parameter increases or decreases total energy investment, letting the model continue to allocate that investment among energy types based on relative cost considerations. Be reminded, however, that the model often "fights" changes that you introduce through such policies. For instance, an increase in energy investment forced by the multiplier **eninvm** could lead to higher production, lower prices, and lower profits to energy producers. That could cause the model's endogenous investment mechanism to begin offsetting the increase in investment you request. Energy investment changes will also affect investment in other sectors of the economy and potentially the growth of GDP.

Not content to let supply and demand patterns determine energy prices, energy exporters, and to a lesser degree energy importers, have in the past attempted to directly control the price of energy. If you set the parameter **enprix** to any nonzero value, it will exogenously set the energy prices **(ENPRI)** for that region. This kind of scenario, like price-fixing in markets, can have perverse consequences (energy prices help equilibrate supply and demand, so exogenous changes in them can upset equilibration).

Regardless of developments on the demand and supply sides of global energy systems, short-term supply disruptions are possible. Both oil importers and oil exporters have the option of setting direct limits on the amount of oil and gas that they import or export.

→ You can examine how an IFs scenario was built to explore a hypothetical OPEC restriction of global oil supply for several years. From the main menu, click on *Scenario Analysis, Quick Scenario Analysis with Tree, Scenario Files, Open, Other, World Integrated Scenarios, NIC, NIC 2020, New Caliphate***.** This set of commands loads a scenario into the tree that was used quite long ago in support of the United States National Intelligence Council's 2020 Project (2004). However, the scenario's timing is always relevant to the changing base year of IFs. Examine the interventions in the tree, particularly the 4-year production cuts in oil output in selected OPEC countries. Run the scenario and compare its results to the Base Case. What happens to energy prices? To GDP? Because the scenario assumes sociopolitical turmoil, in addition to oil supply shortfalls, it may affect GDP more than would a less politically framed scenario of oil supply disruption.

Research Suggestions. For the most effective research use of the model, it is best to begin with a specific question. What might be the implications of slower growth in energy demand for import dependence of a country like Greece that imports much of what it needs? How rapidly might renewable energy forms come to supply 50% of global energy needs? How much damage would a war involving OPEC countries do to the economies of the oil-importing countries? After you clearly identify a question, consider how to implement the scenario(s) needed to investigate (but never definitively answer) it. Always review your results carefully to be certain that the scenario you intended to implement is the one you actually create. And perhaps most important, do not simply accept the results of the model as the answer to your question. Examine them carefully and try to explain why the model produced them. Do they make sense? Why or why not? Use the model as a "thinking tool," not as a crystal ball. Think also about where the model might be inadequate and how to compensate for that inadequacy.

10.4 Conclusion

In this chapter, we investigated the supply and demand of energy. The energy model of IFs is substantial in its size and coverage. Nonetheless, many aspects of the global energy system are highly contentious (for example, the likely change in production costs of renewable energy forms), and no model is capable of conclusively answering all the interesting questions. IFs is structured, however, with enough intervention points that many differing viewpoints can be approximated and explored with model simulations. Although we have so far focused mostly on the economic aspects of the global energy system, the environmental consequences of energy use are also of great concern. We turn to those in the next chapter.

References

Clarke, L., Wei, Y.-M., de la Vega Navarro, A., Garg, A., Hahmann, A. N., Khennas, S., Azevedo, I. M., Löschel, A., Singh, A. K., & Steg, L. (2022). Energy systems. In P. R. Shukla, J. Skea, R. Slade, A. Al Khourdajie, R. van Diemen, D. McCollum, M. Pathak, S. Some, P. Vyas, R. Fradera, M. Belkacemi, A. Hasija, G. Lisboa, S. Luz, & J. Malley (Eds.), *Climate change 2022: Mitigation of climate change. Working Group III Contribution to the IPCC Sixth Assessment Report* (pp. 613–746). Cambridge University Press. https://doi.org/10.1017/9781009157926.008

ExxonMobil. (2025). *ExxonMobil Global Outlook: Our view to 2050.* https://corporate.exxonmobil.com/sustainability-and-reports/global-outlook

International Energy Agency (IEA). (2024). *World Energy Outlook 2024 – Analysis.* IEA. https://www.iea.org/reports/world-energy-outlook-2024

Intergovernmental Panel on Climate Change (IPCC). (2023a). Summary for policymakers. In H. Lee & J. Romero (Eds.), *Climate change 2023: Synthesis report.* Contribution of working groups I, II and III to the sixth assessment report of the Intergovernmental Panel on Climate Change. https://doi.org/10.59327/IPCC/AR6-9789291691647

Intergovernmental Panel on Climate Change (IPCC). (2023b). *Climate change 2023: Synthesis report* (H. Lee & J. Romero, Eds.). Contribution of working groups I, II and III to the sixth assessment report of the Intergovernmental Panel on Climate Change. https://doi.org/10.59327/IPCC/AR6-9789291691647

Moyer, J. D., & Hughes, B. B. (2012). ICTs: Do they contribute to increased carbon emissions? *Technological Forecasting and Social Change, 79*(5), 919–931. https://doi.org/10.1016/j.techfore.2011.12.005

Moyer, J. D., Pirzadeh, A., Irfan, M., Solórzano, J., Stone, B., Xiong, Y., Hanna, T., & Hughes, B. B. (2023). How many people will live in poverty because of climate change? A macro-level projection analysis to 2070. *Climatic Change, 176*(10), 137. https://doi.org/10.1007/s10584-023-03611-3

Moyer, J. D., Verhagen, W., Mapes, B., Bohl, D. K., Xiong, Y., Yang, V., McNeil, K., Solórzano, J., Irfan, M., Carter, C., & Hughes, B. B. (2022). How many people is the COVID-19 pandemic pushing into poverty? A long-term forecast to 2050 with alternative scenarios. *PLoS One, 17*(7), Article e0270846. https://doi.org/10.1371/journal.pone.0270846

Sahadevan, D., Irfan, M. T., Luo, C., Moyer, J. D., Masoc, C., & Beynon, E. (2025). Charged for change: The case for renewable energy in climate action | UNDP Climate Promise. Pardee Institute for International Futures. *Octopus Energy*. https://climatepromise.undp.org/charged-for-change-renewable-energy-climate-action

Solórzano, J., Hughes, B. B., Irfan, M. T., Xiong, Y., Kwasi, S., Ibrahim, A., Makhmatova, S., Hanna, T., & Moyer, J. D. (2024). Protocol for making forecasts exogenous in the International Futures model. *STAR Protocols, 5*(4), Article 103414. https://doi.org/10.1016/j.xpro.2024.103414

U.S. Energy Information Administration (EIA). (2023). *International energy outlook 2023*. https://www.eia.gov/outlooks/ieo/

The Environment

11

Abstract

Treatment of environmental change in International Futures (IFs) focuses on (1) net change in forestation as part of the land-usage module in the agriculture model; (2) usage of renewable and nonrenewable water resources in its water model; and (3) the accumulation of atmospheric carbon dioxide and its impact on global temperatures in the environment model itself. Therefore, representation of environmental change is, like all models in the IFs system, tied closely to other models of the system (especially agriculture and energy, see Chaps. 9 and 10) but is very much linked also to demographics, economics (where temperature change affects economic growth patterns), water, and other models. This chapter provides illustrative forecasts across the century in the three areas of principal focus. It then offers basic information on the dynamics behind those forecasts. It concludes with some attention to scenario analysis with exercises for those who may wish to build future-oriented scenarios.

11.1 Introduction

Since the early 1970s, environmental questions have climbed steadily upward on local, national, regional, and global agendas. In response to this growing concern, the 1992 United Nations Conference on the Environment and Development in Rio de Janeiro (also known as the "Earth Summit") issued a call for what is, in effect, a new global environmental order. The United Nations Environment Programme's (UNEP's) Global Environment Outlook series, a direct result of the 1992 conference, is a good place to find recent information on environmental changes (e.g., UNEP GEO-6 in 2019 and GEO-7 in 2025). Much of the earth's environment on which all life depends has continued to deteriorate with human impact that has expanded and accelerated across a long history. We have come to recognize that the current and projected scale of human activities pose threats to the long-term

B. B. Hughes, *Exploring and Shaping International Futures*, World-Systems Evolution and Global Futures, https://doi.org/10.1007/978-3-032-22084-4_11

Table 11.1 Human land usage and impact on forests and grasslands across time (percentages)

	Forests	Grasslands and Shrubs	Crops	Grazing
BCE8000	57	42	<1	<1
BCE3000	55	44	<1	<1
1700	52	38	3	6
1900	48	27	8	16
1950	44	12	12	31
2023	38	16	15	30

Note Twenty-nine percent of *Earth's* land surface is deserts, ice, etc.; table shows 100 percent use of the other 71 percent

Sources Ritchie (2021: Forest area.) Published online at OurWorldinData.org. https://ourworldindata.org/forest-area#article-citation. Data for 2023 from UN FAO (2024: 312–316, Table 49)

viability of the *land, oceans, and atmosphere* as envelopes for life on earth. We can look in turn at those, beginning with land.

The agricultural revolution started about 12,000 years ago. Over time it led to conversion of land use, including of grasslands to managed grazing and of grasslands and forests to cultivation. The expansion of those uses was quite slow for millennia (see Table 11.1). Yet, by 1700 about 3 percent of the earth's initial forest and grassland area was in crops and about 6 percent was grazed. The subsequent explosion of global population and great improvements in diets pushed up those numbers rapidly. In 2023, they reached 15 percent and 30 percent (plus 1 percent in urban areas) with an approximately 19 percent reduction in forest-area share and 26 percent drop in grasslands share. In total, nearly half of the world's forest and grassland has been converted to crop, grazing, and habitation usage.

Table 11.2 reinforces understanding that ongoing deforestation is strongly linked to the continued rise in population and economic advance toward high-income levels. The bulk of deforestation since 1990 has been in Africa and South America. With some reforestation, there has been much less net change of the total in China, Europe, Japan, and North America across that period. In general, there has been significant deforestation in the tropics, with most reforestation in temperate areas. Tropical deforestation contributes much more to species loss than does temperate deforestation. Given the pressure on the agricultural systems of developing countries for higher production, as well as the desire of many such countries for revenues from hardwood exports, the trends should not surprise us.

Forecasts in this area are complicated in the face of competing pressures for deforestation from timber and agricultural land demand and simultaneously for reforestation with increasing incomes and environmental pressure for increase of the carbon sink. Daigneault et al. (2022) used multiple models and the Shared Socioeconomic Pathway scenarios (SSPs) to conclude that the average projection is a global increase in forest carbon stocks by 10 percent in 2055 and 30 percent by 2105. Nepal et al. (2019) concluded that aggregate global forest area would

Table 11.2 Forest cover and deforestation (million hectares)

Region	Planted forest		Naturally regenerating forest		Total forest		Percent change in total
	1990	2023	1990	2023	1990	2023	
Africa	9.2	11.6	703.8	617.0	713.0	628.6	−11.8
Asia	71.8	137.6	497.7	487.1	569.5	624.7	9.7
Europe	65.2	83.2	943.8	933.8	1,009.0	1,017.0	0.8
North America	23.1	47.9	732.1	704.2	755.2	752.1	−0.4
Oceania	2.8	4.9	181.6	179.9	184.4	184.8	0.2
South America	7.0	21.2	966.6	818.7	973.6	839.9	−13.7
Sum	179.1	306.4	4,025.6	3,740.7	4,204.7	4,047.1	−3.7

Sources Ritchie (2021: Forest area.) Published online at OurWorldinData.org. https://ourworldindata.org/forest-area#article-citation. Data for 2023 from UN FAO (2024: 312–316, Table 49)

increase between 7 and 36 percent by 2100 relative to 2015 across different SSP scenarios.

The International Futures (IFs) Base Case is less optimistic than these other forecasts. Version 8.57 suggests that net reforestation may begin globally only by 2074 and that in 2100, global forest area may be 5 percent less than in 1990 and nearly identical to that of 2020. Figure 11.1 shows the basis for that greater pessimism—namely, continued net deforestation in lower-middle-income and especially in low-income countries across the century.

Turning to human interactions with water environments, the expansion of global human impact into the seas with whaling goes back to at least the seventh century in Japan and to the early 1600s in Britain and the Netherlands. By 1900, the bowhead, gray, and right whales were almost extinct and various agreements to

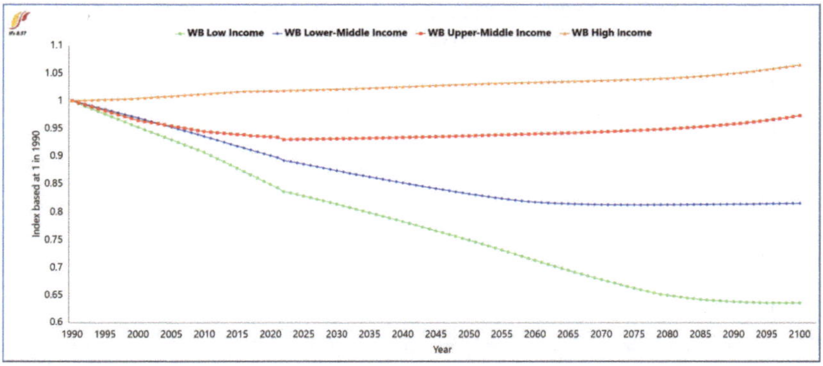

Fig. 11.1 History and IFs forecast of forest-area change. *Note:* Forest area indexed from 1990. *Source* IFs 8.57 Base Case with historical data from UN Food and Agriculture Organization

limit whaling emerged in the following century. Ocean fish catch has expanded across the centuries, reaching about 93 million metric tons in recent years, with nearly half of that in East Asia and the Pacific. Most global fisheries are being fished at or over their long-term sustainable capacity and many stocks are in clear decline (Sampson, 2024).

Some fish stocks, like cod in Eastern Canada, have collapsed and there appears to be peaking and possible general decline in wild catch in the face of declining fish stocks and some international agreements. The expansion of aquaculture has been very rapid and has now passed wild catch. Figure 11.2 shows forecasts of both from the IFs Base Case. The wild catch forecast is, however, very questionable because IFs does not represent the ocean fish stock and the possible or probable resultant decline in the flow variable of annual catch; hence the near stability in the wild catch forecast. Doubts can be raised about the aquaculture forecast also. As well as aquaculture's potential positive impacts, we know it also has some quite negative environmental impacts, including the significant wild fish catch that feeds much farmed fish and the disease and genetic impacts of escaping farmed fish.

Humans also have close interactions with, and significant environmental impacts on, the terrestrial water environment. Chapter 9 (Agriculture) touched on the importance and growing use of irrigation in our expansion of crop production with population growth and as a foundation for improvements of nutrition, efforts that have extended over several millennia. Tied strongly to agriculture is the increasing use and overuse of freshwater from both renewable and fossil sources. For data and analysis in this area, look to the annual *United Nations World Water Development Report* (e.g., UNESCO, 2025) coordinated by UNESCO for the inter-agency UN-Water grouping, including FAO, UN-Habitat, United Nations Industrial Development Organization, and others. Boretti and Rosa (2019) used projections of population and GDP to conclude that global water demand will increase by 20 to 30 percent by 2050, with growth greatest in industrial and domestic use but agriculture remaining the largest user. Given current stresses on that source, they

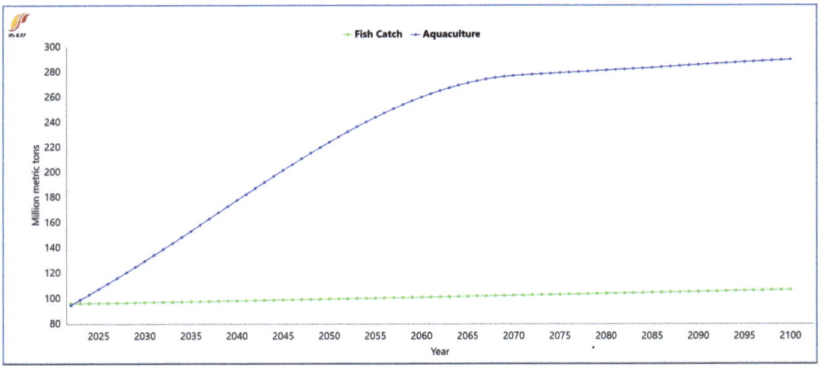

Fig. 11.2 Global fish catch and aquaculture. *Source* IFs 8.57 Base Case

suggest that groundwater extraction will very problematically grow by 39 percent by 2050. Their analysis questioned more optimistic forecasts from the UN's 2018 World Water Development Report.

Other important aspects of our interaction with water systems are extraction of safe water for drinking and disposal of human waste, as discussed in Chap. 8 (Human Development). With respect to both, human efforts (not least those of the Mesopotamians, Minoans, and Romans) go very far back (Abu et al., 2024). Yet, much has changed in recent centuries. In the nineteenth century, the development of cast-iron pipes and understandings about microbes led to separations of and considerable improvements in the two systems.

In terms of human impacts on the environment, those affecting the atmosphere are fairly recent (although changing land use impacts, including dust storms have a long history). Urban air pollution in London rose rapidly in the eighteenth and nineteenth centuries with the increasing use of coal for heating and cooking. One bright spot with respect to the impact on humans on the environment is the improvement in urban air pollution that began in London around 1900, spread to other higher-income countries later in the century, and has begun to occur even in less developed countries in the current century. An important measure of urban air quality, now widely used and represented in IFs, is the quantity of very small particulate matter. In recent decades, that level has been falling in higher-income countries and somewhat more broadly (Fig. 11.3). Bubbles early in the century reflect data inadequacies.

Similarly, massive efforts to improve the cookstoves inside homes, including movement away from the use of biofuels like wood and dung, have already resulted in considerable progress around the world that is likely to continue (Fig. 11.4). Also, water quality in many streams and lakes of more developed countries has improved.

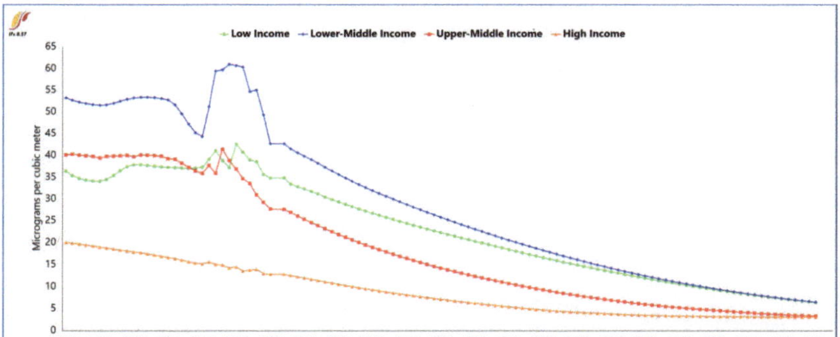

Fig. 11.3 Mean average microgram exposure to particulate matter of 2.5 μm diameter or smaller. *Source* IFs 8.57 Base Case with historical data from the World Bank's World Development Indicators

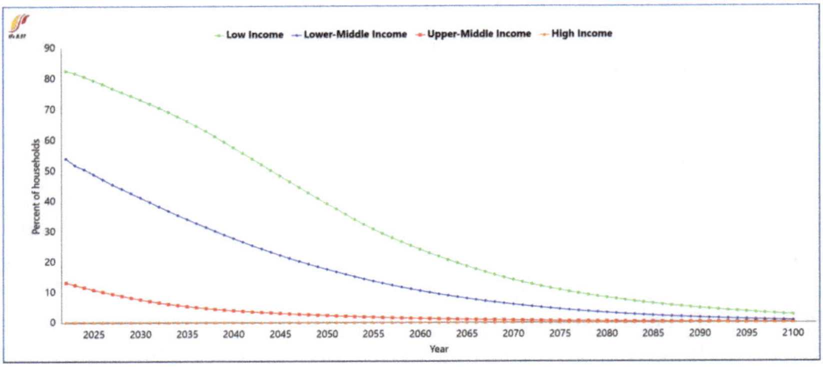

Fig. 11.4 Household use of traditional fuels in cookstoves. *Source* IFs 8.57 Base Case

Turning to the global atmosphere, emissions of chlorofluorocarbons (CFCs) and other substances that were depleting ozone dropped precipitously after nearly universal participation in the Montreal Protocol international treaty of 1987; the hole in the ozone layer above Antarctica that motivated the accord has noticeably diminished in size since 2000.

With respect to the global atmosphere, however, the environmental impact that now appropriately receives the most attention is carbon emission and its contribution to global warming as the primary greenhouse gas. Although some reduction in emissions has begun to emerge in high-income countries, emissions continue to grow in other country-income groups (Fig. 11.5).

IFs forecasts both the flows of carbon emissions and the changing atmospheric stock levels. It computes possible global and country-average temperature change associated with those changes. There is no question that carbon dioxide in the

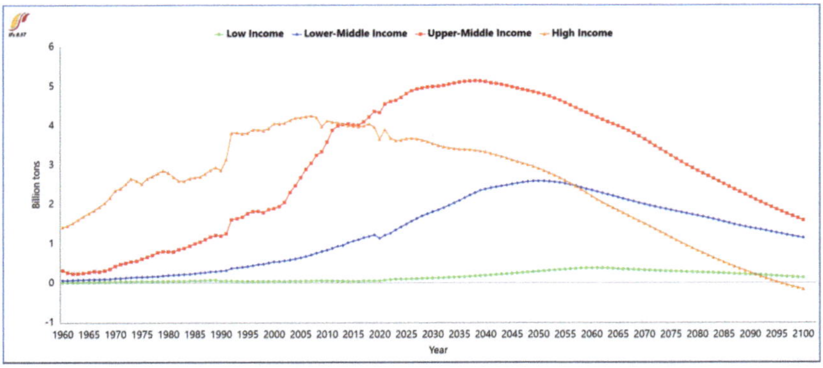

Fig. 11.5 Carbon emissions from fossil fuels. *Source* IFs 8.57 Base Case with data from the Carbon Dioxide Information and Analysis Center at Appalachian State University

atmosphere is increasing and that temperature rise and glacial retreat are related to the phenomenon.

The Intergovernmental Panel on Climate Change (IPCC), drawing on the collective research and insight of hundreds of scientists around the world, has written in its Summary Report that moving average global temperature will most likely be 1.5 °C above 1850–1900 levels in the very near term (IPCC, 2023a: 12). In fact, the annual global temperature reached that level in 2024.

Looking forward, the IPCC Summary Report (2023a: 22) suggests that with currently implemented policies, annual emissions of carbon will rise more than 10 percent before beginning a very slow decline in the last third of the century to levels still above current ones, with a generally similar pattern for total greenhouse gas (GHG) emissions. The IPCC Technical Report (IPCC, 2023c: 58) shows the related global temperature change in 2100 falling between just under 1.5 °C in SSP1 and somewhat above 4.5 °C in SSP5. (See Chap. 1 for elaboration of the SSPs.)

Like IFs, the scenario work of the IPCC has embedded population, economy, and energy scenario considerations into much broader analysis of climate change drivers, emission patterns, impacts, and feedback. Its efforts cut across three working groups and three different categories of forecasting models (Pirani et al., 2024). In Assessment Round 6 (AR6), the three Working Groups (WGI, WGII, and WGIII) used earth system and general circulation models to explore the implications for variables, including temperature, of different patterns of greenhouse gas emissions, including attention to net zero emission futures. They analyzed impacts on biodiversity and on geophysical and broader human and socioeconomic systems. They used SSP scenarios and integrated assessment models (IAMs) to close the loops with earth system models. Much of the work of the IPCC has involved looking at desired patterns of decline in greenhouse gas emissions as well as both the potential for moving onto those paths and the consequences of doing so or not doing so. Our energy chapter discussed the importance of the transition from fossil fuel use to green energy. But Fig. 11.6 shows also the importance of initiatives including demand-side attention to transportation, industry, and buildings and to land use change (to absorb rather than release carbon).

The work of the IPCC across assessment rounds has drawn heavily on the GHG emissions forecasting done by IAMs. Gidden et al. (2019) projected a range of temperature increase by century end between well below 2.0 °C and nearly 5 °C. Their projections of net annual carbon dioxide emissions in 2100 range from about -20 to more than 120 gigatons, with an SSP2 (middle-of-the-road) scenario that peaks in about 2040 and declines toward 10 gigatons (Gidden et al., 2019: 1452).

Such large ranges suggest that a great deal about global warming, its drivers, and its consequences remains uncertain. That includes the rate of transition away from fossil fuel use; the ability of the oceans and other sinks to absorb and retain increased atmospheric carbon dioxide; the contributions that other gases, especially methane, will make to the greenhouse effect; the degree to which changes will be continuous and smooth or will lead to "tipping" in carbon and temperature systems; the change of disease patterns with temperature rise; and the net costs of

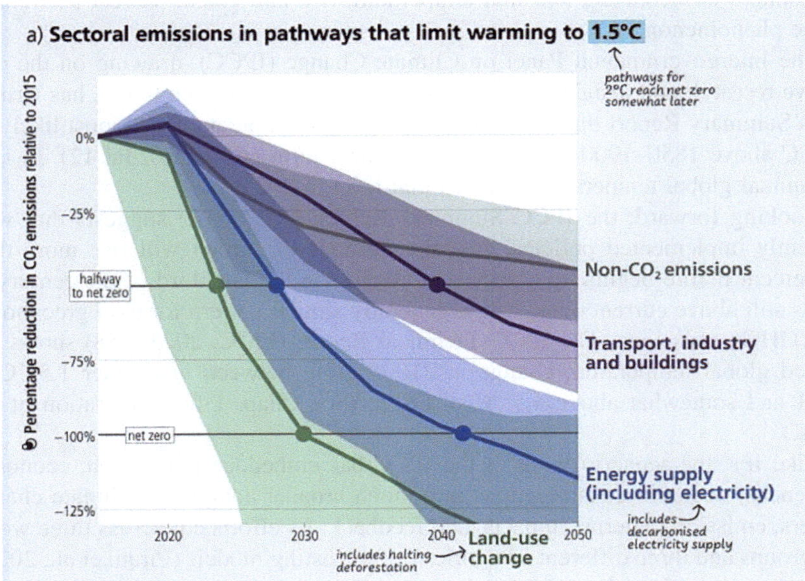

Fig. 11.6 IPCC analysis of policy changes that could potentially mitigate global warming. *Source* IPCC (2023b: 94, Fig. 4.1.a). Available at https://www.ipcc.ch/report/ar6/syr/

global change to human settlements and ecosystems near oceans, to agricultural systems, and to forests.

There are many other important environmental issues that we have not discussed here and that IFs does not represent. For instance, IFs has no representation of species loss, no representation of ocean life stock, and no treatment of soil loss. And there are other important topics that also cannot be explored with IFs, including landfills (covered garbage dumps) and nuclear waste, both of which continue to proliferate.

In short, we face a present with already obvious environmental challenges. And we face a future shaped often by strong trends but very considerable uncertainty on many issues.

11.2 Environmental Dynamics and Leverage

Perhaps the best-known shortcut into thinking about the environment and human leverage is the I = PAT identity (called IPAT), in which environmental impact (I) is defined in terms of the size of the population (P) times the affluence (A), and therefore potential production/consumption level of that population, times the technology (T), represented as the environmental impact per unit of production/consumption (Waggoner & Ausubel, 2002).

Almost all attempts to forecast environmental impact use some explicit or implicit form of this identity and IFs is no exception. Figure 11.7 portrays the relatively simple "tail end" dynamics of selected environmental issues that IFs represents, but the figure leaves out the population, affluence (often tapped with GDP per capita), and technology (PAT) elements that the reader will understand to drive food demand, land and water usage, and energy sources and use in IFs (for instance, Moyer and Hughes [2012] used IFs to analyze the impact of information and telecommunications technology on carbon emissions). Instead, the figure shows only some immediate drivers of water use, forest transformation, and carbon emissions. One of the benefits of IFs, however, is that those immediate drivers do have deeper and complex roots in the model, facilitating the intervention and scenario analysis that the next section will discuss. For information beyond that in this section, see the IFs Wiki, the IFs network diagram, and the Pardee Institute's home page.

Although deforestation has many causes, including the export of timber by cash-starved developing countries and the growth of cities and industrial areas into forest land, the primary driver is the conversion of forest into agricultural

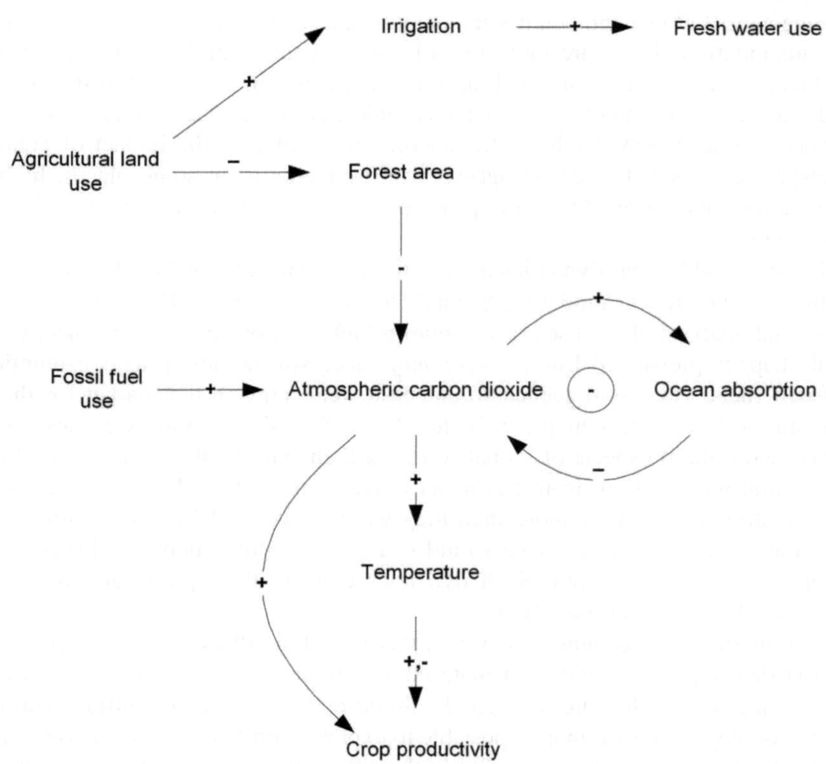

Fig. 11.7 Selected environmental dynamics in IFs

land for grazing or crop production. IFs represents that linkage (as discussed in the Agriculture chapter), including the growth of reforestation in many countries.

IFs computes freshwater needs related to crop irrigation, the single biggest use of water in most countries. More generally, its water model tracks water demand across agricultural, industrial, and municipal categories. On the supply side, it represents surface water withdrawals; groundwater withdrawals from recharging sources; fossil groundwater withdrawals; desalination supplies; and usable treated wastewater.

There are several greenhouse gases, including CFCs that deplete ozone and the methane produced by cattle flatulence and termite activity. Yet, the principal greenhouse gas is carbon dioxide and, while additions of it to the atmosphere have varied sources including the burning of firewood, its principal source is the burning of fossil fuels. IFs tracks carbon emissions from fossil fuels and net emissions from changes in forest area. Some of the carbon dioxide in the atmosphere is absorbed into the oceans (and potentially can be similarly released). Representation of all environmental issues requires tracking both stocks and flows. IFs does this for atmospheric carbon but does not yet represent oceanic stocks of carbon dioxide.

As atmospheric carbon dioxide increases, global temperature rises. IFs uses a structure based on the MAGICC model (Meinshausen et al., 2011) to calculate average annual global and country temperatures, as well as changes in temperature and precipitation. These are indicators of potential future problems ranging from the melting of polar ice caps with accompanying rise in sea levels to the spread of disease-carrying insects into higher latitudes. Although the linkages are complicated, IFs also provides basic frameworks for exploring the impact of greater atmospheric carbon dioxide on agricultural productivity (in some places, higher temperatures and more CO_2 can improve productivity) and of temperature change on economies.

There are additional dynamics around the environment that Fig. 11.7 does not capture but the broader model represents (Hughes et al., 2012). For instance, environmental analysts often discuss the **rebound effect**. Consider, for instance, what would happen globally if Europe were very successful in cutting its consumption of fossil fuels. One consequence would be a weakening in the market for those fuels and a decline in their price. If the United States and China were less concerned about the prospects of global warming than Europe, the lower price of oil and natural gas resulting from European usage constraint might encourage those two countries to consume more than they would have, had Europe continued to consume at historical rates. This would cause some global demand, lowered by Europe's action, to rebound. Such dynamics exist in the equilibrium model of energy in IFs (see again Fig. 10.5).

It is important to recognize, however, that potential affluence of a society need not translate into consumption of material and goods and energy. Consumers can demand less materially-intensive goods. Some refer to this as **dematerialization** of an economy, another dynamic possible to explore with IFs. A broader concept is **degrowth**, which can mean reduction in the size of the economy, not just changes in consumption patterns (Moyer, 2023).

The role for human leverage through environmental decision-making and action is large. Because environmental damage (like depletion of atmospheric ozone) from broad societal action is outside the private economic calculation of most producers and consumers, they frequently pay little voluntary attention to limiting it, even in the most developed countries. Some elements of the environment, such as forests, can be privatized, thus potentially encouraging owners to look beyond immediate economic exploitation and to treat them well over the long run. But air and water systems generally remain public. Thus, in addition to some voluntary and selfless action, it has been the support of non-governmental organizations within and across countries and the pressure of legal and institutional machinery both within and across countries that has led to most progress when it occurs.

Around carbon emissions, there are also intense discussions concerning costs and effectiveness of **mitigation** (reducing emissions and avoiding their impact) and **adaptation** (paying the costs of living in a warmer world with significantly different hydrological and biological patterns). The IPCC (IPCC Summary Report, 2023a) made rough calculations of the mitigation costs of implementing the Paris Agreement on freezing and then lowering emissions. IFs has some but limited structures to allow analysis of this important issue.

At the interstate level, the United Nations Conference on the Human Environment, held in 1972 in Stockholm, was a milestone in general environmental analysis and action. It preceded the creation of the United Nations Environment Programme in 1973 and an explosion of activity protecting domestic and international environments, also incentivized by the Millennium Development Goals and then the Sustainable Development Goals following the United Nations Conference on the Environment and Development in 1992. Global conventions and treaties are proliferating along with non-governmental organizations. Most improvements in environmental quality, or at least reductions in the pace of deterioration, do not just happen as countries become wealthier. Human effort must flow into action consciously targeted at ameliorating the damage of other human activity.

11.3 Interventions and Scenario Analysis

What are the key indicators within IFs for environmental issues? As always, some of these can be found quickly in the *Flexible Displays* of IFs, especially within the *Environment* category. More are in the *Basic Report*. The full set is available through *Self-Managed Display*. In this section, we look again, in turn, at land/forest, water (ocean and terrestrial), and atmospheric environmental issues.

Much scenario analysis on issues of the environment involves working across all interacting aspects of the model and the global system discussed to this point. You can therefore revisit Chaps. 6 through 10 to identify relevant variables and parameters for scenario analysis.

11.3.1 Land Use and Deforestation

One key variable is the amount of the world's land area in forest (**WFORST**). You can also look at regional or country-specific forest areas by examining all uses of land (**LD**). A parameter you can directly manipulate is the multiplier on forest area (**forestm**) and another affects land devoted to crops (**ldcropm**). Given a constraint on total land for crops, grazing, forest, urban usage, and a residual called "other," increases or decreases in either forest or crop area will proportionately affect land available in other categories. Changes in forest area (which stores carbon) will have secondary effects on atmospheric carbon dioxide (**CO2PPM**) and changes in cropland will affect water usage (**WATUSE**).

→ The target growth in land parameter (**tgrld**) controls the initial percentage increase in cropland (most expansion of which comes from forest). Given likely interest in rainforest area (although IFs does not distinguish it from other forest), reduce that parameter to 0.0 for Latin America. Run the model and look at the improvement this makes in the amount of land remaining under forest in Latin America (**LD**) and even in global forest area. Note that there may be no substantial world economic cost to this (look at world **GDP**). But does Latin America incur a cost? Consider GDP per capita at purchasing power parity (**GPDPCP**), human development generally (**HDI**), the portion of the population malnourished (**MALNPOPP**), and ability to raise foreign currency by exporting more agricultural products (**AGX**) than the continent imports (**AGM**).

→ Explore similar scenarios for forest preservation in Africa and South/Southeast Asia, two other regions in which forest cover is decreasing rapidly. When several Southern regions collectively protect their forests, does it affect the North-South economic gap? Do any Northern regions actually benefit by exporting more and higher-priced food to the regions that protect their forests?

Some of the Southern regions have argued that if the richer portion of the world wants to protect Southern rainforests, it will need to share in the costs of doing so. One possibility would be additional foreign aid (controlled by **aiddon**, a parameter discussed in Chap. 7); another would be debt relief (swapping forest preservation for debt), which you could simulate with the parameter for the loan repayment rate, **repayr**. Such side payments by rich countries might help poor countries protect environments that also benefit the rich. Develop what you consider to be some options and examine whether they can provide adequate compensation for the economic losses of the South (you might review economic analysis in Chap. 7).

Another side payment might be improved agricultural technology to compensate for lesser amounts of land under cultivation. Were the rich countries to support more research and development on agriculture, it might lead to the potential for considerable improvements in agricultural yields in the South. You might simulate such improvements via raising the yield multiplier (**ylm**) that we used in Chap. 9.

Food production has managed to stay ahead of population growth in the last four decades largely because of rapid technological progress known as the “Green Revolution.” Instead of an acceleration of such progress, which the Base Case already assumes will continue at a generally comparable rate into the future, there is the possibility that the growth rate will be slower in the future. Slower growth in agricultural yields could lead to compensatory changes in practice, resulting in environmental damage, including the need for more land conversion and water use. The yield multiplier (**ylm**) can simulate such an assumption of slowing technological advance if you reduce its value below 1.0.

11.3.2 Water Usage

Chapter 9 laid out scenario analysis concerning ocean fishing and mariculture. We turn here to extractive uses of water.

The IFs water model forecasts demand in the municipal, industrial, and agricultural categories (**WATDEMAND**). It represents supply broadly, including natural withdrawals (**WATERWITHDRAWAL**) from surface and ground water with attention to the fossil portion of groundwater withdrawal. Total supply (**WATERTOTALSUPPLY**) includes those withdrawals, but also water from desalination (**WATERDELSALINATED**) and recycling of wastewater (**WASTEWATERSUPPLY**). As in many other models of IFs, a price index (**WATPRICE**) sends signals to demand and supply sides to maintain at least approximate balance over time. This uses a proportional-integral-derivative (PID)-type controller similar to that discussed briefly in Chap. 7 (Economics). A water demand multiplier (**waterdemandm**) can change demand in municipal, industrial, and/or agricultural categories.

Global agricultural demand for water, largely for irrigation, is about 70 percent of total human demand. Thus, treatment of irrigation in IFs is important both in terms of analysis of that demand relative to supply and also because it affects agricultural production. In the agriculture model, the yield of irrigated land is several times as great as that of nonirrigated land.

Both the area irrigated and the price of water are important. With respect to the area irrigated, the multiplicative parameter **landirareaactualm** with a default value of “1” allows user direct control (**LANDIRAREAACTUAL**). With respect to the price effect, a country-specific parameter **watpricetoylelas** controls the magnitude of the price effect on irrigated land yield with a variable **YLIMPACTWATPRICE.** Even when **watpricetoylelas** is nonzero, as normally in the Base Case, changes in that potential impact will only affect irrigated yield in the agricultural model if the parameter **watertoyieldswitch** is also turned on (e.g., with a value of 1.0), not typically done in the Base Case.

11.3.3 Atmospheric Environment Other Than Greenhouse Effect

We begin here with issues of more local atmospheric pollution and then move to the global issue of greenhouse effect and climate change.

11.3.3.1 Urban Air Quality

Figure 11.3 showed the micrograms per cubic meter of atmospheric particulate matter with a diameter of 2.5 μm or less in urban areas, using the variable **ENVPM2PT5**. That is responsive primarily to energy type usage (especially coal) but can be controlled with the country-specific multiplier **envpm2pt5m** having the typical Base Case value of 1.0. The health impact in IFs for that pollution is on respiratory infections and diseases plus cardiovascular disease. If the model user does not want to change the particulate level over time, holding it constant can be done with a global hold switch **envpm25hldsw**. (This switch is normally turned off in the Base Case to allow the model to endogenously change the particulate level and its impacts.)

11.3.3.2 Indoor Air Quality

IFs also represents indoor air pollution from traditional and/or solid fuel use. There are two focal points for analysis: (1) fuel use other than modern fuel (e.g., gas or electricity) for cooking alone and (2) solid fuel use for heating and cooking. Those variables affect deaths from respiratory infections, especially of concern with respect to indoor pollution for mothers and children under five years of age and from lung cancer and strokes for older people.

With respect to cooking, **COOKSTOVES** carries information in millions of households on the fuel type in traditional, improved, and modern fuel categories and can be changed with **cookstovesadd**. The percentage of households using each type of fuel can be found in *Flexible Displays* (look in the *Infrastructure* category).

The use of solid fuels (e.g., wood and coal) for both cooking and heating is tracked as a percentage of population with **ENSOLFUEL**. The multiplicative parameter (***ensolfuelm***) can change solid fuel use and a switch (***ensolhldsw***) can hold the percentage of solid fuel use at the level of the first year, which can support counterfactual scenario analysis. Another switch (**hlsolfuelsw**) set to 1 ("on") in the Base Case determines whether solid fuel usage rates will pass through to health outcomes.

11.3.4 Greenhouse Effect and Climate Change

Many IFs users will be especially interested in the issue of climate change. Key indicators include the level of atmospheric carbon dioxide in parts per million (**CO2PPM**) and the percentage increase in atmospheric carbon dioxide (**CO2PER**) relative to estimates of pre-industrial levels. Look also at the annual emissions of carbon from fossil fuel use and net deforestation plus other sources

(**CARANN** and **CARFOROTH**) and the global temperature in degrees centigrade (**WTEMP**). Chapters 7 and 9 identified the implications for the economic and agriculture models of two additional variables computed in the environment model: **CLIMECONIMP** and **ENVYLCHG**, affecting value added (therefore GDP) and crop yields, respectively. Those chapters identified the parameters and logic behind the variable calculation and use; doing so will not be repeated here.

Specific interventions and more elaborate scenarios can be developed using the *Quick Scenario Analysis with Tree.* Some scenarios may draw heavily from interventions catalogued under the *Environmental Uncertainties* branch. These elements tend to focus on unknowns with respect to the future environment (such as the percentage of carbon dioxide in the atmosphere (**co2perm**), not on specific human interventions. For more agent-linked interventions, look under *Interventions and Agent Behavior* at the *Household Actions* and *Government Fiscal Decisions* categories, not just for interventions focused directly on the environment, but also for parameters related to the population, affluence, and technology elements of the IPAT identity.

In the scenario tree, under *World Integrated Scenario Sets* there are several prepackaged integrated intervention scenarios, including some that Chap. 14 will discuss. Examples might be the four scenarios of the IPCC's Third Assessment Round (A1, A2, B1, B2) or the scenarios of UNEP's GEO-3, GEO-4, and GEO-7 (UNEP, 2019) processes. Always useful in wide-ranging scenario analysis are the five SSP scenarios, currently packaged as pre-run scenarios with versions of IFs.

The primary source of increased CO_2 is the burning of fossil fuels (although deforestation also contributes and other gases not currently represented in IFs add to the greenhouse effect). The potentially biggest contribution to addressing both energy and environmental problems lies in the more efficient use of fossil fuels over the near term and movement to renewable energy forms quite rapidly.

→ The energy demand multiplier (**endemm**) allows you to explore the leverage of energy conservation without having to specify exactly how it might come about. Reducing **endemm** for all regions from 1.0 to 0.8 over the first years of a model run introduces a global reduction of energy demand by approximately 20 percent relative to the Base Case. Look at world energy production and price (**WENP** and **WEP**), atmospheric carbon dioxide (**CO2PER**), and world GDP (**WGDP**). Are the effects as great as you might have expected? If not, why not?

Remember still again how the model sometimes "fights" you; this is partly the rebound effect. Energy conservation may lead to energy price reductions (**WEP**). If so, the normal incentive to conserve energy given by higher prices is reduced and that can offset much of the positive effect of the conservation scenario you added.

IFs uses a structure based on the MAGICC model to calculate measures of average annual global (**WTEMP**) and country temperatures (**ENVTEMP**), as well as percentage changes in precipitation (**ENVPRCHG**) since 1990 with or without changes in atmospheric CO_2. Those underlie **CLIMECONIMP** and **ENVYLCHG.**

→ Reductions in energy use with **endemm** are "free." Conservation efforts, either voluntary actions by citizens or government regulation, can sometimes reduce energy usage in ways that have little or no economic cost (sometimes even a savings). Yet, there is a significant debate over the extent to which such "free" reductions can be obtained, for instance by voluntary consumer action. Much analysis of the Greenhouse Effect and political discussion of it focuses on the prospects of a carbon tax (**carbtax**). Such a tax would raise the price of fossil fuels to their users proportionately to their carbon content. That would encourage conservation or switching to less expensive energy forms. The extent of **GDP** loss (or gain) from such taxes is a matter of great debate; Repetto and Austin (1997) provided a very good early analysis. Although IFs does not capture the full range of potential costs or benefits, experiment with different levels of a carbon tax and look at the impact on the energy system, on atmospheric carbon, and on the economy. A carbon tax of about $100 per ton (a value for **carbtax** of 100) would add about $0.20 cents to the cost of a gallon of gasoline. Try ramping up the tax over time and look at the impact on annual emissions (**CARANN**), carbon dioxide levels (**CO2PPM**), and world temperature (**WTEMP**). China produces more carbon emissions than any other country in the world, more than twice the second-place United States. Would a carbon tax in only China have a major effect on global patterns?

Moving beyond conservation to technological advance, another suggestion of environmentalists often is that the world relies much more heavily on renewable energy. This suggestion is often coupled with the argument that government should support more research and development on renewable energy to lower its cost.

→ You can simulate a decreasing cost for renewable energy by lowering the capital cost multiplier for its different forms (**qem**), say from 1.0 to perhaps 0.8 over the next 10 years. Since this multiplier controls all costs for production (operating and capital), a 20 percent reduction over such a period can be very optimistic (given recent patterns, however, the cost of solar cells could very possibly fall more than that). Does this scenario by itself, or in combination with the conservation scenario, significantly lower the production of global fossil fuels (look at **WENP** for fossil fuels compared to the Base Case)? Does it control the growth of atmospheric CO_2? You may need to run the model for many years to answer such questions. And as Chap. 10 discussed, you can directly force the system to increase production and use of renewable energy forms, including hydroelectric, wind, solar, and geothermal power, using **enpm**. Think about whether you want to include nuclear in that set, especially given the possibility for emergence of fusion energy.

You could also look at energy costs of fossil fuels and consider raising them. One way that could happen is if we have overestimated fossil fuel supplies—as we exploit them, the remaining supplies may become too expensive to extract. Lower values of the resource multiplier (**resorm**) can introduce more conservative assumptions, more relevant for oil and gas than for bountiful coal. And some of the costs of nuclear energy, such as disposal of wastes and decommissioning of old plants, have yet to be borne by many nuclear producers; thus, the capital cost of nuclear energy may rise in the future (use **qem** again). You may possibly need to look out across the century to see significant effects.

The model also offers a way of investigating environmental damage to the agricultural system from the growth in atmospheric carbon dioxide that your scenarios almost certainly will fail to stop in the very near future. You can raise the absolute value of an elasticity of agricultural production with growth in atmospheric carbon dioxide (**elasac**) on a region-by-region basis to posit greater damage (leaving the negative sign intact); lowering it simulates less damage and a value of 0.0 would turn off the linkage. Positive values would represent improvements in agriculture from warming, something that may be possible in some cooler geographical areas. This elasticity could also serve you as a proxy for other forms of environmental damage to agriculture (ozone emissions and acid rain are often caused by some of the same industrial and energy practices that generate carbon dioxide).

Even with conservation and technological advance it is challenging to control either deforestation or the increase in atmospheric carbon dioxide. Many observers argue that the rapid growth of human population (even with declines in the fertility rate) in combination with the efforts we make to feed, clothe, and shelter ourselves (maybe even to buy a car) inevitably increases the pressure on the environment. Thus, they argue, one of the most fundamental changes we must make is to further slow and then halt the growth of human population—to bring it into some kind of balance with the environment. You may want to investigate alternative population scenarios again now that you are focusing on environmental quality. Can you reasonably reduce population growth fast enough (with **tfrm**) to have an environmental impact?

Some analysts also suggest that modern high-income economies are extremely wasteful of energy and other raw materials, simultaneously depleting resource bases and contributing to environmental degradation. They propose slower economic growth rates especially for that country set, concentrating remaining economic growth on true improvements in the quality of life. In fact, it is conceivable that the early post-World War II period was extraordinary with respect to general technological advance and that the period since the early 1970s, characterized by slower growth in productivity and economic output worldwide, better indicates the future. If so, the assumptions in the Base Case about technological advance in the leading state (**mfpleadr**) could be too high. You may wish to experiment with more conservative assumptions.

Research Suggestions. How might the world of the future look for an environmentalist? Experiment with a variety of what you consider reasonable assumptions about environmental policies and look at how much difference they make with respect to variables such as GDP per capita, life expectancy, calories per capita, and malnutrition. Develop an integrated scenario for what you consider to be an environmentally friendly but still reasonable future. Remember that it is best to develop scenarios by changing one parameter at a time and examining and understanding the impact of that one before adding changes together.

11.4 Conclusion

This chapter has drawn together many of the issue areas and global dynamics discussed to this point. It is impossible to examine the future of the environment without considering developments in population, economics, human development, food, and energy. The following two chapters shift our attention once again. We turn to developments in domestic and global social and political systems.

References

Abu, T. Z., Achore, M., Irfan, M., Musah, I., & Yussif Azzika, T. (2024). The past, present, and future of Ghana's WASH sector. An explorative analysis. *Water Security, 23*, 100185. https://doi.org/10.1016/j.wasec.2024.100185

Boretti, A., & Rosa, L. (2019). Reassessing the projections of the world water development report. *NPJ Clean Water, 2*(1), 15. https://doi.org/10.1038/s41545-019-0039-9

Daigneault, A., Baker, J. S., Guo, J., Lauri, P., Favero, A., Forsell, N., Johnston, C., Ohrel, S. B., & Sohngen, B. (2022). How the future of the global forest sink depends on timber demand, forest management, and carbon policies. *Global Environmental Change: Human and Policy Dimensions, 76*, 1–13. https://doi.org/10.1016/j.gloenvcha.2022.102582

Gidden, M. J., Riahi, K., Smith, S. J., Fujimori, S., Luderer, G., Kriegler, E., van Vuuren, D. P., van den Berg, M., Feng, L., Klein, D., Calvin, K., Doelman, J. C., Frank, S., Fricko, O., Harmsen, M., Hasegawa, T., Havlik, P., Hilaire, J., Hoesly, R., … Takahashi, K. (2019). Global emissions pathways under different socioeconomic scenarios for use in CMIP6: A dataset of harmonized emissions trajectories through the end of the century. *Geoscientific Model Development, 12*(4), 1443–1475. https://doi.org/10.5194/gmd-12-1443-2019

Hughes, B. B., Irfan, M. T., Moyer, J. D., Rothman, D. S., & Solórzano, J. R. (2012). Exploring future impacts of environmental constraints on human development. *Sustainability, 4*(5), 958–994. https://doi.org/10.3390/su4050958

International Panel on Climate Change (IPCC). (2023a). Summary for policymakers. In H. Lee & J. Romero (Eds.), *Climate change 2023: Synthesis report.* Contribution of working groups I, II and III to the sixth assessment report of the Intergovernmental Panel on Climate Change. Core writing team (pp. 1–34). IPCC. https://doi.org/10.59327/IPCC/AR6-9789291691647.00

International Panel on Climate Change (IPCC). (2023b). *Climate change 2023: Synthesis report.* Contribution of working groups I, II and III to the sixth assessment report of the Intergovernmental Panel on Climate Change [Core writing team (H. Lee & J. Romero (Eds.)]. IPCC. https://doi.org/10.59327/IPCC/AR6-9789291691647

International Panel on Climate Change (IPCC). (2023c). D. Ürge-Vorsatz,: *Climate change 2022: Mitigation of climate change technical summary.* Contribution of working group III to the sixth assessment report of the Intergovernmental Panel on Climate Change. M. Pathak, R. Slade, R. Pichs-Madruga, D. Ürge-Vorsatz, P. R. Shukla, & J. Skea (Coordinating Lead Authors). IPCC. https://www.ipcc.ch/report/ar6/wg3/downloads/report/IPCC_AR6_WGIII_TechnicalSummary.pdf

Meinshausen, M., Raper, S. C. B., & Wigley, T. M. L. (2011). Emulating coupled atmosphere-ocean and carbon cycle models with a simpler model, MAGICC6—Part 1: Model description and calibration. *Atmospheric Chemistry and Physics, 11*(4), 1417–1456. https://doi.org/10.5194/acp-11-1417-2011

Moyer, J. D. (2023). Modeling transformational policy pathways on low growth and negative growth scenarios to assess impacts on socioeconomic development and carbon emissions. *Scientific Reports, 13*(1), 15996. https://doi.org/10.1038/s41598-023-42782-y

Moyer, J. D., & Hughes, B. B. (2012). ICTs: Do they contribute to increased carbon emissions? *Technological Forecasting and Social Change, 79*(5), 919–931. https://doi.org/10.1016/j.techfore.2011.12.005

Nepal, P., Korhonen, J., Prestemon, J. P., & Cubbage, F. W. (2019). Projecting global and regional forest area under the Shared Socioeconomic Pathways using an updated environmental Kuznets Curve model. *Forests, 10*(5), 387. https://doi.org/10.3390/f10050387

Pirani, A., Fuglestvedt, J. S., Byers, E., O'Neill, B., Riahi, K., Lee, J.-Y., Marotzke, J., Rose, S. K., Schaeffer, R., & Tebaldi, C. (2024). Scenarios in IPCC assessments: Lessons from AR6 and opportunities for AR7. *NPJ Climate Action, 3*(1), 1. https://doi.org/10.1038/s44168-023-00082-1

Ritchie, H. (2021). Forest area. *Our World in Data.* https://ourworldindata.org/forest-area#article-citation

Sampson, S. (2024). *FAO Report: Global fisheries and aquaculture production reaches a new record high.* Newsroom. https://www.fao.org/newsroom/detail/fao-report-global-fisheries-and-aquaculture-production-reaches-a-new-record-high/en

United Nations Educational, Scientific and Cultural Organization (UNESCO). (2025). *Mountains and glaciers: Water towers | UN World Water Development Report.* https://www.unesco.org/reports/wwdr/en/2025

United Nations Environmental Program (UNEP). (2019). *GEO-6.* Global Environment Outlook (GEO). http://www.unep.org/geo/geo-resources/geo-6

United Nations Food and Agriculture Organization (UNFAO). (2024). World food and agriculture—Statistical yearbook 2024. *FAO.* https://doi.org/10.4060/cd2971en

Waggoner, P. E., & Ausubel, J. H. (2002). A framework for sustainability science: A renovated IPAT identity. *Proceedings of the National Academy of Sciences, 99*(12), 7860–7865.

Part V

Sociopolitical Impetus

Domestic Sociopolitical Systems 12

Abstract

Previous chapters emphasized that the concept of "sustainable human development" encompasses a broad range of interacting issues across human development and environmental sustainability dimensions. This chapter, the first in Part 5 of the volume, further extends the perspective on sustainable human development to include the dynamics of sociopolitical systems that interact with and significantly shape changes in human development and environmental sustainability. It includes sections on three categories of variables characterizing domestic sociopolitical systems: (1) life conditions (heavily treated in Chap. 8); (2) social organization; and (3) culture and values. The chapter presents International Futures Base Case forecasts of key variables in each category and surveys the model specifications underlying such forecasts. The chapter concludes, as usual, with attention to scenario analysis and exercises for users. Chapter 13 will turn our attention to global sociopolitical dynamics.

12.1 Introduction

It is common to group a wide range of phenomena under the general rubric of sociopolitical characteristics. We will distinguish three general categories. The first is life condition. Into this category we place income and wealth, life expectancy, fertility rate, literacy rate, education level, calorie availability or nutrition level, and access to safe water and air.

The second category is social organization. We can divide that into formal and informal subcategories. In the subcategory of formal aspects, we can place level of democratization and general policy orientations of the government, such as economic liberalization and patterns of spending. In the subcategory of less formal aspects of social organization, we can place the effectiveness of government, level of social trust, the extent of corruption in economic and political life, the social

B. B. Hughes, *Exploring and Shaping International Futures*, World-Systems Evolution and Global Futures, https://doi.org/10.1007/978-3-032-22084-4_12

status of women, and the level of peacefulness or violence in the society. The sixteenth SDG calls for attention to social organization:

SDG 16: "Promote peaceful and inclusive societies for sustainable development, provide access to justice for all and build effective, accountable, and inclusive institutions at all levels."

Elaboration of the goal's focus on "inclusive society" explicitly directs attention to social justice and attention to the vulnerable, connecting social organization to the desire and the ability to improve life conditions.

The third category of sociopolitical characteristics is culture and values, which provide a possible foundation for societies that promote peaceful, inclusive, and effective action. All societies have historical but evolving cultural foundations, patterns of belief and behavior. These foundations shape the character of social organization and its attention to life conditions.

12.2 Life Condition

Three of the life conditions that attract much attention from analysts are income and especially the avoidance of poverty (for which household income is often used as a proxy); life expectancy (which is greatly affected by nutrition); and educational level (for which literacy rates are sometimes used as a quick measure). Earlier chapters, especially those on Human Wellbeing and Economics, have devoted considerable attention to these, but it is useful to review some of that and then bridge to broader sociopolitical analysis.

At the beginning of the twentieth century, the average GDP per capita of Western European countries was only $5,400 (purchasing power parity dollars, in year 2017 prices),[1] about the average in lower-middle-income countries today. Even the rich in those days scored considerably lower on many measures of individual life condition than contemporary countries with much lower-income levels. For instance, life expectancy was approximately 50 years in the richest Western European countries in 1900. In 2025, it reached 67 on average even in low-income countries, up from just 45 in 1960. Chapter 2 pointed out that in 1820 perhaps only 12 percent of adults were literate globally. By 2025 that had risen to 86 percent and even in low-income countries to 61 percent. Social development, at least in terms of individual life condition, has been a remarkable success story.

Yet, global gaps in life condition are wide across countries and large numbers of people live in poverty, hunger, and illiteracy. In the 1990s, the United Nations Development Programme (UNDP) developed a single Human Development Index (HDI) across the income, education, and life expectancy dimensions of social condition and refined the Index in 2010. Its values range from 0 to 1.0. Figure 12.1

[1] Calculated using Maddison (1995, 2001) data and recent values from the World Bank.

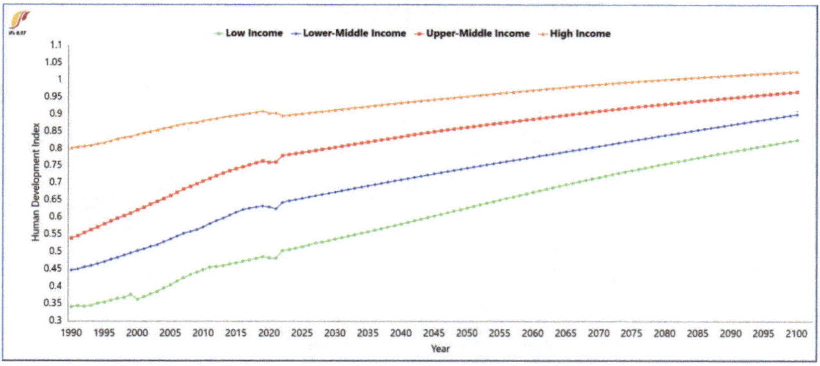

Fig. 12.1 Human Development Index. *Notes* Using the 2010 reformulated index. The transients near 2020 are related in part to COVID-19's impact on life expectancy. *Source* IFs 8.57 Base Case with historical values from various UNDP Human Development Reports that are not completely comparable across time

shows that index across the World Bank's country-income groupings. While it suggests some convergence toward still increasing high-income country levels, that is quite slow in recent data and in the IFs Base Case forecast.

One of the Sustainable Development Goals discussed in Chap. 8 is the universal completion of secondary education and that chapter showed how far we may be from universal enrollment even by 2100. Figure 12.2 shows the educational attainment level of females and males around the world in 2050, by age, in the IFs Base Case. This figure, layering educational attainment onto a population pyramid, is one of the *Specialized Displays* in IFs, and the numbers are generated by the extensive formal education model that Mohammod Irfan has developed. The figure shows the global progress toward universal primary education completion that is likely to have been made by 2050, but even achievement of that by 20–24 year-olds is unlikely to be complete. The world will likely then be far from achievement of universal secondary education.

12.3 Social Organization

Turning from issues of individual life condition to those of social organization, two that attract much contemporary attention are the advance of democracy and the level of sociopolitical instability. An important concern is that democracy's advance has been subject to reversible waves. According to Huntington (1991) who looked back into the1800s, a "third wave" of global democratization began in the late 1970s and accelerated with the collapse of communism in the early 1990s. Figure 12.3 shows that third wave in all World Bank income groups, but also some reversal of it in the early years of this century.

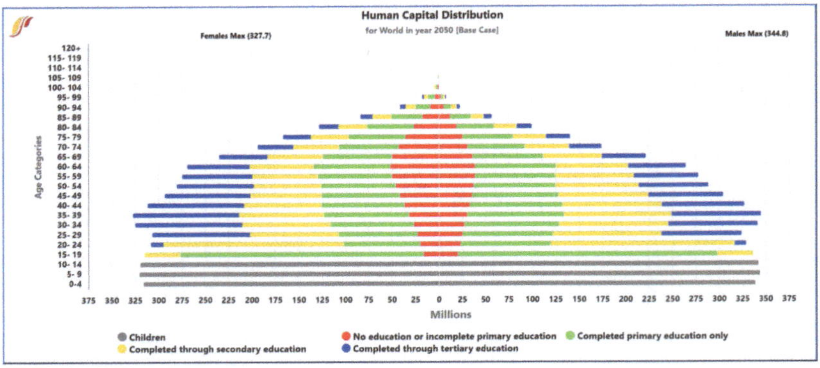

Fig. 12.2 Global educational attainment by age and sex in 2050. *Source* IFs 8.57 Base Case

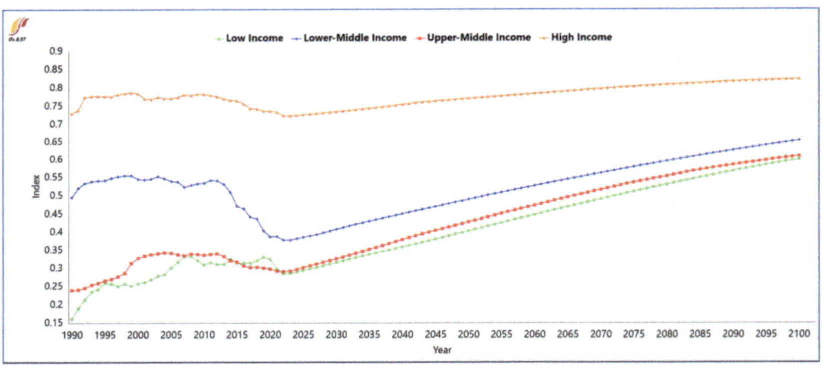

Fig. 12.3 Average polyarchy (electoral democracy) score. *Note* The Electoral Democracy Index runs from 0 (fully autocratic) to 1 (fully democratic). *Source* IFs 8.57 Base Case with historical data from the VDEM project (Coppedge et al., 2025; Teorell et al., 2019)

Although democracy is an important value around much or most of the world, effectiveness of governance in supporting and advancing wellbeing is something that almost all societies crave. Figure 12.4 shows relatively recent values of that using the World Bank's government effectiveness index and IFs forecasts. Given the historical stability of the index for all country groupings, and even some setback early in this century, the IFs forecasts appear optimistic. Yet, the gap between high-income countries and the other three groupings suggests the prospect of some rise, at least in the developing world. This is an area of the model that might be quite important in scenario analysis, especially because government effectiveness, democracy, government corruption, economic freedom, and a broader freedom measure all enter the computation of multifactor productivity in the economic model. They also affect the global system (see the next chapter).

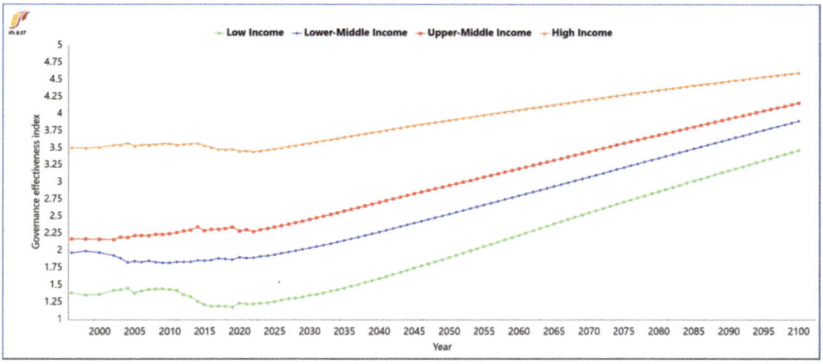

Fig. 12.4 Government effectiveness. *Source* IFs 8.57 Base Case with historical data from the World Bank

In addition to type of governance and its effectiveness, a key issue of social organization is sociopolitical stability. Civil war is almost invariably horrendous and when it is persistent over long periods it can create what have come to be labeled "failed states" (those with persistent abrupt changes in regime and/or revolutions and genocides/politicides). These are very much a concern also for the global system, both for humanitarian reasons and because of concerns about the breeding of terrorism. Figure 12.5 shows the frequency of new and continuing state failure events using historical data from the Uppsala Conflict Data Program (UCDP) (Brosché et al., 2023). Moyer (2023) studied the impact of state failure on poverty using data from the Political Instability Task Force (PITF). Forecasts of state failure in IFs recognize driving variables that include infant mortality, trade openness, and the impact of spillover from intrastate conflicts in neighboring states (Kwasi et al., 2026). Clearly, there is a strong relationship between such internal conflict and economic development, with low-income countries suffering much more from it than even middle-income countries. That suggests that the incidence of it will likely decline across the century. There is no way for IFs to anticipate individual country-based outbreaks or endings of such instability over the long term.

Although democratization, government effectiveness, and social stability take prominent positions, there are many other contemporary issues related to social organization. One is the growth of civil society in countries/regions where it has been weak, including the former communist states. Another very important one is the level of corruption (the IFs database includes series from both Transparency International and the World Bank). The coming section of this chapter on scenario analysis will invite the reader to explore other issues around social organization including corruption. Another sociopolitical issue is the status of women in society. Still another is the character of domestic income and wealth distributions (IFs has an extensive series of Gini indices compiled by the World Bank). Chapter 8 (Human Development) directed considerable attention to the status of women and to income distribution and invited scenario analysis around those issues.

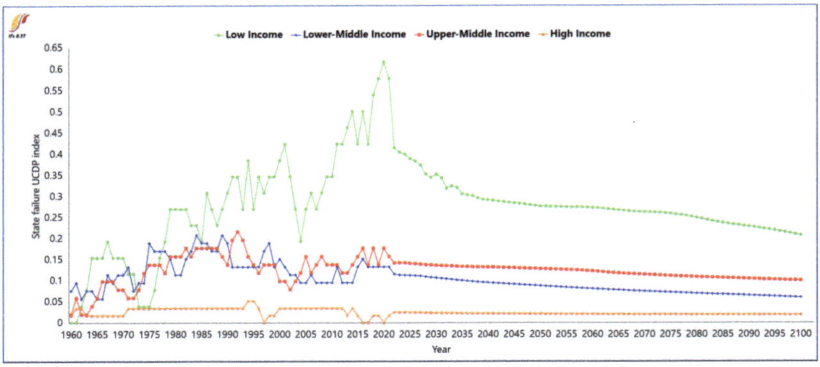

Fig. 12.5 State failure from civil war. *Source* IFs 8.57 Base Case with historical data from the Uppsala Conflict Data Program

12.4 Values and Culture

Values and culture have moved to a prominent position among contemporary global issues, including the rise across many countries in recent years of strong anti-immigrant political forces. At the end of the Cold War, Samuel Huntington (1993) asked what the fault lines of the global system would be as we moved into a new era free of the Western/Communist divide and concluded that they would be along the lines of religious and cultural divisions, what he termed the "clash of civilizations." This is a particularly difficult issue area to represent in a model. IFs uses a very strong source of data, the World Values Survey (WVS) initiated by Ronald Inglehart, which has undertaken an extensive set of surveys since 1980 in societies with the bulk of the world's GDP and population.[2]

Through factor analysis, the WVS found that two orthogonal (uncorrelated) dimensions organized large numbers of responses to their survey questions: (1) a survival vs. self-expression dimension and (2) a traditional vs. secular-rational dimension. Figure 12.6 illustrates a *Specialized Display* in IFs. It shows the responses across age cohorts on the first dimension of those surveyed in sub-Saharan Africa during Wave 6 (2010–2014). Clearly, Africans were heavily focused on the survival end of the scale, with those who were older even more likely to be so oriented. In building a basic forecasting capability representing culture and values, IFs has taken advantage of the strong relationships of these value dimensions to GDP per capita, the cultural traditions that distinguish regions around the world inherent in the WVS data, and the cohort shifts that frequently reshape those dimensions and traditions.

[2] During his life, Ronald Inglehart of the University of Michigan and founder of the WVS was exceptionally generous in providing the data to the IFs project and collaborating on ideas for forecasting of value dimensions.

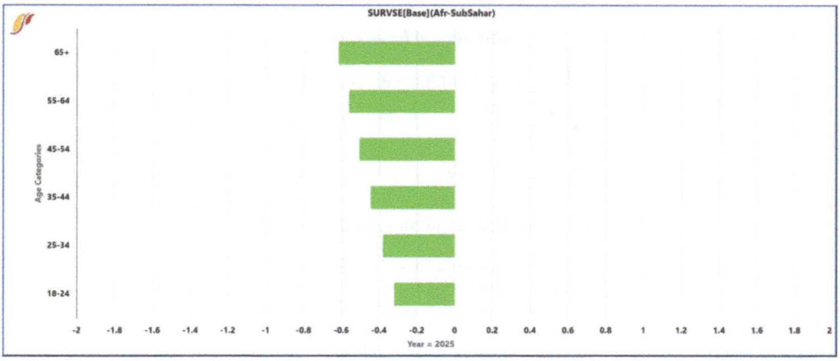

Fig. 12.6 Survival vs. self-expression values in sub-Saharan Africa by age cohort. *Source* IFs 8.54 showing data from the World Values Survey (Wave 6)

There is much that IFs cannot forecast with respect to sociopolitical futures. This is an area that is inherently more complex than demography or even economics. Nonetheless, there is considerable richness to the representation in IFs that can be of use to analysts. We turn to the dynamics of the representation.

12.5 Sociopolitical Dynamics and Leverage

The above discussion of sociopolitical issues identified three general clusters of variables and concerns: life conditions; social organization (formal and informal); and culture/values. Figure 12.7 shows a general conceptualization that has guided the building of representations within IFs. Not all components of it are by any means fully elaborated (the culture/values elements are particularly skimpy), but all components have core representation. Note that each of the three major components of the sociopolitical system not only interacts with each other; they also interact closely with other subsystems in IFs, including the demographic and economic models (Moyer, 2024). Also, the domestic and global sociopolitical systems have important implications for each other (Moyer, 2023). For information beyond that in this section, see the IFs Wiki, the IFs network diagram, and the Pardee Institute's home page.

As is true more broadly in our analysis of possible futures, an understanding of the dynamics around sociopolitical systems can benefit from distinctions between stocks and flows (like population versus births and deaths or capital stocks versus investment and depreciation flows). Sociopolitical systems are, however, often difficult to characterize in those terms. Many of the sociopolitical variables of interest to us (including life conditions, institutional structures, and value systems) tend to have characteristics of stocks, including their tendency to change slowly over time. They don't, however, have some obvious analogue for flows that increment or decrement them over time.

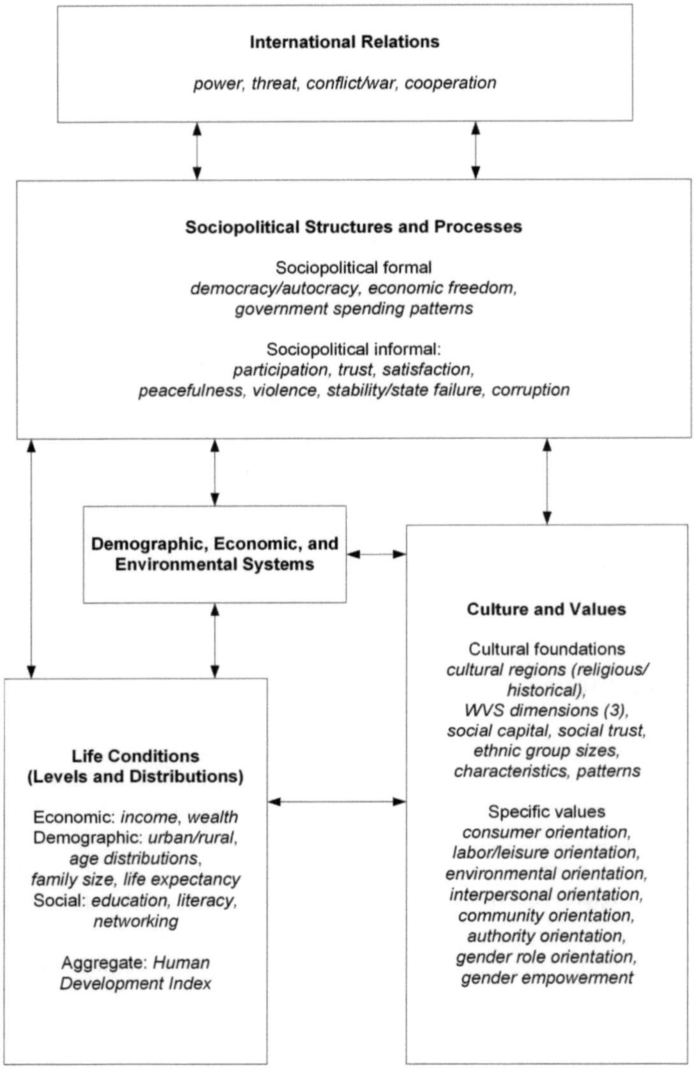

Fig. 12.7 Sociopolitical components of the social fabric in IFs. *Source* Author

Thus, instead of the heavily structural modeling of demographics, economics, health, education, agriculture, and energy, always augmented by statistical and causal analysis, our modeling in this area tends to be more consistently reliant on statistical analysis of changing sociopolitical variables. GDP per capita at purchasing power parity (PPP) is often a useful driver in understanding patterns of social change. It is especially often used in the data preprocessor of the IFs system, which ensures that all countries have values for all variables in the base year of model forecasts. Even when a more complex formulation is used in forecasting, countries missing a data point for the base year will have a computed or estimated value for GDP per capita at PPP (**GDPPCP**) in the base year because GDP and population sizes are two variables for which the IFs system "requires" all countries to have data values.

Other variables are almost always statistically-based elements of forecast formulations in the body of the model proper. Consider, for instance, forecasting government effectiveness (**GOVEFFECT**). In addition to GDP per capita, IFs includes a democracy variable (**DEMOCPOLITY**) and an internal calculation of dependence on energy exports as driving variables. Figure 12.8 shows the degree to which GDP per capita does relate to governance effectiveness across countries in the base year, explaining about 65 percent of variance in it in the model's base year.

Figure 12.9 shows, however, that the VDEM project's measure of democracy also is correlated with governance effectiveness, albeit at a lower level and a different relationship form, as is the Polity project's measure (Marshall & Gurr, 2020). Often those who build and enhance IFs will use statistical estimation packages to discover and add multivariate relationships after exploring the bivariate ones. The statistically estimated formulation used in IFs for governance effectiveness adds a third variable, dependence of the economy of energy exports, to the two discussed here.

As a user of IFs, you have powerful multivariate statistical analysis capabilities within the *Data Analysis* main menu option and the *Analyze Across Countries* suboption (as described in Chap. 4). Further you can do scatterplots and bivariate statistical analysis within *Flexible Displays* when you analyze relationships in the base year, which is tied closely to data values.

Returning to the relationship between GDP per capita and individual life conditions, social organization variables, and values, it is useful to understand some typical characteristics of those relationships. We can speak generally of three

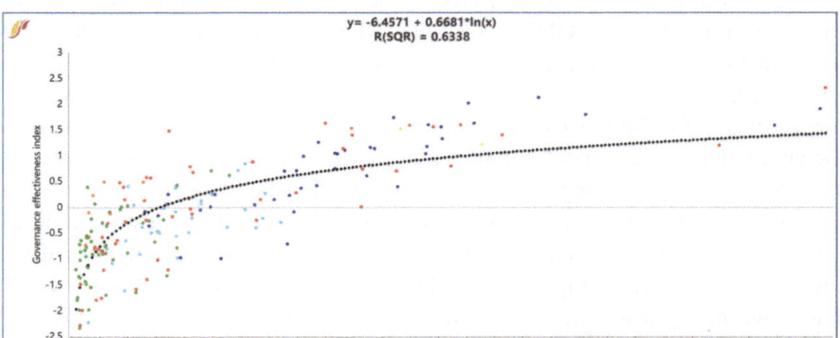

Fig. 12.8 Governance effectiveness as a function of GDP per capita at PPP (2021$). *Source* IFs 8.57 showing most recent data from the World Bank available in 2025

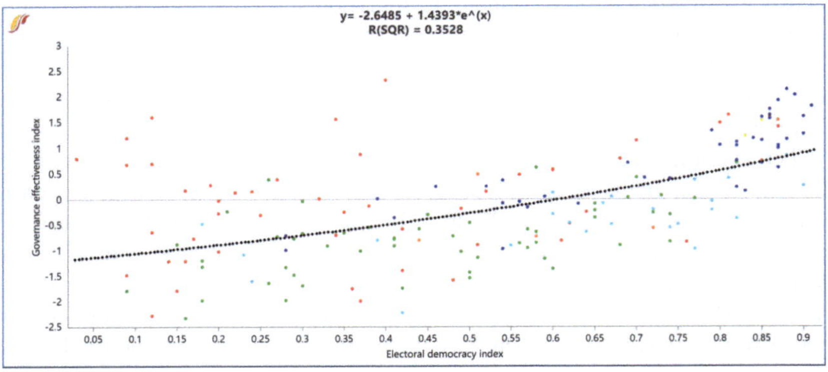

Fig. 12.9 Governance effectiveness as a function of the Electoral Democracy Index. *Source* IFs 8.57 showing data available in 2025 from the World Bank and the VDEM Project (Coppedge et al., 2025; Teorell et al., 2019)

stages that we find, in varying degrees, in most of them (Hughes, 2001). As in Fig. 12.8 and in a great many relationships, we can see a "sweet spot" in development. That is, at low but increasing levels of GDP per capita, sociopolitical conditions improve rapidly. Second, at higher levels, between about $10,000 and $20,000 per capita at PPP in U.S. 2021$, that improvement slows, becoming more of a "steady slog." We would see a similar pattern in the data on literacy or corruption. With respect to infant mortality, however, the movement from "sweet spot" to "steady slog" is at somewhat lower levels of GDP per capita, presumably reflecting its high priority early in the development process.

Third, in many relationships (including those where we see "sweet spots" and "steady slogs"), we also see a "systemic shift." Specifically, over time (even at quite stable levels of GDP per capita) the character of the relationship between GDP per capita and a dependent variable shifts significantly, moving most or all of the curve upward. In other words, global sociopolitical conditions often improve across time regardless of GDP per capita. This obviously indicates that factors other than GDP per capita are at work. For instance, technological change can improve some individual life conditions (like life expectancy or infant mortality) regardless of GDP per capita. Similarly, there can be global changes in ideas or behavior that affect aspects of social organization such as national and global efforts devoted to advancing government quality (Hughes, 2004).

Turning very briefly from driving linkages to forward linkages, sociopolitical variables clearly affect much else in the system. In the next chapter we discuss, for instance, the relationship between democracy level and international conflict. Because we have, however, focused heavily on GDP per capita as a driver, it is only appropriate to stress that sociopolitical variables also have strong forward linkages to GDP growth. It has long been recognized (going back to Solow, 1956) that economic productivity can explain half or more of economic growth; Romer

(1994) and many others have therefore explored the drivers of increased productivity and found many of them in human capital (education and health) and social capital (including governance effectiveness and level of corruption). Chapter 7 discussed this further and one figure presented a *Development Profile* from IFs that showed the impact of socioeconomic variables on economic growth.

As we turn to interventions and scenario analysis, it will not surprise you that we once again pay much attention to life conditions, social organization, and culture and values.

12.6 Interventions and Scenario Analysis

Before you turn to key indicators in forecasts, there are large numbers of data series in the IFs database that can provide fodder for cross-sectional and longitudinal analyses with processes discussed in Chap. 4 (Understanding IFs). Look on the Main Menu under *Data Analysis* and select either analysis type. In the box labeled *Show Variable Name Subset*, you might select the *SocioPolitical* subset and with cross-sectional analysis explore the relationships between GDP per capita at PPP and several sociopolitical variables of interest to you.

Looking at forecasts from the Base Case and pre-run scenarios, there are important indicators available for selected countries and country groupings in the *Flexible Displays* form—look, for instance, under *Human Development*, *Demographics, Economy*, *Education*, *Governance*, *Health,* and *Infrastructure* categories and their sub (meso) categories. There are also, as indicated throughout this chapter, some *Specialized Displays* for sociopolitical variables. These include education levels by age-sex cohorts; the value dimensions of the *World Values Survey*; the *SDG Overview Table*; and the *Development Profile*. Further, there are selected sociopolitical variables available in the *Basic Report* display for both countries and country groupings.

And, of course, there are large numbers of variables available to you via *Self-Managed Display*. There you need to select by variable name rather than description, but *Subsets of Names* (including *Governance* and *Human Development*) can help focus your search. Among the indicators of individual life condition of potential interest to you will be domestic and global literacy rates (**LIT** and **WLIT**); adult educational attainment levels (**EDYRSAG15** and **EDYRSAG25**); domestic and global life expectancy (**LIFEXP** and **WLIFE**); the total fertility rate (**TFR**); the infant mortality rate (**INFMOR**); the percentage of children or the full population who are malnourished (**MALNCHP and MALNPOPP**); and the **GDP**. You might give some special attention to the Human Development Index (**HDI**). See again Chap. 8 on Human Development that identified many of these in the context of exploring progress toward the Sustainable Development Goals.

Among the social organization indicators of interest will be Freedom House's measure of democratization (**FREEDOM**); the Polity Project's measures of democracy (**DEMOCPOLITY**) or VDEM's measure (**DEMOCPOLYARCHY**); a set of measures of state failure probability from the earlier State Failure

Project (**SFINSTABALL** and many others starting with "**SF**"); as well as the **SFCIVILWARUCDP** measure from the UCDP project discussed earlier.

Other social organization variables include Transparency International's measure of government corruption perception (**GOVCORRUPT**); the World Bank's measure of governmental effectiveness (**GOVEFFECT**); the United Nations Development Programme's Gender Empowerment Measure (**GEM**); a domestic Gini index of income distribution (**GINIDOM**); the Frasier Institute's measure of economic freedom (**ECONFREE**); and the A.T. Kearney/*Foreign Policy* measure of globalization (**GLOBALIZ**).

Among measures of culture and values, look for the WVS dimensions of survival/self-expression (**SURVSE**); traditional/secular-rational (**TRADSRAT**); and materialism/post-materialism (**MATPOSTR**).

Turning from looking at relationships to exploring leverage, you can likely benefit, as usual, by some attention to the IFs Wiki. That can help you understand which variables and parameters may be of interest and what forward impacts a focus on specific variables might have.

In the *Scenario Tree,* you will find many drivers of interest (*Parameter Search* may be useful because most often multipliers and additive parameter names include the name of the variable or variables they affect). As you explore individual life conditions, many of the levers for intervention will be found in the population and economic models and you may wish to review those chapters for some of the key ones. Governmental action, particularly expenditure patterns (discussed in the Economic chapter), often changes life conditions. For instance, you would expect higher spending on education to result in higher literacy rates and years of education. Spending that boosts or retards economic productivity could further affect social conditions via its impact of economic growth. Experiment with the multiplier that shifts government fiscal expenditures among categories (**gdsm**), directing more or less money to health, education, and R&D. In addition, there are multipliers on spending at different levels of education (**gdsedm**).

As you move to social organization, there is, in general, less connection in IFs between conditions and economic level or government fiscal choices. For instance, the level of corruption is correlated with GDP per capita, but some very poor countries are much less corrupt than middle-income or even richer countries. In this area of the model, it is especially appropriate to represent many social choices simply with multipliers on the underlying variable. For instance, you can use multipliers on freedom (**freedomm**), democracy (**democm**—look at other parameters beginning with "democ" for parameters to introduce waves), government corruption (**govcorruptm**), government effectiveness (**goveffectm**), and economic freedom (**econfreem**) to directly manipulate social change in these areas. Some, but not all the underlying variables (such as economic freedom and governance quality) will have forward linkages to other variables, so that your manipulation may have broader consequences.

The representation of culture and values in IFs is rudimentary and forward linkages are nearly nonexistent. Because change on the WVS value dimensions is driven by economics and demographics, change in those models will again have impact. And additive factors (**survseadd**, **tradsratadd**, and **matpostradd**) will change levels on the value dimensions.

→ By the time you have reached this chapter, you will likely have developed considerable analytic capabilities with IFs. And as with environmental change, sociopolitical change is heavily influenced by other issue areas of the model system, particularly demographic and economic growth. So, the exercise here is more open-ended than many earlier ones.

- Explore the Base Case to get a clear sense of what the trajectory and magnitude of sociopolitical change may be in the three categories that were the focus of this chapter.
- Then explore scenarios of faster or slower economic growth to see how much impact they have. Follow that by exploring interventions via government expenditures on education and health.
- Finally, explore using some of the direct multipliers noted above to see their impact, especially on social organization. In working with multipliers like these, it is important to recognize that changing variables such as the level of corruption or economic freedom is very difficult in any society. Do not assume that your interventions, if they have positive consequences for your scenario, are somehow a "free lunch." Think seriously about whether they are reasonably attainable and what social and other costs, perhaps not represented in the model, there might be.
- Although we want to explore scenarios that improve sociopolitical systems and therefore positively affect much within society, it is also possible that those systems deteriorate. The parameter **sfintlwaradd** can increase levels/probabilities of internal conflict. It is worth exploring the implications of that happening in a country of interest to you.

Partly because the bodies of work that empirically and theoretically describe the dynamics of sociopolitical systems are less well developed and more contentious than those in many other issue areas, it has proven especially challenging to develop formulations for them in the IFs system. Ideally, equation structures, involving functional forms and the specification of variables included (not just parameters), should be open to change by users throughout the model. Some steps have been taken toward this end by identifying a small number of bivariate and multivariate functions that are accessible, primarily with respect to sociopolitical systems. You can find those under *Scenario Analysis, Change Specific Functions.*

12.7 Conclusion

This chapter emphasized the degree to which domestic and international sociopolitical changes are interrelated. Nonetheless, there was value in focusing separately on the domestic elements in this chapter. The next chapter takes us primarily to the relationships among societies rather than the change within them.

References

Brosché, J., Nilsson, D., & Sundberg, R. (2023). Conceptualizing civil war complexity. *Security Studies, 32*(1), 137–165. https://doi.org/10.1080/09636412.2023.2178964

Coppedge, M., Knutsen, C. H., Gerring, J., Lindberg, S. I., Teorell, J., Altman, D., Angiolillo, F., Bernhard, M., Cornell, A., Fish, M. S., Fox, L., Gastaldi, L., Gjerløw, H., Glynn, A., Good God, A., Grahn, S., Hicken, A., Kinzelbach, K., Krusell, J., … Ziblatt, D. (2025). *V-Dem country-year dataset v15*. Varieties of Democracy (V-Dem) Project. https://doi.org/10.23696/vdemds25

Hughes, B. B. (2001). Global social transformation: The sweet spot, the steady slog, and the systemic shift. *Economic Development and Cultural Change, 2*(January), 423–458. https://www.journals.uchicago.edu/doi/abs/10.1086/452510?journalCode=edcc

Hughes, B. B. (2004). Regimes and social transformation. In A. Underdal & O. Young (Eds.), *Regimes and social transformation:* Kluwer Academic Publishers. https://link.springer.com/chapter/10.1007/978-1-4020-2208-1_13

Huntington, S. P. (1991). *The third wave: Democratization in the late twentieth century*. University of Oklahoma Press.

Huntington, S. P. (1993). The clash of civilizations. *Foreign Affairs, 72*(3), 22–49.

Kwasi, S., Xiong, Y., Meisel, C. J., Morgan, C., & Moyer, J. D. ([Manuscript in Preparation]). SDG 1: Reassessing the relationship between civil war and poverty. In *SDGs and Peace.*

Kwasi, S., Xiong, Y., Meisel, C.J., Morgan, C., & Moyer, J. D., "SDG 1: Reassessing the relationship between civil war and poverty," in The Elgar Companion to Peace and the Sustainable Development Goals edited by Döring, S., Petrova, K., & Hegre, H., Elgar Publishing, forthcoming 2026, https://www.eelgar.com/shop/usd/the-elgar-companion-to-peace-and-the-sustainable-development-goals-9781035329427.html.

Maddison A. (1995). *Monitoring the world economy 1820–1992*. OECD.

Maddison, A. (2001). *The world economy: A millennial perspective*. OECD.

Marshall, M. G., & Gurr, T. R. (2020). *Polity5: Political regime characteristics and transitions, 1800–2018*. Dataset Users' Manual. Center for Systemic Peace. https://www.systemicpeace.org/inscr/p5manualv2018.pdf

Moyer, J. D. (2023). Blessed are the peacemakers: The future burden of intrastate conflict on poverty. *World Development, 165*, Article 106188. https://doi.org/10.1016/j.worlddev.2023.106188

Moyer, J. D. (2024). A double-edged sword into a plowshare: Analyzing geopolitical implications of alternative socioeconomic development pathways. *One Earth, 7*(2), 336–347. https://doi.org/10.1016/j.oneear.2024.01.002

Romer, P. M. (1994). The origins of endogenous growth. *Journal of Economic Perspectives, 8*(1), 3–22. https://www.jstor.org/stable/2138148

Solow, R. M. (1956). A contribution to the theory of economic growth. *Quarterly Journal ofbr/ Economics, 70*(1), 65–94. http://links.jstor.org/sici?sici=0033-5533%28195602%2970%3A1%3C65%3AACTTTO%3E2.0.CO%3B2-M

Teorell, J., Coppedge, M., Lindberg, S., & Skaaning, S.-E. (2019). Measuring polyarchy across the globe, 1900–2017. *Studies in Comparative International Development, 54*(1), 71–95. https://doi.org/10.1007/s12116-018-9268-z

The Global Sociopolitical System 13

Abstract

Chapters 2 and 3 emphasized that understandings of global change generally are rooted in a combination of three things: (1) analysis of historical trends; (2) causal analysis exploring relationships of variables statistically; and (3) dynamic analysis that identifies feedback loops and shapes understandings broadly with world views. Chapter 3 identified realism and liberalism as two such world views and they help frame this chapter. International Futures (IFs) explores some interactions of states (countries) within the broader system without identifying individual partner/target states (monadically); it also explores some interactions as specifically state-to-state (dyadic or bilateral). It represents both conflict and cooperation in the global system. Realist analysis in IFs computes multiple power measures, both monadic and dyadic, looking at variables such as systemic polarity, power transition, and threat of conflict. Liberal analysis in IFs looks at variables such as democracy extent and explores many mostly cooperative interactions across countries, including trade, foreign investment, and foreign aid. The chapter facilitates scenario analysis with exercises that could emphasize a single world view (realism or liberalism) or enable exploration of global dynamics that reflect a combination.

13.1 Introduction

Whether one views international interactions from the perspective of conflict and immiserating exploitation or that of peaceful and cooperative diffusion of human development, there is no question that the interactions between countries and

Collin Meisel, Pardee Institute Director of Analysis, is a co-author on this chapter. Caleb Petry, Pardee Institute Model Development Research Associate, contributed much to the chapter's work.

B. B. Hughes, *Exploring and Shaping International Futures*, World-Systems Evolution and Global Futures, https://doi.org/10.1007/978-3-032-22084-4_13

groups of countries affect movement toward or away from goals for sustainable human development. The final Sustainable Development Goal (SDG) calls explicitly for the second pattern of interaction:

SDG 17: "Strengthen the means of implementation and revitalize the global partnership for sustainable development."

The elaboration of that goal in targets and with indicators calls for foreign aid to be increased to 0.7 percent of donor country GDPs, for foreign direct investment (FDI) to rise, for environmentally protective technology transfers, and for open and nondiscriminatory trade with developing country export increases. It also identifies other aspects to strengthen global partnerships, all of which are consistent with a world view of intercountry interaction that Chap. 3 and Fig. 3.5 identified as liberalism (sometimes called globalism or even idealism).

Yet, in that same chapter, Fig. 3.4 similarly identified some of the dynamic elements of the security dilemma as part of what is often called the realist world view. That perspective draws our attention to both historical and contemporary interaction of states as very frequently conflictual and often negative in impact even when it does not involve overt conflict.

History has shown us both sides. It often highlights the long history of warfare. Major conflicts are often seen by scholars and observers as due to changing power distributions, including those in which a globally leading or "hegemonic" power is challenged by a rising power (Allison, 2017; Chan, 2007; Organski, 1958; Tammen et al., 2017; Thucydides, 431 B.C.E). Examples often include the Peloponnesian War when Athens' rise threatened Sparta and the two world wars of the twentieth century as German power came to rival that of the United Kingdom.

There have also been long periods of relative stability and much lower levels of warfare and lost life. Not least are the years after World War II, during which the great power of the United States, manifested in part by very cooperative action including the Marshall Plan and other foreign aid and by strong support for open trade and investment flows as well as by very strong military capabilities, contributed to avoidance of direct great-power war.

The International Futures (IFs) system includes variables and dynamics that draw on both world views, consistent with the notion that global dynamics are exceedingly complex and that we can learn from investigating multiple possible futures. Subsequent sections in this chapter will discuss those futures, often but not always clearly framed by the realist or liberal perspectives.

13.2 Monadic and Dyadic Representations of Global Dynamics

Before turning to the discussion of realist and liberal perspectives in IFs, however, it is important to address the issue of monadic versus dyadic (or bilateral) variables and dynamics in IFs. In previous chapters, there has been very limited discussion of the latter. Full understanding of demographics and economics (including elaborated agriculture and energy models) benefits from attention to international flows, including those of migrating people with associated flows of remittances as well as of trade and finance. Historically, IFs has represented such international interactions with a monadic pooled approach; that is, all outflows from countries were summed, all inflows were summed, the two sums were reconciled (most often averaged), and country values for both were normalized to the global total.

The first departure from this approach was in the representation of threat between countries with specific country-to-country or bilateral values. More recently, bilateral representations have been added for trade and many other variables.[1] The computational intensity of those led to the introduction of a parametric switch in IFs that by default leaves all international calculations except threat in monadic mode. (Important Note: That switch can be moved to bilateral mode in the downloaded version of IFs, but not in the web-based version. Explanation of doing so will be given in the subsequent discussion of bilateral power.)

Briefly, the bilateral mode draws heavily on the gravity model of trade initially developed by Tinbergen (1963) and elaborated by many others including Armington (1969), Anderson (1979), and Santos Silva and Tenreyro (2011). In simple terms, analogously to gravity's physical relationship to mass and distance, the gravity model of bilateral trade dynamically anticipates change in trade mass of the trading partners to be related to GDP or size of specific economic sectors of the trading partners. Unlike physical distance with physical gravity, distance in economic trade is shaped by changing technology, changing complementarity of trading profiles, and tariffs or other trade barriers. Shorter-term economic forecasting can compute the trade between dyads and sum them to obtain country totals. In longer-term forecasting with IFs, this method proved often to create unstable country totals. Therefore, the forecasting of most variables in IFs with dyadic representation involves separate calculation of country total (monadic) values and constraint of dyadic values to those monadic totals.

More generally, bilateral forecast variables in IFs can be broken down along two dimensions (see Table 13.1). The first is that variables can either be constrained by a separate calculation of a core IFs monadic variable or unconstrained and therefore determine the monadic calculation in IFs. The second is undirected vs directed flows. Undirected flows are those, like membership or lack thereof, in which the actor-to-partner and the partner-to-actor share a value, such as a shared IGO membership. On the other hand, directed flows, such as trade exports and imports, are those in which the actor-to-partner and the partner-to-actor values generally differ.

We later cover usage of those bilateral variables of special importance to both the realist and liberal perspectives on global system dynamics.

[1] For much more detail on the bilateral representations in IFs, see the IFs Wiki documentation of the economic model, especially Chapter 5 on "Bilateral Elaboration of Trade and Other Interstate Flows" from Hughes (2015).

13.3 Global System Overview

Keeping in mind the distinctions between monadic (pooled) and bilateral treatment of interactions among countries and the alternative modes of bilateral interaction laid out in Table 13.1, we can survey treatments in IFs through the lenses of realism and liberalism.

13.3.1 The Power of Realism

Power is a key variable in realism. What is power and how do we know it when we see it? Fundamentally, it is the ability to influence outcomes. In measuring it, however, analysts typically turn to the capabilities that underlie that ability and much debate surrounds identifying and weighting those capabilities. Early work in projects like Correlates of War (Small & Singer, 1982) tended to focus on three such variables: population size; economic size measured as industrial capabilities or as GDP; and military size, often tapped by spending levels or occasionally by military personnel counts.

13.3.1.1 Four Monadic Capability-Based Power Measures in IFs

The representation of power within IFs has roots in that history but has gone beyond it and provides multiple measures to the user in addition to the ability to create one's own index. Four measures in IFs represent country power: **POWERHH**, **POWERHHMI**, **POWERGPI**, and **POWER**. Table 13.2 shows the variables in IFs that make up each of the four measures. Only the **POWERHH** measure can be changed easily by IFs users because its components are multiplicatively weighted by values in the parameter ***wpwghtpow***. The other three are useful for analysis, but the contributions from the 26 terms in the ***wpwghtpow*** vector are currently hard-coded.

The idea of a Technology Power (TechPower) index was suggested for the IFs project around 2000 by Evan Hillebrand and Paul Herman and became **POWERHH**. That measure is an improvement of the then widely used Singer-Small index because it recognizes the importance in the modern world of technological sophistication; it uses the product of GDP and GDP per capita as a proxy for that.

Table 13.1 Alternative modes of bilateral representation for variables in IFs

	Constrained	Unconstrained
Directed	Arms trade and transfers Foreign aid Level of representation (embassies) Migration Trade (default approach)	Trade (alternative approach)
Undirected/Shared	Shared IGO memberships	Shared alliance memberships

Source Original version from David Bohl, updates from Caleb Petry

Table 13.2 Variables used in power measures in IFs and their multiplicative weightings

	POWERHH (using **wpwghtpow**)	POWERHHMI	POWERGPI	POWER/DiME
		Hard-coded	Hard-coded	Hard-coded
Population	0.8	0.8		
GDP at PPP	1.1	1.1	0.2	
GDP at MER	0.0			
GDPP*GDPPCP (Technology proxy)	0.3	0.3		0.2
GDP*GDPPC	0.0			
Govt Spending	0.0			
Mil Spending	0.9	0.9	0.2	0.2
Conv Power	0.0			
Nuclear Power	0.0			
Nuclear Warheads	0.0	0.25	0.05	0.2
Energy Net Export (Prod in code)	0.0		0.05	
Trade Total	0.0		0.05	0.2
FDI Inflows	0.0		0.05	
Aid Net	0.0		0.05	
HumCapWkPopHDI	0.0			
Govt Revenues	0.0			
Innovation R&D	0.0		0.05	
Connectivity ICT Production	0.0		0.05	

(continued)

Table 13.2 (continued)

	POWERHH (using **wpwghtpow**)	POWERHHMI	POWERGPI	POWER/DiME
		Hard-coded	Hard-coded	Hard-coded
Diplo Embassy	0.0	0.3		0.2
Diplo IGO	0.0	0.2		
Diplo Treaty	0.0	0.1		
Diplo DPI (index of 3 above and GOVREV))	0.0		0.05	
GDPxGDPPCP	0.0			
Effective Human Capital	0.0		0.05	
GDPPC	0.0			
GDPPCP	0.0			

Source Author

Beckley subsequently popularized that product's importance, computing it in his book using market exchange rate values (Beckley, 2018a) and in an article using purchasing power parity values (Beckley, 2018b).

Paul Herman helped identify the original Base Case weightings for **POWERHH** in IFs. **POWERHHI** (I for index) is a variation adding population and military spending. Jonathan Moyer created the **POWERHHMI** variation by further adding nuclear power and diplomatic strength. The POWERGPI measure differs by adding several economic flow variables (Moyer & Markle, 2018); it was used by Heim and Miller (2020).

Table 13.2 shows another measure added to the evolving set within IFs, simply named **POWER**, but known in the published world as the DiME (Diplomacy, Military, Economy) index, including for study of the rise of China relative to the United States (Moyer et al., 2023). **POWER** involves five equally weighted terms. The table indicates that IFs users have control over those weightings for the **POWERHH** measure via changes in the **wpwghtpow** parameter vector.

In the world of power politics, a key "leverage" question is: does military spending level strengthen power and the security position of a country? Your first reaction might be that the answer is obviously "yes." Yet, the causal loop diagram of the security dilemma in Chap. 3 (see Fig. 3.4) outlined a reason for caution before assuming that higher military spending always enhances security. Arms spending threatens the security of other states and often gives rise to increased spending by the threatened states in a dynamic known as action-reaction.[2]

In addition, spending can have economic costs that burden an economy. While in the last century the United States spent five percent or more of its GDP on defense for nearly forty years, Japan spent one percent or less, and the Europeans spent about three percent. There are many reasons for the great post-World War II economic success (and power) of Japan and Europe, and their lower military spending may be an important one. Thus, military spending may translate to enhanced power, but the dynamics are devilishly complex.

[2] Lewis Fry Richardson, a British meteorologist, devoted a substantial portion of his life to studying arms races in balances of power. Early in the twentieth century, Richardson formulated a pair of equations that, with variations, we still rely on to understand arms races (Richardson, 1960; see the chapters in Zinnes and Gillespie (1976) for variations in presentation of the equations).

$$\text{XSPENDING}^{t+1} = k * \text{YCAPABILITIES}^{t} - a * \text{XCAPABILITIES}^{t} + g$$

$$\text{YSPENDING}^{t+1} = l * \text{XCAPABILITIES}^{t} - b * \text{YCAPABILITIES}^{t} + h$$

The first equation says that military spending of country X at time $t + 1$ responds via a "reactivity" coefficient "k" to capabilities of country Y in the last time period but is reduced by a portion "a" of country X's own capability. Finally, a third coefficient "g" is called the "grievance" term and sets a basic level of spending independent of the dynamic terms. The second equation says that country Y sets its spending levels in the same fashion, although the parameters will normally differ. The structure of IFs reproduces the essential logic of the Richardson equation system. The key difference is that in IFs some additional structure elaborates the major terms of those equations.

13.3.1.2 Bilateral Power in IFs: Formal Bilateral Influence Capacity (FBIC) Index

The international politics literature often conflates the terms power and influence, defining both as country A's capacity to have country B do X. Measurement of power typically focuses on resources that would underlie such capacity. It is useful, however, to combine measurement of resources with consideration of the relational context of states. That consideration benefits from the movement of measurement from monadic to dyadic (or bilateral). Potentially, two pairs of two states could have the same relative power resources but in the context of their relationship have different relative influence patterns.

The FBIC measure in IFs (**INFLUENCE**) specifies the relational context in terms of (1) bandwidth, the extent of A and B's interaction across economic, political, and security dimensions; and (2) dependence, the relative dependence of the states on each other for economic prosperity and security. Although bandwidth measures are symmetrical in the dyad, dependence measures are not. Moyer et al. (2021: 7) laid this out in terms like those of Table 13.3. Meisel et al. (2023) analyzed the current and future relationship of the United States and China using the FBIC Index.

As indicated earlier, the FBIC measure and other dyadic measures are not yet available in the web-based version of IFs. They are available to compute when using the downloaded version. To do so, rebuild the model base (all initial conditions) using *Extended Features/Rebuild Base* with bilateral data after ticking the Using Bilateral Data box; that will then automatically use bilateral data from the *IFs/Data* folder. After rebuilding the base, the model needs to be run making sure that the bilateral switch (**bilateralsw**) is set to 1 (on) and a new scenario saved if a user wishes to have regular access to a scenario with bilateral forecasts enabled. Even after this process is done, scenario analysis requiring access to bilateral forecast variables must have this switch set to 1. Most IFs users will not

Table 13.3 FBIC Index taxonomy and variables

Dimension	Economic	Security	Political
Bandwidth	Total goods trade	Total arms transfers	Level of diplomatic representation
	Trade agreements	Military alliances	Shared intergovernmental organization membership
Dependence	Goods trade, % of total Goods trade	Arms import stock, % of total arms import stock	
	Goods trade, % of GDP	Arms import stock, % of military spending stock	
	Aid, % of Total aid		
	Aid, % of GDP		

Source Moyer et al. (2021: 7, Table 1). Reprinted with permission of the Atlantic Council

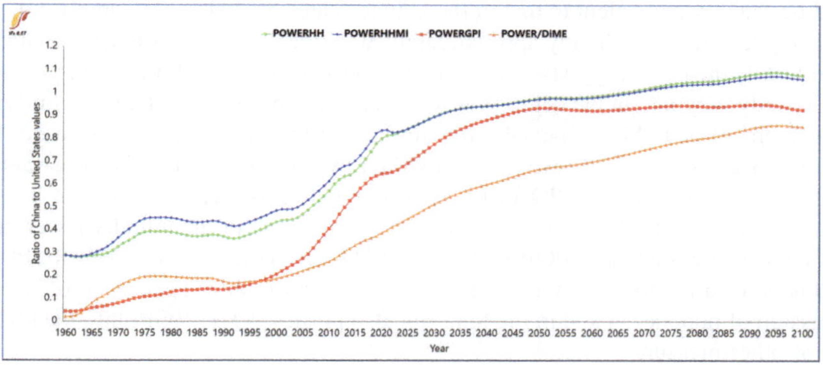

Fig. 13.1 Rise of Chinese power relative to that of the United States. *Note* Figure uses 5-year moving average to smooth data irregularities. *Source* IFs 8.57 Base Case with historical data from multiple sources

undertake bilateral analysis because of the complexity in setting it up.[3] However, it is important for anyone undertaking extended global sociopolitical/geopolitical analysis.

13.3.1.3 Polarity and Power Transition

The bipolar world order centered on the Cold War conflict between East and West organized almost all aspects of the global sociopolitical system after World War II. That order ended with the collapse of communism in the Soviet Union and the disintegration of the U.S.S.R. into fifteen independent countries in 1991. The end of the Cold War led to discussion of a new world order, the characteristics of which are much more difficult to define than those of the old order and are still changing.

Figures 13.1 through 13.4 provide some insight on power in the current and future world and take advantage in doing so of the four monadic measures of power in IFs. Power forecasts change with updates of data and changes in the model so it is worth exploring whether the version of the model you are using produces similar or changed patterns to those shown below.

The first of the graphics shows the possible rise of Chinese power relative to that of the United States. Three of the four measures suggest that Chinese power

[3] A further complication occurs when setting the bilateral switch to 1: Tariff revenues calculated from the bilateral trade forecasts then automatically feed into government revenues. This feeds through the IFs economic model and slightly alters IFs forecasts for variables as central to the model as GDP, and hence GDP per capita. Users wishing to use scenarios with the bilateral switch on should always compare their results to a "bilateral base case" (i.e., a scenario where the only parameter intervened upon is the bilateral switch that is set to 1). This will allow them to understand to what extent their other scenario interventions are "doing the work" aside from the changes inherent in the bilateral base case.

will be roughly equivalent to that of the United States by 2045–2050 and two of the four suggest that it is already approximately 80 percent of that in the United States. The DiME measure suggests a much lower ratio currently and a reaching of near parity only very late in the century. A look back at Table 13.2 explains the outlying measurement of DiME, especially the absence in that measure of population, but the absence also of GDP (except as modified by GDP per capita) is important as well. The absence also in POWERGPI of population (except as modified by age structure, income level, and development) explains some of the initially lower ratio in that measure relative to POWERHH and POWERHHMI, but in POWERGPI the inclusion of a number of more external connection variables appears to offset the omission of population and in the long run also pushes POWERGPI into the higher set of other measures.

The relative rise of China in all power measures suggests the near inevitability of a return to an increasingly bipolar global system. Yet, Fig. 13.2 shows that India is poised to rise even relative to China and that the rise is likely to accelerate in the second half of this century. Interestingly, while two of the measures (POWERHH and POWERHHGPI) suggest rough end-century parity of India with China and therefore something more of a tripolar system, the two measures that do not include population (POWERGPI and POWER/DiME) leave India still somewhat below China.

One-fourth of the way through this century, many observers of the global sociopolitical system would name the Russian Federation as the third most powerful country of the world, after the United States and China, or even, given its especially great nuclear and other military strength, as the second. Figure 13.3 compares its power on all four measures relative to India. Although the DiME measure of power places it above India, even it shows the relative power shifting quite rapidly in favor of India.

Finally, it is of interest to consider the FBIC bilateral influence measure and to look with Fig. 13.4 at its historical path since 1960. The solid lines in that

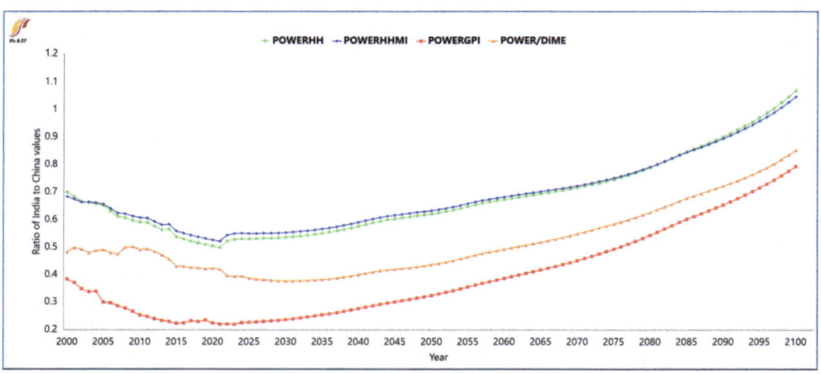

Fig. 13.2 Fall and rise of Indian power relative to that of China. *Source* IFs 8.57 Base Case with historical data from multiple sources

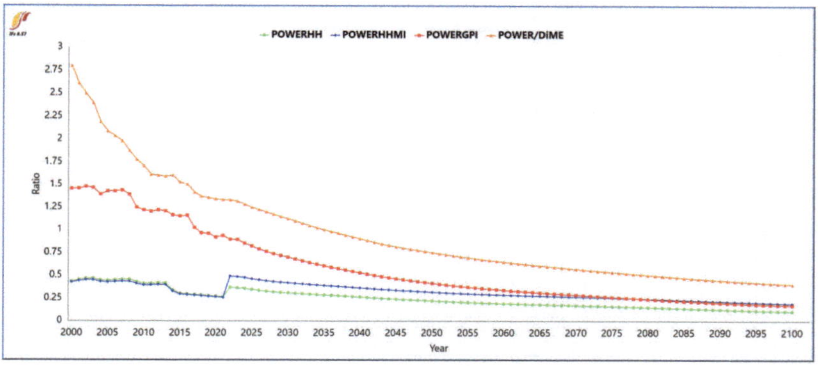

Fig. 13.3 Fall of Russian power relative to that of India. *Source* IFs 8.57 Base Case with historical data from multiple sources

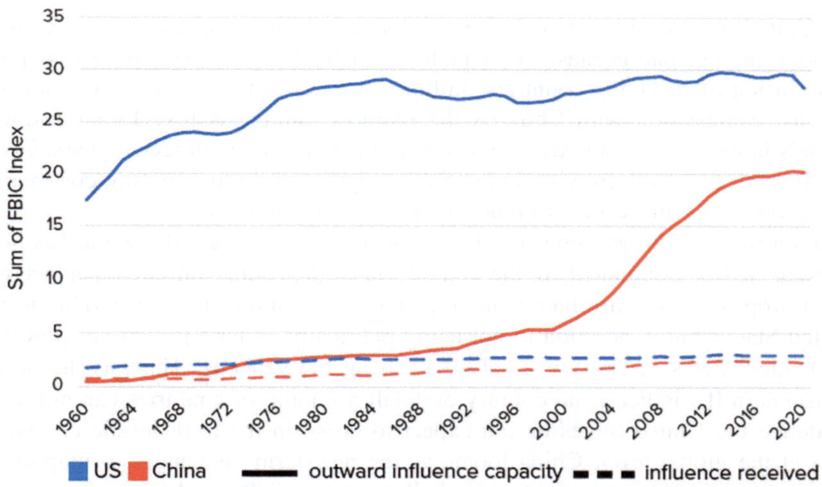

Fig. 13.4 Global sum of outward and inward FBIC Index scores for China and the United States. *Source* Moyer et al. (2021: 11 [Fig. 4]). Reprinted with permission of the Atlantic Council

figure show the global sums of outward influence capacity for the United States and China and suggest the rapidity of the increase in Chinese influence capacity since the beginning of the twenty-first century, greatly narrowing the gap with the United States. In short, the emergence of a bipolar world is evident there as well as in the monadic power measures. Influence received by both countries from around the world is dramatically lower than the outward capacity, but there has been some growth in it as well.

An open question is whether we will soon see decline in influence of the United States and continued growth for China. Worldwide, IFs forecasts suggest

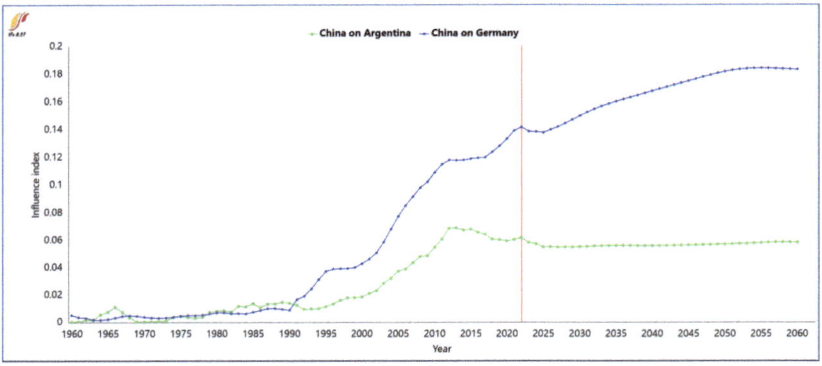

Fig. 13.5 FBIC Index scores for China's influence on Argentina and Germany. *Note* The vertical line is at the year between data and forecast. *Source* IFs 8.55 Base Case with historical data from multiple sources

the United States will retain the most global influence capacity in the coming decades, in large part because of the inclusion in the FBIC measure of arms exports as a portion of receiving countries' military stock. But for selected countries and regions, contestation with China is the likeliest future. As Fig. 13.5 illustrates, China's influence on both Argentina and Germany has risen in recent years. Given the impact of its still growing industrial strength, that influence may continue to rise, especially with respect to other industrialized countries.

In terms of traditional power politics, for most of a century the world has been unipolar, led or dominated by the United States depending on one's perspective. The European Union has had economic and demographic size comparable to the United States, but in addition to not being politically unitary, its ratio of power to the United States in 2020 was only 0.57 using POWERHH. Thus, as the power measures in IFs indicate, no country or small grouping of countries can militarily challenge the United States or can expect to do so in the next decade or two, at least at the global level. China looms as the near-term challenger to unipolarity in most assessments and could pose challenges especially in Asia, such as in the Taiwan Strait, at any time. One of the key questions about the future is what will happen if China gains military and/or economic parity with the United States and then moves into an even more powerful position. Will that power transition happen peacefully or not? Unfortunately, historical power transitions have been almost invariably marked by wars (Heim & Miller, 2020).

13.3.1.4 Threat of Conflict

IFs computes the threat of conflict (**THREAT**) within each dyad (pair) of states in the system as a function of power, polarity, and power transition, but also as a function of many other influences. The Wiki in the IFs Help system describes threat in more detail (see also Fig. 13.6). When the dyadic version of the model is used, a parameter switch (**wpindsw**) with multiple settings can turn on new calculations of threat and conflict.

As a foundation for thinking about threat of conflict, it is important to understand and measure conflict itself. The IFs team looks to multiple projects and relevant threads for understanding it. One thread is Diehl et al.'s (2021) measure of rivalry, which helps us understand when the conditions for conflict are present. Another thread comes from the Uppsala Conflict Data Program and Peace Research Institute of Oslo's (UCDP-PRIO) Armed Conflict Dataset (Davies et al., 2024; Gleditsch et al., 2002) which provides an undirected measure of military conflict. The third thread, upon which our statistical analysis especially builds, provides a directed measure of interstate conflict and comes from Gibler and Miller's (2024)Militarized Interstate Events (MIEs) dataset. The MIEs build on the older

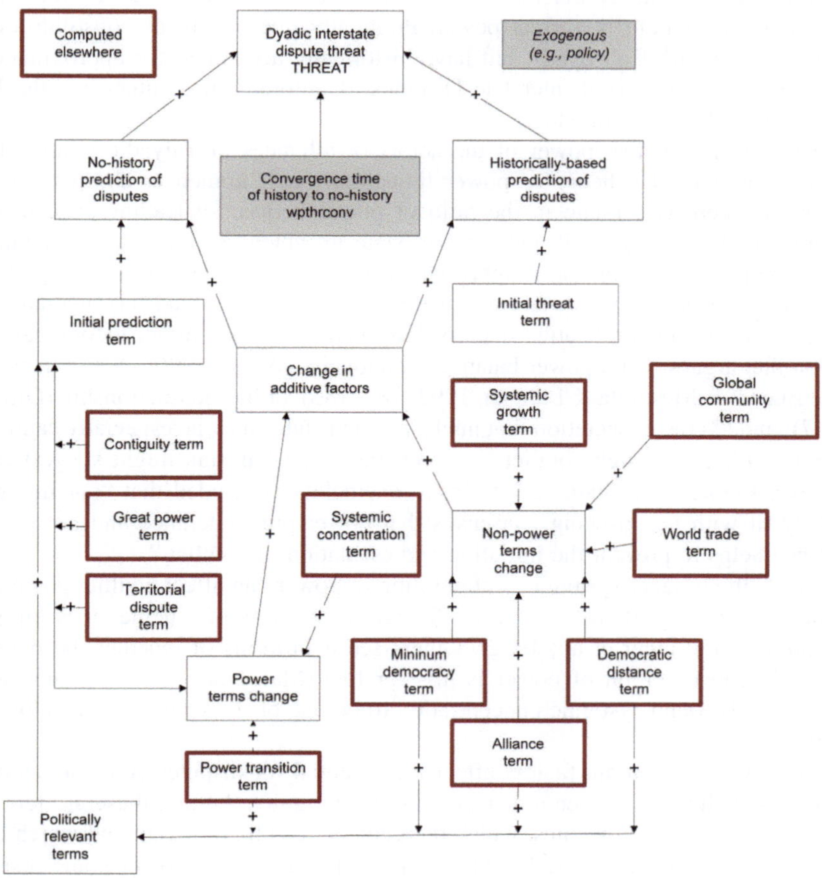

Fig. 13.6 Threat of interstate dispute. *Source* Author

and often used Militarized Interstate Disputes (MIDs) dataset (Bennett & Stam, 2004).

As of this writing, the IFs project has been reworking the forecasting of threat (**THREAT**) and of interstate conflict (Moyer et al., 2026). This effort will be ongoing with new data and analysis also from others. A parameter (**wpindsw**) allows switching between newer and older formulations for it. Ongoing model development and analysis takes advantage of the representations of power discussed above, but it also looks to many other variables.

There are three major ways in which power and its polarity are often understood to affect the threat of conflict. The first is through the absolute power of the actors, especially whether they are great or major powers in the global system. Great powers have been considerably more conflict-prone than other systemic actors. In fact, analysts of conflict often distinguish two types of dyads, calling one of them "politically relevant," conferring that label when the countries are contiguous and/or when one or both is a great power. Politically relevant dyads constitute only about one-eighth of all dyads, but have historically accounted for approximately 75 percent of Militarized Interstate Disputes. (Contiguity also enters into the IFs calculation of conflict threat.)

Second, the relative power of the actors or alliances in a dyadic relationship affects conflict. The theory of power transition (see Tammen et al., 2017) suggests that there is a range in the ratio of power between a leading state in the system (global or regional) and a rising state in which the potential for conflict rises sharply. That range is roughly between 0.8 and 1.2. We saw in Fig. 13.1 that many measures of power suggest that the China-U.S. dyad has entered that range. Somewhat complicating matters, however, are (1) a contrary theoretical tradition that argues that a power balance is conducive to less conflict between states (see Bueno de Mesquita & Lalman, 1992: 182–186), or has no relationship (Gibler, 2017); and (2) the expectation that nuclear armed states may be especially cautious about conflict (although conflicts between India and Pakistan might suggest otherwise). Bennett and Stam (2004: 164) nonetheless concluded that "our findings clearly fit with the growing consensus that power preponderance and not power balance helps to prevent the initiation and escalation of conflict."

Third, the broader systemic configuration of power can affect conflict potential. Some see the importance of this configuration in terms of whether it is bipolar or multipolar (Singer et al., 1972). Others see it in terms of whether the overall systemic concentration of power is high or low (Mansfield, 1994). Bennett and Stam (2004) found systemic concentration to be one of the strongest predictors of conflict.

Many other important factors affect the potential for disputes and wars among countries including, as mentioned, proximity/contiguity. Among these are territorial or other resource disputes, which theoretically can be resolved, and which can also arise very quickly. The IFs project looked to Huth (1996) for some insight and data here.

Much more beyond power, polarity, power transition, and contiguity affects the threat of conflict. There is another important perspective on the global political future that might also be considered realist, while shifting some of our attention from states to other actors. Advance of military technology, to the point where small and potentially rogue states, groups, and even individuals can kill thousands and perhaps millions of people, is a new global phenomenon. The destruction of New York's World Trade Center in 2001 and concerns about the movement of weapons of mass destruction into the hands of nonconventional global actors and terrorists suggest one possible character of system transformation. Further, in addition to individuals and groups, the last chapter noted the argument that cultural regions have become important globally, not just locally.

Discussion to this point has focused almost entirely on factors that tend to increase the threat of conflict. Yet, much analysis of global sociopolitical change focuses on factors that may decrease it. Mentioning those is a logical bridge to the discussion in the next section of liberalism. In general terms, observers at least as far back as Immanuel Kant have called for and identified a relationship between increased prosperity and global peace.

There is quite widespread agreement that level of democracy (more common among countries with higher income) affects the potential for conflict between states (Oneal & Russett, 1997; Ray, 1995). Analysis increasingly partitions the effect into two pieces (as recommended to the IFs project by Paul Senese). More democratic pairs, often measured in terms of the democracy level in the less democratic state of the pair, experience less conflict. Also, reduced difference in the political regimes of two states may reduce conflict (Leeds, 1999; Peceny et al., 2002).

The preponderance of empirical analysis further supports the proposition that increased trade (and perhaps financial flows) reduces conflict (Bennett & Stam, 2004), contributing with joint democracy to enhanced peace among states in the manner that Kant posited long ago. Most studies focus on trade specific to the dyad, generally using dyadic trade over GDP as a measure of trade dependence, and often focusing on the less dependent of the two trading partners (Oneal & Russett, 1997). Mansfield (1994) found that the *systemic* level of trade over GDP is also inversely and strongly related to war, and IFs incorporates a relationship tied to that finding. Another important force reducing conflict may be human development as discussed broadly across its facets in Chap. 8.

How does one pull together all this disparate work on the foundations of conflict? The preferred approach of many in the field of international relations is to do an estimation of all the factors and develop the best fitting, albeit complex, relationship. Relationships based solely on statistical analysis may, however, be badly behaved in long-term forecasting. Further, as we have discussed, and as Fig. 13.6 also suggests, the number of factors that shape threat and their interaction is incredibly complex. The approach in IFs has therefore been one that combines extensive statistical analysis with attention to a more algorithmic methodology that looks for what economists often call "stylized facts,"[4] or what also can be competing theories. Many questions surround past and ongoing development of IFs. How should conflict episodes be discounted in the more distant past?[5] Are there threshold points at which driving variables become important or their impacts saturate (or change direction)? What is the shape of relationships with variables individually and then collectively? Using information from many studies and quantitative analysis done specifically for IFs,[6] the project has built, and will keep refining, formulations for forecasting the threat of conflict based on statistics and more structural understandings or stylized facts from a broad range of research with respect to each of many drivers.

13.3.2 The Appeal of and Potential for Liberalism

As incredibly important as the study of power, power transition, and threat of conflict is in thinking about the future, much of this book and the work with IFs focuses on the possibilities and hopes for those more prosperous and peaceful scenarios of global change exemplified by the SDGs. Let's return to the potential for positive change, including the reduction of conflict threat among states.

Unlike the central role that power plays in understanding threat and potential for conflict in IFs representation of the elements of the global sociopolitical system that the realist perspective emphasizes, there is no single variable that does the same in the liberal perspective. There are, instead, many variables that collectively advance the prosperous and peaceful scenarios of the global future discussed in earlier chapters and called for in SDG 17. These include:

- The advance of democracy, because it provides a potential avenue for the broader population to shape societies in support of the general good. This was discussed in Chap. 12 (Domestic Sociopolitical Systems) and is used in threat formulations. So does advance in the status of women, also identified there and in Chap. 8 on Human Well-being.
- As agents for progress and prosperity, there is much that democratic (and also non-democratic societies quite frequently) can do, including domestic spending on education, health, R&D, and infrastructure.

---

[4] Stylized facts are rough-magnitude insights, put together using multiple studies, to enable reasonable statements about absolute and relative strength of variables. They are transparent and easily changeable and therefore often suitable for formulations in forecasting models.

[5] Based on historic MIDs data, each dyad of states was assigned an initial condition for conflict probability (sometimes called a "prior"). Declining weight is placed on older data (Crescenzi & Enterline, 2001); Crescenzi helped prepare the initial conditions in a process akin to the exponential smoothing often used in forecasting.

[6] The IFs team received advice and/or other help from Stuart Bremer, Mark Cresenzi, Doug Lemke, Edward Mansfied, and Paul Senese, none of whom can be blamed for any inappropriate use of it; Bennett and Stam graciously provided a pre-publication manuscript.

- In interaction with other states, national governments can provide foreign aid (see Chap. 7) and interact peacefully via the creation of embassies, the signing of treaties, and the participation in intergovernmental organizations. Although this chapter has noted that these actions also enhance power and that some are explicitly recognized in power measures (including POWERHHMI, POWERGPI, and POWER [DiME]), they can and have also facilitated peaceful interactions.
- Regional and global organizations, including the European Union and the United Nations, have become stronger. Moreover, there is a grouping of countries globally that constitute a "zone of peace." There has been no military conflict since 1945 among the rich and democratic countries of the European Union, North America, and Asia (Australia, Japan, and New Zealand).
- Within the formal security sphere, the North Atlantic Treaty Organization (NATO) progressively expanded its current 32-country membership by reaching out to former communist countries of Central and Eastern Europe and by recently adding Finland and Sweden. NATO has an agenda that extends beyond military issues to human rights and interstate cooperation. At the end of the Cold War, it reached out broadly with its Partnership for Peace, including to the Russian Federation. Also, the Organization for Security and Cooperation in Europe has a still larger membership of 57 states including the Russian Federation in another much less successful effort to bridge Europe, Central Asia, and North America.
- The global expansion of trade and foreign direct investment (FDI) was much more rapid than the growth of GDP for many decades (including through the first decade in this century), increasingly tying countries together and potentially making them more adverse to conflictual disruptions of those important linkages. Figure 13.7 shows the growth in global trade relative to GDP historically and the very uncertain IFs Base Case forecast for the rest of the century.

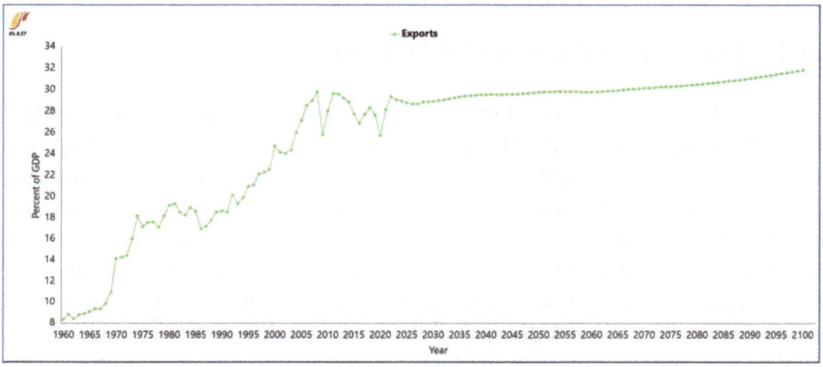

Fig. 13.7 Global trade as a percentage of GDP. *Source* IFs 8.57 Base Case with historical data from the World Bank

- With or without the cooperation among states, migration among them and the financial remittances that newer migrants often send back to former homes can be very beneficial for remittance- or returning migrant-receiving countries. (See the Population and Economics chapters on this.)
- States in interaction with each other, as well as states individually and households, are also often agents with power to protect the environment and support longer-term prosperity and well-being (see the Environment chapter on this). International non-governmental organizations (INGOs, including those focused on the environment) have supplemented and partnered with states and intergovernmental organizations (IGOs) in advancing well-being internationally.

Collectively, all such forces have, in fact, often protected peace and advanced prosperity globally. It is important that they are represented in IFs. Another very important dynamic has been the movement from divergence of economic growth rates in richer and poorer countries to convergence as those growth rates accelerated in developing countries. Although Chap. 2 pointed out some interruption in convergence, it now appears a likely long-term pattern.

13.4 Interventions and Scenario Analysis

As always, we begin discussing scenario analysis by identifying important indicator variables and then investigate our leverage with respect to them. The first subsection below focuses on the use of IFs for scenario analysis of changing power patterns globally, followed by a second subsection that turns to the issue of threats that countries pose to each other and the dangers of war. The final subsection returns to the potential for positive global sociopolitical change. Improving global futures is the primary thrust of most analyses for which IFs was created and of this book on its use. For information beyond that in this section, see the IFs Wiki, the IFs network diagram, and the Pardee Institute's home page.

13.4.1 Power Capabilities and Dynamics

As discussed earlier regarding Table 13.2, power (**POWERHH, POWERHHMI, POWERGPI, and POWER (DiME)**) is both an indicator of interest and a key determinant of other variables, including threat levels and warfare probability, which in turn drive government spending on the military and thus affect government finances more broadly. In *Flexible Displays*, the *Power* meso-category within the *International Relations* macro-category allows you to explore forecasts with all four variations of measurement. The ***wpwghtpow*** parameter allows you to change the formulation of **POWERHH** very flexibly across the driver variables identified in Table 13.2. Should you want to change those weightings, go into the *Scenario Tree*, do *Parameter Search*, *Load* ***wpwghtpow***, and pick the dimension for which you wish to specify a value.

→ One of the common debates about power measurement is the weight to assign population. Table 13.2 showed that population is one of the drivers of power in the **POWERHH** and **POWERHHMI** measures and discussion pointed out that with ***wpwghtpow*** you have control over it in the calculation of **POWERHH**. Some analysts even suggest that higher population, especially in poorer countries, should be considered a burden, not a contribution to power (Beckley, 2018a, 2018b). With attention to **POWERHH,** experiment with the weighting assigned to population and see how much difference it makes in the relative power of the United States and China and/or of India and Europe.

Perhaps the most analyzed power-related change in the contemporary global sociopolitical system is the growing power of China. Chinese leaders speak of the "peaceful rise of China," but many others are less sanguine.

→ Take time to explore in the Base Case scenario the relative power capabilities (economic, demographic, military, and aggregate) of China and other countries. As was done in Figures 13.1 through 13.3, undertake your own graphical analysis over a long period of time (at least until 2050) of the power of the United States, China, and India for one more of the measures of power in IFs. What is your perception about when power transitions are likely to occur?

→ How fast is the European Union overtaking the United States? When might China pass Japan in power? Could these be "trick" questions?

→ Table 13.2 showed that military spending is a component of the power calculation in all four measures in IFs. The only other variable with such ubiquity is some version of GDP or GDP per capita. Create an intervention for increased Chinese power based upon higher military spending (increase spending by 50 percent relative to the Base Case over five years and then hold it there). The parameter **gdsm** is a multiplier (Base Case values are 1.0) on all government expenditures and can be used in your scenario. Run the model for at least 20 years. The increase in power that China achieves in your scenario comes at the expense of other countries and regions (to see this you could compare the power [**POWER**] of other actors with their power in the Base Case). What implications does your scenario have for power transition?

→ Turn attention to the other driver of power common to the measures in IFs, namely GDP. Experiment with interventions to obtain a sense of the implication that alternative GDP growth futures have with respect to when power transitions might take place. The Economy chapter elaborated how to build alternative economic growth scenarios. For example, you may want to develop scenarios using the parameters **mfpleadr** or especially **mfpadd**. Again, population growth is a driver of **POWERHH** and **POWERHHMI**. You may wish also to explore alternative assumptions about its growth, for instance by using **tfrm** and/or **mortm**.

13.4.2 Threat and War

It bears repeating that power distributions and changes in them, the advance of democracy, and other system transitions and transformations are fundamentally important topics for analysis. At the same time, analysts of international politics look to them in significant part because they also affect levels of international threat and war. As difficult as it is to model threat and war, and as shaky as formulations in IFs or other models might be, it is important to think also about their dynamics.

13.4.2.1 Threat of Conflict

One foundational question in forecasting threat is the degree to which past conflicts should determine the initial probabilities versus relying on the drivers and their parameters alone to determine those values. For instance, analyzing the drivers of threat alone might suggest that the probability of conflict between the United States and Cuba was 5 percent or less. Yet, the historically based probability of conflict in the dyad, rooted in the Cold War and the continued hostility of the United States toward a communist regime, could suggest something considerably higher. Similarly, tying the model's base year value of threat to drivers in a formulation, not to history, could lead to a forecast of significant probability of conflict between the United States and Canada, in part because the literature on drivers says that contiguous states are most likely to fight. Further, using drivers alone might create a high forecast of conflict probability between the United States and the United Kingdom, because of their high power levels.

Ideally perhaps, the initial conditions for possible conflict might better be set by experts familiar with the countries in each dyad. In IFs, this would mean setting 188 times 187 values and is totally impractical, so historical conflict data are used as a default. The representation of the bilateral variable **THREAT** in IFs begins by recognizing the importance of recent and base year values of it for all country dyads. Initialization has moved from the MIDs dataset to MIEs data. In the forecasts of IFs, values over time slowly converge from that initial condition to the one predicted by the drivers, representing a supposition that history fades over time. The parameter to control the speed of this default rule is ***wpthrconv***. Setting ***wpthrconv*** to 1 means that driver-predicted values immediately replace historical ones. (Anyone interested in a particular dyad should look at its initial condition for **THREAT** in both directions [e.g., United States-Canada and Canada-United States] and could set values they feel appropriate.)

Moving to dynamics across time, our earlier discussion of both realist and liberal perspectives identified a great many interacting drivers of change in **THREAT** across time in IFs. Go to the *Scenario Tree* to change the parameters that determine the weighted contribution of those threat drivers. Most of those parameters begin with "**wp**" for world politics and a *Search* within *Self-Managed Display* will call them up. Illustratively, they include ***wpally***, ***wpcommun***, ***wpcon***, ***wpcontiguity***, ***wpdemdist***, and ***wptrade***.

In overview, the terms that contribute to **THREAT** and are controlled by these ***wp…*** parameters fall into the two general categories of the sociopolitical variables discussed in this chapter and across the volume more broadly. Those categories and their component elements as represented in model code are DeltaPowerTerms and DeltaNonPowerTerms.

In the DeltaPowerTerms category for each state in a bilateral relationship are (1) national power itself; (2) a term related to power transition in the bilateral relationship; (3) a territorial dispute term; (4) a term representing the degree of power in its contextual concentration among the great powers; and (5) an exogenous term with which the user can specify the extent of global community (***glocom***). In the DeltaNonPowerTerms category are (1) the democracy level of the less democratic country in the pair; (2) the democratic distance between them; (3) a trade level term; (4) a term specifying whether the states are allies; and (5) a GDP growth term.

→ Enhance or decrease the role of power transitions (use ***wppowtran1*** and ***wppowtran2***) in changing threat levels. The former controls the ratio (e.g., 0.9) of power between leading and challenging states at which the transition begins to especially augment threat of war. The latter increases and decreases the magnitude of the effect. How does your change affect levels of threat among China, India, Japan, and the United States? Assuming that the **THREAT** levels change, explore the implications for conventional war probability (**CWARPB**), for military and other governmental spending (**GDS**), and for other distal impacts of interest to you. Does this element of geopolitical relationships have broad impacts across the SDGs? If so, how?

13.4.2.2 Legacy Representation: Conventional and Nuclear War Probability

Warfare is, unfortunately, a part of global sociopolitical dynamics. Recognizing this, very early work on IFs made some effort to represent its probability (as discussed in this subsection) and even its horrific implications (as discussed in the next subsection). In recent decades, all development and IFs project usage of these representations ceased. The code remains in IFs, however, so these two subsections briefly explain their usage, while inveighing against significant use.

Warfare's primary driver in IFs is the threat of such conflict, as discussed above. War also, however, has other drivers, often in the form of shorter-term triggers, such as crises generated by the action of particular leaders or by unanticipated events. IFs, as a long-term model heavily tied to continuous change, cannot pretend to forecast such crises. Therefore, while useful in modeling the increased or decreased pressures for conflict, IFs is not appropriately used to study the actual likelihood of international conflict. Nonetheless, a parameter exists to turn such conflict on (***waron***). When the possibility of warfare is turned on within IFs, it will occur randomly, probabilistically based on the underlying levels of bilateral **THREAT**. When parametrically on, the model computes values for conventional and nuclear war probability (**CWARPB** and **NWARPB**) and its actual occurrence (**CWAR** and **NWAR**). The only significant value of the latter two variables, however, is to remind us that we do not want to risk such horrific outcomes of international interaction.

The probability of conventional war (**CWARPB**) increases in IFs with the threat level (**THREAT**) of militarized disputes between two actors. It is also affected by whether the two states are contiguous. **CWARPB** is expressed as a probability (0–1) instead of a percentage. Historically, there has been about one war per three disputes and the parameter ***wpthrwar*** translates **THREAT** into **CWARPB** with such a ratio.

13.4.2.3 Legacy Representation: Conventional and Nuclear War

No wars will occur in the IFs Base Case regardless of how high **THREAT** and **CWARPB** might be because the parameter that controls them (***waron***) is set to zero. It is turned off in the Base Case in part so that repeated runs of the base or other scenarios will always generate the same results, making them much easier to analyze. Because war is a probabilistic or stochastic event, even an unchanging probability of war across multiple model runs will generate different profiles of warfare when ***waron*** is activated. To activate ***waron***, set it to "1" in the tree. (Once ***waron*** is set at one, another optional parameter [***cwarf***] can be used to force a global war if it is set at 100 or more.)

When war is turned on, positive values of **CWARPB** will lead to incidences of conventional war (**CWAR** = 1 as opposed to **CWAR** = 0 for no war) among actors and their allies through the functioning of a random number generator. For example, values for **CWARPB** of 0.1 should generate about one war every ten years. When nuclear weapons are present, conventional war may also probabilistically escalate to nuclear war (**NWAR**), depending on **NWARPB**, as controlled by parameter ***nwarf***. The user can also change war probability with an exogenously specified baseline shift in the probability of war (***cwarbase***). In the Base Case, all the ***cwarbase*** shifts in war probabilities are set to 0.0; therefore the probability of conventional war in the Base Case and in all scenarios for which you specify no baseline probability is driven solely by the level of **THREAT**.

Having explained the possibility of turning on actual representation of overt warfare in IFs, it bears repeating: as important as studying the **THREAT** of overt conflict is, the manifestation of that threat in the form of war is not something we believe we can truly forecast. We have put no real effort into doing so. The only real value of overt warfare parameters and variables is to remind us that *we do not want to risk such horrific outcomes of international interaction.*

13.4.3 Positive System Transformation and Interactions

It is, of course, not just power and changes in its distribution that shape the global sociopolitical landscape. The character of the countries of the world and their broader patterns of interaction are extremely important. The previous chapter on

Domestic Sociopolitical Systems directly discussed the first of these, including via the chapter's attention to the growth over time, albeit not without setbacks, of democracy. Yet, all earlier chapters, including those treating demographics, economics, human well-being, and the environment, and also Chap. 2 reviewing historical change, addressed the dramatically changing character of countries across time.

What about the interactions of these dynamic societies? Those, too, have been focal points for much previous attention. That includes discussion of migration in demographics; of trade, foreign aid, and international investment in economics; as well as the collective impact of societies on the environment. IFs and its use are fundamentally focused on change across almost all aspects of societies and their interactions. What was not discussed earlier, namely the implications of societal change for national security, peaceful interactions, and global cooperation have been the topic of this chapter, with its focus on power, power transition, threat, and conflict—but also on cooperation—and their implications in turn for societal change. The liberal perspective throughout the post-World War II period has placed much emphasis on the advances of democracy, cooperative global economic interaction, and, most recently, an even broader collaboration on a still wider range of issues like those identified in the SDGs.

We should at least comment, however, on how the countries of the world might collaborate directly on the issue of threat reduction. One venue for collaboration that draws attention from both liberal and realist perspectives and that we have not discussed is arms control, including collective attention to military spending, which can be prominent in such collaboration. For instance, in the late 1980s the Soviet Union attempted to defuse or even reverse its arms race with the United States by unilaterally announcing some measures of restraint, including a ban on nuclear testing. The United States and its NATO allies initially displayed distrust and did not react. The reactivity parameter of the real world remained high with respect to conflict but low with respect to cooperation (the parameter in IFs is **reac**). Eventually, however, the NATO members began to perceive the efforts to be sincere, even while they interpreted them as self-interested attempts to reduce strain on a weakening Soviet economy. Gradually, NATO began to respond in kind and a certain amount of trust emerged. Unfortunately, in more recent years military actions by the Russian Federation have reversed these dynamics.

→ Explore a scenario involving a broad-scale, even global process of arms reduction. In the Base Case, display **GDS** for military spending (the amount of government spending on the military) for key countries and groupings within IFs. Use the multiplier parameter on that spending (**gdsm**) to significantly reduce such spending, at least for key countries. It is probably easier just to reduce spending for the World. Look at results in many areas. How does it affect economic performance (**GDP**)? How does it affect power balances (**POWER**)? And how does it affect **THREAT** levels globally and in particular dyads?

→ Even if drivers of power do not change, it is possible that the reactivity of countries to the power of other countries might change and thereby affect the threat levels. The **reac** parameter allows you to raise or lower such reactivity relative to the base value of 1.0. Try lowering (or raising) the parameter. How greatly does it affect not only **THREAT** but also military spending and much else in the global future?

13.5 Conclusion

No chapter in this volume can stand alone—all the issues have been related. In this chapter, we have made specific references to the ways in which demographic and economic issues affect the global sociopolitical system (for instance, through the contributions that demographic size makes to power or threat and to the manifold impacts of economic performance on the system). We could easily have talked about the importance of food supply, energy (both as an input to economies and as a source of potential conflict among societies), and the environment.

Through the accumulated exploration of multiple issues that cry out for interacting system analysis, this and earlier chapters have appropriately led us to a point that encourages stepping back and looking at the system in a fully integrated fashion. The last exercise in this chapter could logically have been the development of an integrated scenario across issue areas, perhaps a normative one in which we attempt to describe the world we would like to see. That is, however, a large exercise. Thus, instead of doing it here, we have reserved the next and final chapter for precisely that activity.

References

Allison, G. (2017). *Destined for war: Can America and China escape Thucydides's trap?* HarperCollins.

Anderson, J. E. (1979). A theoretical foundation for the gravity equation. *The American Economic Review, 69*(1), 106–116. http://www.jstor.org/stable/1802501?origin=JSTOR-pdf

Armington, P. S. (1969). A theory of demand for products distinguished by place of production. *Staff Papers (International Monetary Fund), 16*(1), 159–178. https://www.jstor.org/stable/3866403

Beckley, M. C. (2018a). *Unrivaled: Why America will remain the world's sole superpower*. Cornell University Press.

Beckley, M. C.(2018b). The power of nations: Measuring what matters. *International Security, 43*(2), 7–44. https://doi.org/10.1162/isec_a_00328

Bennett, D. S., & Stam, A. C. (2004). *The behavioral origins of war.* University of Michigan Press. The authors were kind enough to provide a pre-publication manuscript in 2001 and page references are to that manuscript.

Bueno de Mesquita, B., & Lalman, D. (1992). *War and reason.* Yale University Press.

Chan, S. (2007). *China, the US and the power-transition theory: A critique.* Routledge. https://doi.org/10.4324/9780203940662

Crescenzi, M. J. C., & Enterline, A. J. (2001). Time remembered: A dynamic model of interstate interaction. *International Studies Quarterly, 3*(September), 409–431. https://www.jstor.org/stable/3096085

Davies, S., Engström, G., Pettersson, T., & Öberg, M. (2024). Organized violence 1989–2023, and the prevalence of organized crime groups. *Journal of Peace Research, 61*(4), 673–693. https://doi.org/10.1177/00223433241262912

Diehl, P. F., Goertz, G., & Gallegos, Y. (2021). Peace data: Concept, measurement, patterns, and research agenda. *Conflict Management and Peace Science, 38*(5), 605–624. https://doi.org/10.1177/0738894219870288

Gibler, D. M. (2017). State development, parity, and international conflict. *American Political Science Review, 111*(1), 21–38. https://doi.org/10.1017/S0003055416000514

Gibler, D. M. & S. V. Miller. (2024). The Militarized Interstate Events (MIE) dataset, 1816–2014. *Conflict Management and Peace Science 41*(4). https://doi.org/10.1177/07388942231182674

Gleditsch, N. P., Wallensteen, P., Eriksson, M., Sollenberg, M., & Strand, H. (2002). Armed conflict 1946–2001: A new dataset. *Journal of Peace Research, 39*(5), 615–637. https://doi.org/10.1177/0022343302039005007

Heim, J. L., & Miller, B. M. (2020). *Measuring power, power cycles, and the risk of great-power war in the 21st century*. RAND Corporation.

Hughes, B. (2015). *IFs economic model documentation*. Frederick S. Pardee Institute for International Futures, Josef Korbel School of Global and Public Affairs, University of Denver. https://korbel.du.edu/pardee-resources/ifs-economic-model-documentation/

Huth, P. (1996). *Standing your ground: Territorial disputes and international conflict*. University of Michigan Press.

Leeds, B. A. (1999). "Domestic political institutions, credible commitments, and international cooperation. *American Journal of Political Science, 43*(4), 979–1002. https://doi.org/10.2307/2991814

Mansfield, E. D. (1994). *Power, trade, and war*. Princeton University Press.

Meisel, C. J., Petry, C. L., Moyer, J. D., & Burrows, M. J. (2023). *Spheres of influence in the coming decades: Four alternative scenarios*. Stimson Center.

Moyer, J. D., & Markle, A. (2018). *Relative national power codebook* (SSRN Scholarly Paper No. 4712180). Social Science Research Network. https://doi.org/10.2139/ssrn.4712180

Moyer, J. D., Meisel, C. J., & Matthews, A. S. (2023). Measuring and forecasting the rise of China: Reality over image. *Journal of Contemporary China, 32*(140), 191–206. https://doi.org/10.1080/10670564.2022.2071879

Moyer, J. D., Meisel, C. J., Matthews, A. S., Bohl, D., & Burrows, M. J. (2021). *China-US competition: Measuring global influence* (SSRN Scholarly Paper No. 3941475). Social Science Research Network. https://doi.org/10.2139/ssrn.3941475

Moyer, J. D., Petry, C. L., Meisel, C. J., & Solórzano, J. (2025). Green wars: Will a renewable energy transition cause international conflict? *(Under Review)*.

Moyer, J.D., Meisel, C.J., Petry, C.L., & Solórzano, J. (2026). The Geopolitics of Decarbonization: Renewable Energy Transitions and International Conflict *(Under Review)*.

Oneal, J. R., & Russett, B. M. (1997). The classical liberals were right: Democracy, interdependence, and conflict, 1950–198. *International Studies Quarterly, 2*(June), 267–294. https://www.jstor.org/stable/3013934#:~:text=https%3A//www.jstor.org/stable/3013934

Organski, A. F. K. (1958). *World politics*. Knopf.

Peceny, M., Beer, C. C., & Sanchez-Terry, S. (2002). Dictatorial peace? *The American Political Science Review, 96*(1), 15–26. https://www.jstor.org/stable/3117807

Ray, J. L. (1995). *Democracy and international conflict*. University of South Carolina Press.

Richardson, L. F. (1960). *Arms and insecurity*. Quadrangle Books.

Santos Silva, J. M. C., & Tenreyro, S. (2011). Further simulation evidence on the performance of the Poisson pseudo-maximum likelihood estimator. *Economics Letters, 112*(2), 220–222. https://doi.org/10.1016/j.econlet.2011.05.008

Singer, J. D., Bremer, S., & Stuckey, J. (1972). *Capability distribution, uncertainty, and major power wars, 1820–1965* (B. Russett, Ed.). Sage.

Small, M., & Singer, J. D. (1982). *Resort to arms*. Sage Publications.
Tammen, R. L., Kugler, J., & Lemke, D. (2017). Foundations of power transition theory. In *Oxford research encyclopedia of politics*. https://doi.org/10.1093/acrefore/9780190228637.013.296
Thucydides. (431 B.C.E.). *The history of the Peloponnesian war*. The Gutenberg Project. https://www.gutenberg.org/files/7142/7142-h/7142-h.htm
Tinbergen, J. (1963). Shaping the world economy. *The International Executive, 5*(1), 27–30. https://doi.org/10.1002/tie.5060050113
Zinnes, D. A., & Gillespie, J. W. (Eds.). (1976). *Mathematical models in international relations*. Praeger.

Part VI

Integrated Scenario Analysis

Preferred (and Other) Futures 14

Abstract

We began this volume by identifying three questions. What kind of future would we prefer? Where do current changes appear to be taking us? And what leverage do we have to bring about a future we prefer? Building blocks for helping address those have been (1) identifying current trends and emphasizing causal analysis of global dynamics; (2) elaborating the structure and capabilities of International Futures (IFs) as a tool for analysis of global dynamics and possible strategic intervention points; (3) progressively analyzing human development, sustainable development, and sociopolitical change. This final chapter turns to more fully integrated analysis of global change. It includes a very brief survey of ongoing strengthening of the foundations for analysis with the evolution of data, trend analysis, and models. It then turns to some of the many projects that have developed and elaborated issue-integrated scenarios, both dystopian and desired. Key elements of thinking and analysis that shape such scenarios are discussed. As a final exercise, users are invited to elaborate their own integrated scenarios of where global change appears to be taking us and how, with reasonable but possibly aggressive strategic interventions, we collectively might help bring a preferred future into being.

14.1 Introduction

In thinking about the future, it is possible to ask an overarching question "What might the future of human development, environmental, and sociopolitical systems be throughout the twenty-first century? An obvious answer to that comes quickly:

Jonathan Moyer, Director of the Pardee Institute for International Futures, is a co-author on this chapter.

B. B. Hughes, *Exploring and Shaping International Futures*, World-Systems Evolution and Global Futures, https://doi.org/10.1007/978-3-032-22084-4_14

"No one truly knows." That simple answer creates a serious challenge. Personal and collective actions, even decisions not to act, will affect the future in potentially very significant ways, perhaps not moving it toward ones that we would like to see unfold.

It therefore makes sense to improve our thinking about the future as much as possible. It helps to decompose the question into three somewhat more manageable ones: What kind of future would we **prefer**? Where do current **patterns** appear to be taking us? What **leverage** do we have in pursuing our preferred future?

Chapter 1 set the stage by discussing the earlier Millennium Development Goals (MDGs) and the contemporary Sustainable Development Goals (SDGs), which collectively define a preferred global future of sustainable human development. It also quickly described some of the major patterns and transformations reshaping our world. In doing so it stressed that, while we can usefully look at and extrapolate trends, we also need to understand causality. Finally, it introduced the International Futures (IFs) system and our usage of it both to strengthen understanding of causality and to explore the leverage that we and our societies might have in pursuit of preferred futures.

Chapter 2 added to our understanding by quantitatively surveying long-term patterns of change and their apparent direction. Chapter 3 turned to the importance of our values and goals and of dynamic, integrated understandings concerning the future, while Chaps. 4 and 5 elaborated the International Futures modeling system. Subsequent chapters explored causality and leverage with respect to global change, focusing on specific issue areas and progressively expanding the capability of IFs users to analyze the integrated interactions among them.

The purpose of this, the final chapter, is to turn back to integrated analysis of the prospects for sustainable human development. It first looks very briefly at some of the progressive strengthening of foundations for forecasting and foresight in data enhancement, trend analysis, and integrated global modeling. It then reviews some of the scenario development that has unfolded over time. Those scenarios have tended to be primarily focused on either environmental sustainability or sociopolitical change. Yet some scenario analysis has considered the big picture of scenario development, looking broadly at sustainable human development.

Following this review of some foundational work by others in Sects. 14.2 and 14.3, we turn in Sects. 14.4 and 14.5 to your own possible use of IFs in wide-ranging analysis of possible and preferred futures. You can, if desired, jump directly or at any time to that use.

14.2 Foundations for Understanding Global Change

We have explored two primary foundations for understanding global change and where it might be taking us. The first involves studying historical data and looking at trends and patterns, recognizing that they may have a variety of shapes going forward (Chap. 2). The trends very often alert us to issues of importance, both positive change and change that concerns us. The second involves digging into causality that drives those trends and their interactions. The rise of integrated assessment models (IAMs) like IFs has given us important toolkits for that. Some brief commentary on the progressive strengthening of those two foundations is useful before we turn to scenario analysis built on them.

14.2.1 Data, Trends, and Identification of Issues

There is a strong foundation for forecasting in regular appraisals of challenges facing, and progress being made in, the push for sustainable human development. Critical elements of that foundation lie in the explosion of data series and derivative indices that cover effectively all countries of the world across all the concepts and variables that underlie goal sets such as the SDGs and modeling such as that of IFs. Most of that did not exist prior to the end of World War II and the collapse of major empires and their colonialism. The International Monetary Fund began publishing its great many series on domestic and international finance in 1948. It is no accident, however, that many data series, including many of the approximately 1,600 time series constantly updated by the World Bank in its World Development Indicators, begin around 1960 with the independence of African countries.

Building on such data foundations, trend analysis has proliferated, especially with respect to economics. For instance, the chief economists of the United Nations (United Nations Economist Network) system came together in 2020 to identify and analyze (but not forecast) "five human-made megatrends which … dominate and frustrate global efforts…" The five are climate change; demographic shifts (especially aging); urbanization; emergence of digital technologies (which now must include artificial intelligence); and high and often rising inequalities (United Nations Economist Network, 2020).

Other UN analysis across multiple issue areas is sometimes issued under the rubric of "Critical Trends." For instance, the Department of Economic and Social Affairs (UN DESA) published *Global Challenge, Global Opportunity: Critical Trends in Sustainable Development* in support of the 2002 Johannesburg Summit. DESA also issues an annual World Social Report. In 2025, in collaboration with the United Nations University's World Institute for Development Economics Research (UN-WIDER), the report emphasized both the progress made over the previous three decades and the remaining challenges related to (1) insecure livelihoods even for those now living on as much as $6.85 a day; (2) persistent inequality with social, economic, and political exclusion; and (3) declining trust and weakening social cohesion with increasing polarization (United Nations, 2025).

Several intergovernmental organizations (IGOs) often issue very useful issue-focused forecasts that are generally short term. For example, the World Bank produces regular *Global Economic Prospects.* The International Monetary Fund provides its *World Economic Outlook* normally twice a year, with quantification of economic prospects a few years out that the IFs system draws upon. The Organisation for Economic Co-operation and Development similarly provides its *Economic*

Outlook with nearer-term horizons. Some reports, like those of the United Nations Human Development Report Office, look further out and those of the United Nations Population Division peer out across the century.

Outside of the IGO community, think tanks exist that look regularly at possible global futures. One preeminent organization is the Millennium Project, linking some 70 "nodes" of experts around the world. It produced its twentieth *State of the Future* report in 2024, focusing on 15 global challenges (Glenn et al., 2024). That edition continued the tradition of reporting on all those challenges across sustainable development dimensions. It also provides updated quantification of the organization's State of the Future Index, with projection to 2035. The 2024 edition of the volume focused special attention on the great challenges posed by rapid and largely uncontrolled advance in artificial intelligence. Jerome Glenn has put much effort into generating national and international policy toward protection of humanity from its dangers.

Massive datasets have also emerged across sociopolitical variables. For instance, the Inter-university Consortium for Political Research (ICPSR), based at the University of Michigan, was founded in 1962. It links more than 800 academic and research bodies and archives hundreds of thousands of research files in the social and behavioral sciences. Similarly, the Harvard Dataverse includes more than 200,000 data series across a much wider set of disciplines. Much more data-gathering and data-making exists around the world, including the Pardee Institute's own creation of data and indices, which are accessible in the Dataverse. The multiple projects within the Institute's Diplometrics program include (1) Country and Organization Leader Travel (COLT), cataloguing international visits by heads of government since 1990; (2) United Nations voting since 1946; and (3) Diplometrics Diplomatic Representation, also since 1946. Data visualizations and data access for these and other Pardee data projects are at https://korbel.du.edu/pardee/data-visualizations/.

Such foundations have made global quantitative trend analysis and statistical causal analysis possible. They have been a significant impetus to the development earlier of the MDGs and more recently of the SDGs and to global modeling. For instance, IFs includes nearly 7,000 data series, available to all users.

14.2.2 Causality: Global Modeling

With the proliferation of datasets and the identification of concerns about some of the trends in them, the foundation and the incentives have grown for undertaking analysis and developing computer models looking across multiple issues and providing longer-term forecasts.

Given the great significance of ongoing global transformations for coming decades and even centuries and given also the widespread recognition of the interactions among multiple changes, one might expect that large numbers of scholars and policy organizations would devote attention to studying them and forecasting their unfolding in an integrated fashion. Surprisingly, however, there are relatively few who attempt to undertake longer-term global forecasting across multiple interacting issue areas.

Chapter 1 and this volume's Appendix 2, summarizing the history of IFs, point to how the earliest such models in the 1970s, popularized by Club of Rome reports, demonstrated the benefits of building and using such systems. The *Limits to Growth* volume (Meadows et al., 1972) used the World3 model. *Mankind at the Turning Point* (Mesarovic & Pestel, 1974) used the World Integrated Model (WIM) developed under the leadership of Mihajlo Mesarovic. The limits identified were fundamentally environmental, including both food and fossil fuel energy, in the face of rapid population and economic growth.

Another early and separate thrust in global modeling was focused heavily on sociopolitical systems. Reaching back to the lifetime work of Lewis Fry Richardson (1960) on equations representing arms races and to Harold Guetzkow's (1959) development of the computer-assisted Inter-Nation Simulation, Karl Deutsch put in place a team at the Science Center in Berlin to develop the GLOBUS model in the 1979–1988 period (Bremer, 1987). GLOBUS was able to draw on large databases of international interactions, including work at the University of Michigan (Small & Singer, 1982) and at the Science Center. That project linked global sociopolitical change to demographic and economic change. The economic model of GLOBUS (Hughes, 1987) drew on WIM and was a very early precursor to that in IFs.

On the more purely environmental side, the First Assessment Report of the Intergovernmental Panel on Climate Change (IPCC) was released in 1990 with a **Business-as-Usual** scenario. It drew on largely physical system models of atmospheric general circulation, the oceans, and the carbon cycle.

Motivated significantly by such concerns about global climate change and other environmental issues, such as deforestation and species loss, a significant number of complex models with attention to the drivers of such physical change (including demographics and economics) were developed in the following decades and became known as integrated assessment models (IAMs). Those efforts led in 2007 to the creation of the IAM Consortium. Early efforts by its members provided a foundation for the development by 2008 of four scenarios focused on the long-term future of global and more local environments. Called **Representative Concentration Pathways** (RCPs), these scenarios began to be used by the IPCC in its Fifth Assessment Report (AR5) during 2014.

The movement of large-scale global modeling into treatment of human development issues such as poverty, health, and education is more recent (Lutz et al., 2008). As this volume has emphasized, the MDGs and SDGs have given great additional attention to and impetus for modeling of human, sociopolitical, and environmental development analyzed broadly.

At this point, however, global models that integrate human development with environmental and/or geopolitical analysis are still rather scarce. Much effort within the IAM community, including the SHAPE project (Sustainable development pathways achieving Human well-being while safeguarding the climate and

Planet Earth) that looked across the SDGs, has relied on multiple models rather than an integrated one. Bjoern Soergel helped make such multiple model analysis happen in SHAPE (Soergel et al., 2024).

14.3 Long-Term Integrated Global Scenario Analysis

As suggested already, integrated scenario analysis has evolved and built on dataset creation, trend analysis, issue identification, and model development. It is useful to understand some of the scenario analysis built on those foundations.

14.3.1 Early Projects

Given that the attention of the early Club of Rome models and reports was heavily on environmental challenges (especially food supply and fossil energy in combination with rapid population growth), the scenarios focused very heavily on economic and demographic drivers of those challenges. The *Limits to Growth* study, with its representation of the world as a whole and limited variable elaboration, reported on 12 scenarios that explored aggregate interventions affecting resource depletion, population growth, industrial output, food production per capita, and a pollution variable. *Mankind at the Turning Point* included four scenarios addressing economic drivers (with attention to reducing global economic gaps across 12 world regions) and seven scenarios focused on the timing and degree of population policy. A chapter also addressed (but did not model) the issue of global confrontation versus cooperation.

Given the importance of the Cold War at the time of its work, the GLOBUS project devoted considerable scenario analysis to the extent of hostility in the U.S.-U.S.S.R. relationship. One elaboration of such analysis explored the impact of trade protectionist policies on the global political climate. Other scenarios explored the impact of population growth and the implications of technological progress for living standards.

Near the same time, using increasingly sophisticated modeling, much attention in scenario analysis remained heavily on environmental issues. Conceived in 1991, the PoleStar project and its model (the PoleStar System) had roots in the 1987 Brundtland Commission report *titled Our Common Future* (United Nations, 1987) and put environmentally sustainable development at the center of model development. Nonetheless, its scenario analysis drew in geopolitical and even some human development elements, especially cross-country inequality. There was a strong relationship between PoleStar and the Global Scenario Group (GSG) organized by the Stockholm Environment Institute in 1995. One can see the linkages of environmental and sociopolitical elements in its scenarios as elaborated under long leadership by Paul Raskin in *Great Transition: The Promise and Lure of the Times Ahead* (Raskin et al., 2002). Briefly, the six GSG scenarios, earlier discussed by Hammond (1998), fall into three general categories:

- **Conventional Worlds (with Market Forces and Policy Reform).** Rapid technological innovation, economic reform, and economic growth. Integrated world markets. Widespread prosperity, peace, and stability, but also pervasive inequality and growing environmental problems.
- **Barbarization (with Fortress World and Breakdown).** Failure of market-led growth to redress social wrongs and environmental disaster. Inward retreating within and across countries and regions. Widening inequality, conflict, and instability.
- **Great Transitions (with Eco-communalism and New Paradigm).** Power is more widely shared and social coalitions work at grass-roots level. Markets are effective but do not substitute for deliberate social choices. Fundamental change for the better in politics, social institutions, and the environment.

The PoleStar Project was in many ways ahead of its time in its efforts to somewhat bridge the human development, environmental sustainability, and sociopolitical dimensions of global futures analysis. Nonetheless, in the first quarter of the twenty-first century much world modeling and scenario analysis have remained primarily focused on environmental or sociopolitical issues. These analyses have recognized the implications of human activity for issues of their central focus and also the general implications of changes in the environment and sociopolitical conditions for humanity. They have, however, generally not elaborated aspects of development in the human condition, the focus of the first set of SDGs (discussed in Chap. 8).

14.3.2 Scenarios Heavily Focused on Environmental Analysis

The United Nations Environment Programme (UNEP) began issuing Global Environment Outlook reports (GEOs) in 1995. Even GEO-1 used model analysis across seven global regions, building on the modeling and scenarios of the Global Scenario Group. Work of the GSG and the PoleStar Project influenced many other environmental analyses, including the World Water Council's Water Vision report in 1999–2000, the OECD Environmental Outlook in 2001, and the Millennium Ecosystem Assessment in 2005.

Other modeling projects undertaken by members of the IAM Consortium mostly illustrate the ongoing and even increasing importance of environmental analysis. As indicated, members of the IAMC released Representative Concentration Pathways (RCPs) scenarios around 2008. Those map out four alternative futures for atmospheric concentrations of greenhouse gases. They began to be used by the IPCC with the Fifth Assessment Report (AR5) in 2014.

Analyses with the PoleStar model[1] of the Stockholm Environmental Institute together with the IMAGE model (Integrated Model to Assess the Global Environment) at PBL Netherlands Environment Assessment Agency and other models were used in support of UNEP's third Global Environment Outlook (GEO-3), to provide some integration across the multiple issue areas of that analysis. The GEO-3 analysis focused on four scenarios:

- **Markets First.** Trust is placed in globalization and economic liberalization processes. Governments are unable to cope with social and environmental issues.
- **Security First.** Great disparities across countries and environmental pressures give rise to conflict; the rich and powerful separate themselves from the broader society.
- **Policy First.** Governments take decisive initiatives toward environmental and social goals, but the steps are not fully adequate in addressing the problems.
- **Sustainability First.** A new paradigm emerges. New values and institutions support changes in behavior and policies on social and environmental issues.

UNEP's GEO-4 process built on the same four scenarios and the IFs system helped provide drivers and an integrated structure for analysis.[2] Versions of IFs at that time included prerun versions of these four scenarios with its installations, just as versions as of this writing do with the Shared Socioeconomic Pathways (SSPs). Future versions of IFs will almost certainly continue to have prerun versions of these or other scenarios. (Should you wish to run them now, the.sce (parameter) files for the GEO-4 scenarios can be found under *IFs/Scenario Small/World Integrated Scenario Sets/UNEP GEO/GEO4.*)

GEO-6 (UNEP, 2019) used a combination of top-down (model-based) and bottom-up (more qualitative) analysis for scenarios. For example, its chapters on "Outlooks and Pathways to a Healthy Planet with Healthy People" used multiple models and qualitative analysis to address potential progress toward many of the SDGs. IFs was used in looking, for instance, at under-5 mortality (Moyer & Hedden, 2020). The scenarios used were mostly those of the Roads from Rio + 20 project (van Vuuren et al., 2012): **Global Technology**, **Decentralized Solutions**, and **Consumption Change** with some attention to the SSPs.

The earlier Millennium Ecosystem Assessment (MA) project of the United Nations focused heavily on the environment while also being attentive to the role of human activity in environmental change and the importance of environmental services for humans. It produced its final reports in 2005. Although the

[1] See http://www.seib.org/polestar/Projects.html; January 16, 2004.

[2] The World Business Council for Sustainable Development (1997, 1999, 2000) also provided additional integrated scenarios with a heavy environmental emphasis. See IEA (2003: 179–185) for a good review.

work emphasized assessment, it also included forward-looking scenarios that overlap significantly with those of the IPCC and the GEO-3 and GEO-4 processes. Specifically, those were:

- **Global Orchestration**. A combination of a globalized world and one that is attentive to environmental issues, equity, and human development concerns.
- **Order from Strength.** Another variation of the GEO-3 **Security First** (or **Fortress World**) scenario sketched earlier in this section.
- **Adapting Mosaic**. Local developments with contextually-appropriate, environmentally-sensitive approaches.
- **Technogarden.** Global connections with a technological focus on sophisticated approaches to environmental management and strong provision of ecosystem services.

Supporting the MA's final report were multiple soft-linked (i.e., separately used with exchanges of quantitative analyses) IAMs: IMAGE 2.2; IMPACT; WaterGap; AIM; and Ecopath.

More generally, the integrated assessment modeling community, with its models and scenarios (the SSPs prominent among them), leads a large-scale global effort by those within the Integrated Assessment Modelling Consortium to elaborate a preferred future with environmental sustainability. Prominent among those models has been IMAGE, with work led by Detlef van Vuuren (van Vuuren et al., 2012, 2015, 2019). Also greatly important in those efforts has been the work of Elmar Krieger and others at the Potsdam Institute for Climate Impact Research (Kriegler et al., 2012) and that of Keywan Riahi and others at the International Institute for Applied Systems Analysis (IIASA), see Riahi et al. (2017). Other modeling projects of importance do now exist around the world. Those efforts made possible the development of the SSP scenarios, with leadership by Brian O'Neill (O'Neill et al., 2013), which this volume has repeatedly referenced and used.

14.3.3 Scenarios Heavily Focused on Sociopolitical Change

In the sociopolitical arena, scenarios can have many elements of the environmentally-oriented ones above, but they typically much more heavily emphasize variability/uncertainty in the roles of key potential actors, especially states. Analysis with GLOBUS was an early exemplar. Scenario analysis in this arena usually draws less heavily on global models.

The Hart-Rudman Commission, more formally the United States Commission on National Security/twenty first Century (1999), began in 1998 to develop reports to accomplish what it described as a comprehensive government-sponsored review of U.S. national security. Although most of the Commission's reporting looked at quite specific issues and geographic regions, and although it argued that the future will likely have a patchwork of elements, it concluded its 1999 Phase I report on *The New World Coming* with a brief elaboration of four scenarios:

- **A Democratic Peace.** Democratic norms predominate, there is no sharp ideological conflict, and political cooperation among states is high. The scenario is built on assumptions of moderate economic growth and a continuing information revolution but also posits global inequalities and other challenges.
- **Protectionism and Nationalism**. Global economic crisis or other factors lead to the rise of nationalism and of regional power blocs. The developing world suffers economic decline in the face of decreasing trade and financial flows. Nationalism, ethnicity, and fundamentalism direct some politics and even governments.
- **Globalization Triumphant**. Rapid progress in technology and global growth in economies allows developing countries to create larger economies than those in the developed world. There is widespread commitment to more equitable income, to development of human capital, to peaceful conflict resolution, and to sustainable environmental policies.
- **Division and Mayhem**. Diffusion of dangerous technologies, breakdown of globalization processes, or environmental crisis could give rise to this world. Internationalism breaks down, with restrictions on trade and on financial and information flows. Many countries fragment ethnically, culturally, or religiously. Terrorist attacks on U.S. cities reorient U.S. policy toward preventing more attacks and away from broader global engagement, giving rise to a downward spiral.

Also largely in the sociopolitical arena, the Forward Studies Unit of the European Commission (1999) put forward five possible futures with a focus on Europe. These include **Triumphant Markets** (another victory of globalization scenario); the **Hundred Flowers** (less successful and inegalitarian globalization with rising dissatisfaction and divisiveness); **Shared Responsibilities** (success of the European model on the continent and more generally); **Creative Societies** (reduced governmental footprint in Europe, ultimately leading to more creative governance and a renewed society); and **Turbulent Neighbourhoods** (ethnic wars, terrorism, and criminality around the world). One can see similarities with the Hart-Rudman scenarios.[3]

Also in the sociopolitical arena, the U.S. National Intelligence Council (U.S. NIC), providing integrated strategic thinking for the head of U.S. intelligence and the broader U.S. government, prepared a report on *Global Trends 2010* (U.S. NIC, 1997). It followed that through 2021 with six more *Global Trends* reports on a four-year cycle with scheduling roughly planned to prepare them in time for a new or returning president. These reports, not continued in 2025, quite routinely

[3] The Millennium Project of the United Nations University has developed and published many scenarios over time, including the following: Cybertopia (free trade, globalization, and strong technology); the Rich Get Richer (global inequalities and environmental problems); a Passive Mean World (barriers to trade and other connections); and Trading Places (countries like China surge forward).

included several scenarios with a heavily geopolitical orientation but with attention also to much broader underlying trends.

It can be of interest to contrast the scenarios identified in the third NIC report, *Global Trends 2020: Mapping the Global Future* (U.S. NIC, 2004) with those in the seventh and perhaps final such report (U.S. NIC, 2021), titled *Global Trends 2040: A More Contested World*. The third report identified demographics, natural resources and the environment, science and technology, the global economy and globalization, national and international governance, future conflict, and the role of the United States as key drivers. But it identified globalization processes as a meta-driver. The scenarios of the project were:

- **Davos World.** A heavily globalization-driven and mostly optimistic scenario, although identifying stresses such as inequality.
- **Pax Americana.** U.S. technological and more general leadership; a generally optimistic world.
- **A New Caliphate.** The rise of a charismatic leader in the Middle East and more generally across the Islamic world, giving rise to significant system turmoil.
- **Cycle of Fear.** Global insecurities rooted in economic stress plus terrorism and other violence give rise to global divisions rather than cooperation.

In contrast, the seventh NIC report, seeing much the same driving forces but looking 20 years further into the future and inevitably also building on patterns visible during its writing, identified scenarios that suggest a more contested world. In the words of that report (U.S. NIC 2021: 109):

- **Renaissance of Democracies.** The United States leads a resurgence of democracies.
- **A World Adrift**. China is the leading but not globally dominant state.
- **Competitive Coexistence**. The United States and China prosper and compete for leadership in a bifurcated world.
- **Separate Silos.** World in which globalization has broken down and economic and security blocs emerge to protect states from mounting threats.
- **Tragedy and Mobilization.** Bottom-up, revolutionary change on the heels of devastating global environmental crises.

Input from personnel of the Pardee Institute using the International Futures model provided support for Global Trends 2020, 2025, 2030, and 2040, generally including quantification of driving trends and/or scenarios. Files facilitating the running of scenarios for Global Trends 2020, 2025, and 2030 can be found through use of the *Scenario Tree/World Integrated Scenarios*.

14.3.4 Scenarios Ranging Across the SDGs

Notably absent from much of the discussions above of both global modeling and long-term integrated scenario analysis is extensive focus on human development issue areas. Quantitative analyses looking primarily at environmental and/or sociopolitical futures suggest the importance of inequalities and of social conditions in general terms but have not given comparable attention to issues like poverty, hunger, gender inequity, health, education, or basic infrastructure.

This limitation has come to be recognized in the IAM world. That recognition helped motivate the SHAPE project that was active from 2019 to 2023. Co-funded by the European Union with many other organizations, SHAPE identified attention to the full set of SDGs as central to its work and focused on the interactions between mitigating climate change and the pursuit of the SDGs. Coming from its work, which combined use of selected IAMs and more specialized models, Soergel et al. (2024) elaborated three primary Sustainable Development Pathway scenarios with clusters of policy emphasis: **Economy-driven Innovation**; **Resilient Communities;** and **Managing the Global Commons**. These built in part on the SSP scenarios. One limitation of the scenario analysis was use of exogenously-specified variables for population, GDP, and education, drawing from SSP quantifications generated by other models, rather than representing those endogenously; thus, they could not be dynamically impacted by alternative forecasts of both environmental and human development variable trajectories.

Although its treatment of environmental issues is considerably more limited than that of many IAMs, the IFs system is largely unique in its ability to support scenario analysis across human development, environmental sustainability, and sociopolitical variables (Hanna et al., 2024; Hughes, 2025). Jonathan Moyer is now leading that effort (Moyer & Hedden, 2020).

14.4 Integrated Scenarios: Common Elements

Surveying the scenario families of the above studies suggests some categories of key uncertainties that often shape integrated scenario analysis, with or without global models.

1. Key Framing Scenario Elements
   - *Pace of Technological Change and/or Economic Growth***.** This is close to a universal pair of uncertainty framing categories, because essentially all scenario groupings represent variation with respect to either or both. That is not surprising: futurists as diverse as Karl Marx, Herman Kahn, and Alvin Toffler have always been influenced by technological and/or economic shapings. Alternative artificial intelligence (AI) futures fit here. And empirically, it is difficult to find sociopolitical or environmental variables that are not correlated with, and significantly driven by, technology and GDP per capita.

- *Environmental Uncertainties.* Framing assumptions with respect to the environment are also prevalent in longer-term analyses but are generally treated as endogenously dependent on other variables rather than exogenously emerging. They can be seen in the scenarios of the Global Scenario Group, in IPCC assessments and GEO reports, and in the work of IAMs. Many scenarios focus heavily on the material intensity of the economy and the character of the energy system as drivers.

2. Global, Regional, and Country Patterns and Behavior
   - *Level of Globalization (Possibly in Juxtaposition with Maintenance or Strengthening of Cultural Differences).* As with technological change and economic growth, and very much related to both of those, nearly all scenario groupings make differing assumptions across scenarios with respect to the extent of globalization. A smaller set looks also to the role of regional cultural differences à la Friedman's (1999) juxtaposition in *The Lexus and the Olive Tree*. The scenario set of the Millennium Ecosystem Assessment has among the clearest separations between emphasis on the former with **Global Orchestration** and on the latter with **Adaptive Mosaic.**
   - *Global power configurations.* Some scenarios look at global politics, explicitly differentiating possible future power distributions
   - *International Cooperation or Lack of Cooperation.* Many scenario sets introduce variations in behavior of countries in interaction, including the United States, China, and those of the European Union, considering the extent of their unilateralism or multilateralism in policy.
   - *Income Distribution.* This dimension of underlying variation in scenarios is also quite common across sets; the distributional attention can be interstate, intrastate, or both. In many analyses, it is treated as a logical extension of assumptions about globalization (globalization often either creates greater inequalities or enhances convergence of economic levels, depending on the systemic perspective of the scenario builder).
   - *Nationalism and other isms.* Some global political scenarios look at the intensity of national, religious, cultural, or other tribal identities and/or ideologies. This can be part of a globalization/regionalization scenario set but can also be separate.
   - *Demographic patterns.* Environmentally-focused scenario sets normally define alternative demographic futures. There has, however, come to be substantial convergence in assumptions and therefore forecasts across demographic analysts.
   - *Governance Success.* In some scenarios, the success or quality of governance is explicitly variable and in some analyses the size of government also varies.
3. Agent Behavior
   - There tend not to be specific agent-class assumptions in most integrated scenario sets (e.g., attention to governments, firms, and households). That is unfortunate because consideration of agent-class behavior is highly relevant to policy analysis around specific issues. As integrated scenario development increasingly becomes connected with policy analysis, and it will be,

attention to agent-class behavior will become more common both in building integrated scenarios and in exploring leverage across them. Financial flows including government finance, as represented in the social accounting matrix, are one key element of agent behavior.

4. Wild Cards
   - The central character of some scenarios (commonly in fiction books and movies) is a focus on improbable but theoretically possible events, from plagues through meteor collisions with earth to specific technological breakthroughs. In short, and appropriately, wild cards are "all over the place." Implications of artificial super intelligence could fall into this category, as could other elements of technological change.[4] These are not common in model-based scenarios.

This quick analysis of common components in integrated scenario sets generally reinforces the structure of the IFs *Scenario Tree* for creating scenarios (see again Chap. 5, especially Fig. 5.1, for discussion of it). It also overlaps with the structure of the folders of many prepackaged scenario components available for loading into the *Scenario Tree* of IFs (they are in.sce files). See Fig. 14.1, which uses Windows Explorer to show the Scenario Small folder that is installed as a sub-folder under IFs. With the *Add Scenario Component* option in the *Scenario Tree*, pre-built scenarios with one or more parameter changes can be loaded in preparation for model runs. Prior to a run it is possible to add multiple parameter assumptions or sets of them as desired. In addition to retrieving such pre-built scenarios or scenario elements from the directories, the user of the *Scenario Tree* in the stand-alone version of IFs (not the web version) can store new scenario files built with the *Tree* into folders for future elaboration or use.

Figure 14.1 shows the IFs Scenario Small folder structure available to users. Note that the folders and subfolders correspond somewhat to the categories of the *Scenario Tree* and even more closely to the components of the scenarios that have been discussed above: framing components around the economy, technology and the environment; interventions at the agent behavior level for households, governments and firms; and surprises and wildcards.

In addition, however, a folder called *World Integrated Scenario Sets* includes subfolders (on the right-hand side of Fig. 14.1) with scenario sets from many of the projects that IFs has undertaken with IGOs, NGOs, national governments, and think tanks. Exploration of some of these and associated reports and publications identified above might be useful in support of your own thinking about scenarios.

For instance, there is a folder related to a project done with the UN Women group (*World Integrated Scenario Sets/UN Women*) focused on enhancing the status and well-being of women; for related scenario discussion see Andrijevic et al. (2025). Another folder (*World Integrated Scenario Sets/COVID/SDG Analysis)* has scenarios prepared with the United Nations Development Programme around the implications of the COVID-19 pandemic for progress on the SDGs (Hanna et al.,

[4] Barry Hughes has pointed in his writing to the importance of the Whipperwillian effect.

Fig. 14.1 Directory Structure of IFs Pre-defined Scenario Elements and Integrated Scenarios. *Note* Folders in left-hand column are in *Scenario Small* (not shown) and folders in the right-hand column are found in *World Integrated Scenario Sets* (shown in the left-hand column). *Source* Screen capture by author of IFs 8.57 folder structure within the IFs folder

2024). Other scenarios focus directly on elements of human development including poverty, education, longevity, women, the human development index (HDI), and SDG pathways.

Still another folder in Fig. 14.1 contains.sce files related to scenarios prepared for UNEP GEO-4 (World Integrated Scenario Sets/UNEP GEO/GEO4). You will also be able to find folders related to scenarios of the IPCC, the UNEP GEO project, and the NIC reports. For example, clicking on the IPCC exposes four scenario files named IPCCA1Sce, IPCCA2Sce, IPCCB1Sce and IPCCB2Sce. These files contain specifications for the scenario tree that roughly parallel the four scenarios developed by the third round of the IPCC studies. If the IPCCA1Sce file were selected and opened, it would populate the scenario tree with a set of appropriate drivers. Then you could select the Explain Scenario option from the Scenario Tree's menu for a verbal explanation of the scenario. Other folders relate to projects and published work undertaken by the Pardee Institute itself. Again, they can be combined as desired.

14.5 Creating Futures by Making Choices

The insistence by participants in the Rio conference in 1992 on simultaneous consideration of environmental and socioeconomic development issues symbolized the contemporary recognition that all issues on the world agenda are linked. Until this chapter, however, we have largely separated consideration of issues centering on human condition progress from those on environmental sustainability and from those on sociopolitical variables. This chapter suggests, however, that use of the IFs system offers unique possibilities for integrated analysis of global futures, including pursuit of multiple SDGs.

It is therefore time for you to reflect on the extensive interaction of these issues and to ask yourself about the comprehensive shape of the futures you expect and wish to see evolve in the coming decades. This book leaves you not with a vision but with an exercise:

→ Take time to sketch stories—to build two scenarios.

The first is for the world you think it is likely you will see in place at the end of 40–50 years (if you are a university student, that may be roughly the span of your working life). Where do you believe global changes are taking us? Is it dystopian, utopian, or some mix? Consider the differences between the IFs Base Case scenario and your own insights, mapping out a scenario with IFs that represents your own current "best guess" with respect to the developments of the coming years. That is, you may want to create (and save) your own Base Case.

The second story is for the world you would like to see at the end of that horizon. What kind of future do you want to see? What do you value and what are your goals with respect to those values? What leverage do humans have in bringing about the future you would prefer? What actions need to be taken and what might be the costs of taking them?

Here are some things you might want to explore. Economically, do you anticipate and/or desire seeing the strengthening of free trade and financial flows, a stabilization of those connections, or a movement of the world to greater autarky in countries or regions? Put another way, would the world be better served if the globalization of trade and financial flows were strengthened or reversed? Demographically, will and should global population peak well before the turn of this century or not?

In terms of the environment, do you foresee increased environmental destruction or the technological and behavioral redress of multiple environmental problems? Do you prescribe substantial global environmental regulation, country-specific approaches, or sponsorship of certain technological advances? What might be, and what do you think should be, the rough size of the world population? Will food and energy be adequate or will prices rise sharply and supply disruptions result?

Can the Sustainable Development Goals (or their successors, likely to be similar) be met and what actions would you like to see toward facilitating their attainment? Can advances in education, health, and poverty reduction be maintained and perhaps accelerated?

You can use the IFs system to help you undertake the exercise of this chapter. There is much that the simulation does not contain that will require either that you elaborate the narrative of your scenario(s) beyond IFs or that you use the model system very creatively. (IFs does not, for example, allow you explicitly to introduce new global or regional institutions with respect to security, economics, or the environment.)

What IFs does allow you to do, as we have seen in the preceding chapters and this one, is to introduce a wide range of alternative assumptions and policy strategies and therefore examine the world as an integrated whole across its many geographic divisions and disparate human and physical systems.

A variation on the exercise would be to put together your own Base Case followed by two extreme scenarios—the most pessimistic and the most optimistic scenarios about the future that you believe have any credibility. Identify clearly the policy foci and variations in causal understanding that differentiate the two scenarios. Your most optimistic scenario should help you identify interventions (or at least intervention points) that you would recommend; your most pessimistic scenario may help spot those to avoid. If you are careful in this process, you may find that, while building your optimistic scenario, some interventions achieve positive results in one issue area but have negative consequences elsewhere. Think about trade-offs and costs of interventions.

You cannot help positively shape the future you want, either your own personal future or that of the broader world around you, unless you first envision it and begin mapping a path to it. Choice under uncertainty will never be easy. In the face of any personal or collective challenge, we can only work to define our values, identify important leverage points, and thereby make our choices somewhat more likely to move us toward our goals. There is no way to avoid challenge: individually and collectively we do have a future to shape.

References

Andrijevic, M., Zimm, C., Moyer, J. D., Muttarak, R., & Pachauri, S. (2025). Representing gender inequality in scenarios improves understanding of climate challenges. *Nature Climate Change, 15*(2), 138–146. https://doi.org/10.1038/s41558-024-02242-5

Bremer, S. A. (Ed.). (1987). *The Globus model: Computer simulation of worldwide political and economic developments*. Routledge.

European Commission Forward Studies Unit. 1999. *Scenarios Europe 2010*. European Commission https://librarysearch.lse.ac.uk/discovery/fulldisplay/alma9963266510302021/44LSE_INST:44LSE_VU1.

Friedman, Thomas L. 1999. *The Lexus and the olive tree*. Anchor Books.

Glenn, J. C., Gordon, T., & Florescu, E. (2024). *State of the future 20.0*. The Millennium Project.

Guetzkow, H. (1959). A use of simulation in the study of inter-nation relations. *Behavioral Science, 4*(3), 183–191. https://doi.org/10.1002/bs.3830040302

Haas, E. B. (1990). *When knowledge is power*. University of California Press.

Hammond, A. (1998). *Which world? Scenarios for the 21st century*. Island Press.

Hanna, T., Hughes, B. B., Irfan, M. T., Bohl, D. K., Solórzano, J., Abidoye, B., Patterson, L., & Moyer, J. D. (2024). Sustainable Development Goal attainment in the wake of COVID-19: Simulating an ambitious policy push. *Sustainability, 16*(8), 3309. https://doi.org/10.3390/su1608 3309

Hughes, B. B. (1987). Domestic economic processes. In S. A. Bremer (Ed.), *The Globus model* (pp. 39–158). Routledge.

Hughes, B. B. (2025). Analysis of integrated global SDG pursuit: Challenges and progress. *Sustainability, 17*(15), 6672. https://doi.org/10.3390/su17156672

International Energy Agency (IEA). 2003. *Energy to 2050: Scenarios for a sustainable future.* IEA. http://library.iea.org/dbtw-wpd/textbase/nptoc/EnergyTo2050TOC.pdf.

Kriegler, E., O'Neill, B. C., Hallegatte, S., Kram, T., Lempert, R. J., Moss, R. H., & Wilbanks, T. (2012). The need for and use of socio-economic scenarios for climate change analysis: A new approach based on shared socio-economic pathways. *Global Environmental Change, 22*(4), 807–822. https://doi.org/10.1016/j.gloenvcha.2012.05.005

Lutz, W., Cuaresma, J. C., & Sanderson, W. (2008). The demography of educational attainment and economic growth. *Science, 319*(5866), 1047–1048. https://doi.org/10.1126/science.1151753

Meadows, D. H., Meadows, D. L., Randers, J., & Behrens, W. K., III. (1972). *Limits to growth.* Universe Books.

Mesarovic, M. D., & Pestel, E. (1974). *Mankind at the turning point.* E.P. Dutton & Co.

Moyer, J. D., & Hedden, S. 2020. "Are we on the right path to achieve the sustainable development goals?" *World Development, 127*(104749). https://doi.org/10.1016/j.worlddev.2019.104749

O'Neill, B. C., Kriegler, E., Riahi, K., Ebi, K. L., Hallegatte, S., Carter, T. R., Mathur, R., & van Vuuren, D. P. (2013). A new scenario framework for climate change research: The concept of shared socioeconomic pathways. *Climatic Change, 122*, 387–400. https://doi.org/10.1007/s10 584-013-0905-2

Raskin, P., Banuri, T., Gallopin, G., Gutman, P., Hammond, A., & Swart, R. (2002). *Great transition: The promise and lure of the times ahead.* Stockholm Environment Institute. http://www.tellus.org/seib/publications/Great_Transitions.pdf

Riahi, K., van Vuuren, D. P., Kriegler, E., Edmonds, J., O'Neill, B. C., Fujimori, S., Bauer, N., Calvin, K., Dellink, R., Fricko, O., Lutz, W., Popp, A., Crespo Cuaresma, J., Kc, S., et al. (2017). The Shared Socioeconomic Pathways and their energy, land use, and greenhouse gas emissions implications: An overview. *Global Environmental Change, 42*, 153–168. https://doi.org/10.1016/j.gloenvcha.2016.05.009

Richardson, L. F. (1960). *Arms and insecurity.* Quadrangle Books.

Small, M., & Singer, J. D. (1982). *Resort to arms: International and civil wars, 1816–1980* (2nd ed.). Sage. https://cir.nii.ac.jp/crid/1970586434801124649

Soergel, B., Rauner, S., Daioglou, V., Weindl, I., Mastrucci, A., Carrer, F., Kikstra, J., Ambrósio, G., Aguiar, A. P. D., Baumstark, L., Bodirsky, B. L., Bos, A., Dietrich, J. P., Dirnaichner, A., Doelman, J. C., Hasse, R., Hernandez, A., Hoppe, J., Humpenöder, F., … Kriegler, E. (2024). Multiple pathways towards sustainable development goals and climate targets. *Environmental Research Letters, 19*(12), 124009. https://doi.org/10.1088/1748-9326/ad80af

United Nations Department of Economic and Social Affairs. (2025). *World social report 2025: A new policy consensus to accelerate social progress.* https://desapublications.un.org/publications/world-social-report-2025-new-policy-consensus-accelerate-social-progress

United Nations Economist Network. (2020). *Shaping the trends of our time: Report of the UN Economist Network for the UN 75th anniversary.* https://www.unep.org/resources/report/Shaping-Trends-Our-Time-Report

United Nations Environment Programme (UNEP). (2019). *Global environment outlook 6.* Cambridge University Press.

United Nations World Commission on Environment and Development. (1987). *Our common future.* Oxford University Press.

U.S. Commission on National Security/21st Century (the Hart-Rudman Commission). (1999, September 15). *The new world coming: American security in the 21st century, supporting research and analysis.*

U.S. National Intelligence Council (U.S.NIC). (1997). *Global trends 2010.* https://www.dni.gov/index.php/who-we-are/organizations/mission-integration/nic/nic-related-menus/nic-related-content/global-trends-2010

U.S. National Intelligence Council (U.S. NIC). (2004). *Global trends 2020: Mapping the global future.* https://apps.dtic.mil/sti/html/tr/ADA497321/index.html

U.S. National Intelligence Council (U.S. NIC). (2021). *Global trends 2040: A more contested world.* https://www.intelligence.gov/publics-daily-brief/publics-daily-brief-articles/national-intelligence-council-releases-global-trends-report-on-the-more-contested-world-of-2040

van Vuuren, D. P., Bijl, D. L., Bogaart, P., Stehfest, E., Biemans, H., Dekker, S. C., Doelman, J. C., Gernaat, D. E. H. J., & Harmsen, M. (2019). Integrated scenarios to support analysis of the food–energy–water nexus. *Nature Sustainability, 2*(12), 1132–1141.

van Vuuren, D. P., Kok, M., Esch, S. van der, Jeuken, M., Lucas, P., Prins, A. G., Alkemade, R., Berg, M. van den, Biermann, F., Grijp, N. van der, Hilderink, H., Kram, T., Melamed, C., Pattberg, P., Scott, A., Stehfest, E., Vries, B. de, Velde, D. W. te, & Wiggens, S. (2012). *Roads from Rio+20. Pathways towards global sustainable development targets for 2050.* PBL Netherlands Environmental Assessment Agency. https://dspace.library.uu.nl/handle/1874/281315

van Vuuren, D. P., Kok, M., Lucas, P. L., Prins, A. G., Alkemade, R., van den Berg, M., Bouwman, L., van der Esch, S., Jeuken, M., Kram, T., & Stehfest, E. (2015). Pathways to achieve a set of ambitious global sustainability objectives by 2050: Explorations using the IMAGE integrated assessment model *Technological Forecasting and Social Change 98*, 303–323. https://doi-org.du.idm.oclc.org/10.1016/j.techfore.2015.03.005

World Business Council for Sustainable Development (WBCSD). (1987). *Exploring sustainable development: Global scenarios 2000–2050.* World Business Council for Sustainable Development.

World Business Council for Sustainable Development (WBCSD). (1999). *Energy 2050: Risky business.* World Business Council for Sustainable Development.

World Business Council for Sustainable Development (WBCSD). (2000). *The wizard of US.* World Business Council for Sustainable Development.

Appendix 1

The Philosophical Approach in IFs to Modeling

Given the goal of sustainable human development, how do we represent the complex, interacting systems that shape prospects for attaining it in a formal, computer-based model such as International Futures (IFs)?

A general conceptual approach helps frame the IFs system. Specifically, global human systems consist of classes of agents and larger structures within which those agents interact. Over time, agents and the larger structures evolve in processes of mutual influence and determination. Two concepts are foundational to building model structures and processes: *stocks* and *flows*. Stocks exist across time and can be incremented or decremented. Examples are population, financial and physical capital, energy resources, atmospheric carbon level, and knowledge. Flows are what increment or decrement those stocks across time. Examples are births, deaths, economic investment or capital depreciation, and carbon emissions. Key or dominant relationships in the models of IFs very often determine the magnitude of such flows, often representing the behavior of agent classes and therefore identifying their potential leverage within the subsystem.

Households, firms, and governments are primary agent classes.[1] Humans as individuals and in households are parts of larger demographic systems or structures. IFs adopts the typical approach to representing the stocks in such demographic structures via age-sex cohort distributions. Critical demographic decisions of households are related to births, migration across countries, and labor force participation. Financially, households make decisions concerning consumption versus savings and what is consumed.

Government behavior greatly influences financial flows within countries and across borders through their search for income and targeting of transfers and direct

[1] IFs is not agent-based in the sense of models that represent individual micro-agents following rules and generating structures through their behavior. Instead, IFs represents both existing macro-agent classes and existing structures with complex historical path dependencies, attempting to show the dynamics of how behavior by those agents can change and how the structures can evolve.

B. B. Hughes, *Exploring and Shaping International Futures*, World-Systems Evolution and Global Futures, https://doi.org/10.1007/978-3-032-22084-4

expenditures in multiple categories including military, education, health, research and development, and infrastructure. While IFs does not represent as many decision and action points for firms as for households and governments, it does include the role of firms in shaping levels and targets of investment.

Social accounting matrices (SAMs) are structural forms that integrate representation of non-market-based financial transfers among such agents with exchanges in a market system. IFs uses a SAM structure to account for inter-agent flows generally. Financial asset and debt stocks, and not just flows, are also critical to maintain as part of this structural system because they make possible and also motivate the behavior of agent classes. For instance, high government debt levels eventually require changes in revenue raising and/or expenditure behavior.

Obviously, not all agent-class behavior is financial. Governments interact with each other within interstate systems that frame the pursuit of security and cooperative interaction. Potential behavioral elements include spending on the military, but also such actions as joining of alliances. Modeling of interstate behavior usefully includes representation of action-reaction dynamics that are sensitive to power relationships across the actors. IFs represents changing power structures (which again have stock and flow elements) and interstate threat. It also represents governance characteristics such as domestic democracy and government effectiveness, which can in turn influence spending patterns.

Still further, human actor classes interact with the broader environment. In so doing, important behavior includes technological innovation and use, as well as resource (like water and forest land) extraction, and emissions release. The structures of IFs track both stocks of resources and selected flows, including emissions (for instance, greenhouse gases). Many behaviors of households (such as energy consumption), of government (such as carbon taxes), and of firms (such as investment in renewable energy production) affect those environmentally related flows and stocks. They can reflect values as well as financial motivations.

The general approach described here is a guide to IFs development, not a straitjacket, and some systems cannot be represented well within it (for instance, we make no attempt to represent culture in terms of stocks and flows). In representing the behavior of agent classes and the structures of systems, IFs draws upon large bodies of insight in many theoretical and modeling literatures and a huge collection of data. Although IFs often breaks new ground with respect to specific subsystems, its strengths also lie significantly in its integration and synthesis of bodies of earlier and ongoing work of others.

Appendix 2

Generations of International Futures (IFs)

The first generation, developed in 1978–1979, had deep roots in the world models of the 1970s, including those of the Club of Rome. In particular, IFs drew on the Mesarovic-Pestel or World Integrated Model (Mesarovic & Pestel, 1974), to which the creator of IFs contributed. IFs consciously also drew on the Leontief World Model (Leontief et al., 1977), the Bariloche Foundation's world model (Herrera et al., 1976), and the Systems Analysis Research Unit Model (1977), following comparative analysis of those models by Hughes (1980). That generation was written in FORTRAN and available for use on mainframe computers through CONDUIT, an educational software distribution center at the University of Iowa. It represented the world in 10 regions.

The second generation of International Futures moved to early microcomputers in 1985 using the DOS platform. It was a very simplified version of the original IFs without regional or country differentiation.

The third generation, first available in 1993, became a full-scale microcomputer model for DOS systems with a Windows-based interface and representation of 162 countries. That generation improved earlier representations of demographic, energy, and food systems and added foundational environmental and sociopolitical content. It built upon collaboration with the GLOBUS project and adapted the economic submodel of GLOBUS (developed by Hughes, 1987). GLOBUS was created with the inspiration of Karl Deutsch and the leadership of Stuart Bremer (1987) at the Science Center in Berlin. The third generation produced three editions or major releases of IFs, each accompanied by a book also called *International Futures* (Hughes, 1993, 1996, 1999). The creation of a data "preprocessor" fully integrated into the IFs system's interface facilitated regular updates and enhancements of the model's base year initialization.

The fourth generation began taking shape in early 2000. It was heavily influenced by increasing interest in the model as a policy analysis tool by several organizations. General Motors commissioned a specialized version of IFs named

B. B. Hughes, *Exploring and Shaping International Futures*, World-Systems Evolution and Global Futures, https://doi.org/10.1007/978-3-032-22084-4

CoVaTrA (Consumer Values Trends Analysis) with updated and extended demographic modeling and representation of value changes. An alliance was established with Ronald Inglehart (1997), founder of the World Values Survey. The Strategic Assessments Group (SAG) of the Central Intelligence Agency supported a specialized version named *IFs for SAG.* The work involved in preparing *IFs for SAG* greatly extended and enhanced the sociopolitical representations of the model, both domestic and international, with substantial help from Evan Hillebrand (Hughes & Hillebrand, 2006). The European Commission sponsored a project named TERRA, which led to a specialized version named *IFs for TERRA*. And the RAND Pardee Center began sponsorship of a project to explore the potential for substantially reducing global poverty and developing a global social safety net. Across the fourth generation, there was heavy emphasis on enhanced usability. For example, ideas from Robert Pestel in the TERRA project led to the creation of a tree structure for scenario creation and management.

The fifth generation emerged across the first decade of the 2000s. Up until this point, IFs was only available as a stand-alone Visual Basic system on DOS machines. Thanks to support from the National Intelligence Council's 2020 Project, a central aspect of the fifth generation was the development of a web-based version. That generation also increased transparency of the system via an extended Help system and the ability for users interested in particular variables/sub-areas of the model to access flow charts, equations, and even computer code on demand. During that decade, financial support from Frederick S. Pardee made possible the creation of the Frederick S. Pardee Center for International Futures (now the Frederick S. Pardee Institute for International Futures) within the Josef Korbel School at the University of Denver.

The sixth generation unfolded roughly between 2010 and 2013 and involved expansion of the model system with education, health, infrastructure, and governance system representations in support of the five-volume Patterns of Potential Human Progress series funded by a gift from Fred Pardee. Also during this period the web-based system was strengthened.

The seventh generation was approximately in the 2014–2022 period and was significantly motivated by a variety of projects that roughly have fallen into human development and geopolitical categories, with increasing connections between them. The system's capabilities for supporting provincial/state/district breakdowns of countries were extended, the addition of significant bilateral linkages and relationships among countries to the monadic representation of them began to move ahead, and the interface system was strengthened. During this period, the directorship of the Center transitioned from Barry Hughes to Jonathan Moyer. That was a very important step not only in assuring continuity of the Center and the IFs project's work, but also because Moyer's very active and successful efforts led to new areas of work with new partners and a concomitant growth in the core Pardee team.

The eighth and current generation began in 2023 with the migration of IFs from Visual Basic (VB6) to Visual Basic (.Net) to assure software continuity in the face of Microsoft's decision to freeze development of the former. The migration has enhanced the free and open presence of IFs on the web. During this period, the IFs project's ties with the Integrated Assessment Model community around the world have been strengthened. Further, following the death of Mr. Pardee in 2022, his estate provided significant additional endowment resources to the Pardee Center. That support, combined with Dr. Moyer's continued thoughtful efforts to significantly expand externally funded research and to lead the Center's maturation and growth as well as that of the IFs system, resulted in an official redesignation by the University of Denver as a research institute rather than a center. Throughout this period, refinement and extensions of the model structure, including on the international relations/Diplometrics side, and regularization of data updates have continued.

The next generations. Refinement and extensions, responding to the needs of partners and to changes in the world and our understanding of it, will drive continued evolution in IFs and its usage. It will always be freely open for use by others.

References

Bremer, S. A. (Ed.). (1987). *The GLOBUS model: Computer simulation of world-wide political and economic developments*. Westview.

Herrera, A. O. (1976). *Catastrophe or new society? A Latin American world model*. Fundación Bariloche.

Hughes, B. B. (1980). *World modeling*. Lexington Books.

Hughes, B. B. (1987). Domestic economic processes. In S. A. Bremer (Ed.), *The Globus model* (pp. 39–158). Routledge.

Hughes, B. B. (1993). *International futures: Choices in the creation of a new world order* (1st ed.). Westview Press.

Hughes, B. B. (1996). *International futures: Choices in the creation of a new world order* (2nd ed.). Westview Press.

Hughes, B. B. (1999). *International futures: Choices in the face of uncertainty* (3rd ed.). Westview Press.

Hughes, B. B., & Hillebrand, E. E. (2006). *Exploring and shaping international futures* (4th ed.). Paradigm Publishers.

Inglehart, R. (1997). *Modernization and postmodernization*. Princeton University Press.

Leontief, W., Carter, A., & Petri, P. (1977). *The future of the world economy*. Oxford University Press.

Mesarovic, M. D., & Pestel, E. (1974). *Mankind at the turning point*. E.P. Dutton & Co.

Systems Analysis Research Unit. (1977). *SARUM 76 global modeling project*. Departments of the Environment and Transport, 2 Marsham Street. London, 3WIP 3EB.

Index

B. B. Hughes, *Exploring and Shaping International Futures*, World-Systems Evolution and Global Futures, https://doi.org/10.1007/978-3-032-22084-4

M

P

MIX
Papier aus verantwortungsvollen Quellen
Paper from responsible sources
FSC® C105338

FSC
www.fsc.org

If you have any concerns about our products,
you can contact us on
ProductSafety@springernature.com

In case Publisher is established outside the EU,
the EU authorized representative is:
Springer Nature Customer Service Center GmbH
Europaplatz 3, 69115 Heidelberg, Germany

Printed by Libri Plureos GmbH
in Hamburg, Germany